Lecture Notes in Computer Science 2228

Edited by G. Goos, J. Hartmanis, and J. van Leeuwen

T0216292

Springer

Berlin
Heidelberg
New York
Barcelona
Hong Kong
London
Milan
Paris
Tokyo

Burkhard Monien   Viktor K. Prasanna
Sriram Vajapeyam (Eds.)

High Performance Computing – HiPC 2001

8th International Conference
Hyderabad, India, December 17-20, 2001
Proceedings

Springer

Series Editors

Gerhard Goos, Karlsruhe University, Germany
Juris Hartmanis, Cornell University, NY, USA
Jan van Leeuwen, Utrecht University, The Netherlands

Volume Editors

Burkhard Monien
University of Paderborn, Department of Mathematics and Computer Science
Fürstenallee 11, 33102 Paderborn, Germany
E-mail: bm@uni-paderborn.de

Viktor K. Prasanna
University of Southern California
Department of EE-Systems, Computer Engineering Division
3740 McClintock Avenue, EEB 200C, Los Angeles, CA 90089-2562, USA
E-mail: prasanna@usc.edu

Sriram Vajapeyam
Independent Consultant, c/o Infosys Ltd.
"Mangala", Kuloor Ferry Road, Kottara, Mangalore 575006, India
E-mail: sriram_vajapeyam@yahoo.com

Cataloging-in-Publication Data applied for

Die Deutsche Bibliothek - CIP-Einheitsaufnahme

High performance computing : 8th international conference ; proceedings /
HiPC 2001, Hyderabad, India, December 17 - 20, 2001. Viktor K. Prasanna ...
(ed.). - Berlin ; Heidelberg ; New York ; Barcelona ; Hong Kong ; London ;
Milan ; Paris ; Tokyo : Springer, 2001
    (Lecture notes in computer science ; Vol. 2228)
    ISBN 3-540-43009-1

CR Subject Classification (1998): C.1-4, D.1-4, F.1-2, G.1-2

ISSN 0302-9743
ISBN 3-540-43009-1 Springer-Verlag Berlin Heidelberg New York

Springer-Verlag Berlin Heidelberg New York
a member of BertelsmannSpringer Science+Business Media GmbH

http://www.springer.de

© Springer-Verlag Berlin Heidelberg 2001
Printed in Germany

Typesetting: Camera-ready by author, data conversion by PTP-Berlin, Stefan Sossna
Printed on acid-free paper   SPIN: 10840999   06/3142   5 4 3 2 1 0

Message from the Program Chair

I am pleased to introduce the proceedings of the Eighth International Conference on High-Performance Computing to be held in Hyderabad, India from December 17-20, 2001. The technical program consists of five sessions, which are arranged in one track, without parallel sessions. We also have five keynote speeches, an invited session, an industrial track session, a poster session, eight tutorials, and three workshops.

The topics of the technical sessions are: Algorithms, Applications, Architecture, Systems Software, and Communication Networks. For each topic we had a program committee headed by a vice-chair: Bruce M. Maggs, Horst D. Simon, Allan Gottlieb, Cauligi S. Raghavendra, and Guang R. Gao. The program committees consisted of a total of 55 members, who did all the hard work of reviewing the submitted papers within the deadlines. We received 108 papers for the technical sessions. Our goal was to attain a high-quality technical program. Twenty-nine papers were accepted, which meant an acceptance rate of 27%. Each paper was reviewed by up to four members of the program committee and one program vice-chair. The accepted contributions came from 18 different countries. This is an indication of the true international flavor of the conference.

I would like to thank the five program vice-chairs who worked very closely with me in the process of selecting the papers and preparing the technical program. Viktor K. Prasanna provided excellent feedback about the technical sessions in his role as general co-chair. I would like to express special thanks to all members of the program committee for their high-quality and timely review work.

Finally, I would like to mention the immense support provided by my assistants Marion Bewermeyer, Sven Grothklags, Georg Kliewer, and Thomas Thissen. Without their constant help I could not have performed the tasks required of the program chair. They did all the work related to the collection of the papers, coordinating the review process, preparing the acceptance decision of the program committee, and notifying the authors about the acceptance/rejection decision.

I would like to welcome all the participants to the conference in Hyderabad and hope that they will enjoy the excellent technical program.

September 2001   Burkhard Monien

Message from the General Co-Chairs

It is our pleasure to welcome you to the Eighth International Conference on High Performance Computing and to Hyderabad, a leading center for Information Technology in India. This message pays tribute to many volunteers who made this meeting possible.

We are indebted to Professor Burkhard Monien for his superb efforts as program chair in organizing an excellent technical program. We would like to thank the vice chairs, Guang Gao, Allan Gottlieb, Bruce Maggs, Cauligi Raghavendra, and Horst Simon, for their efforts in putting together a strong program committee, who in turn reviewed the submissions and composed an excellent technical program. We would like to thank Cauligi Raghavendra for his efforts in organizing the invited session on networks.

Many volunteers have been associated with HiPC over the years, and they have continued their excellent service in organizing the meeting: Vipin Kumar invited the keynote speakers and coordinated the keynotes, Sartaj Sahni handled the poster/presentation session, Sandhya Dwarkadas and Tulika Mitra handled publicity, Nalini Venkatasubramanian and her assistant Mario Espinoza interfaced with the authors and Springer-Verlag in bringing out these proceedings, Manav Misra put together the tutorials, Ajay Gupta did a fine job in handling international financial matters, Dheeraj Sanghi administered scholarships for students from Indian academia, Sudheendra Hangal handled the exhibits with inputs from R. Govindarajan, Sajal Das acted as vice general chair, C. P. Ravikumar and Vinod Sanjay, though not listed in the announcements, managed publicity within India.

We would like to thank all of them for their time and efforts.

Our special thanks go to A. K. P. Nambiar for his continued efforts in handling financial matters as well as coordinating the activities within India.

Major financial support for the meeting was provided by several leading IT companies. We would like to thank the following individuals for their support:

N. R. Narayana Murthy, Infosys
Karthik Ramarao, HP India
Kalyan Rao, Satyam
Shubhra Roy, Intel India
Amitabh Srivastava, Microsoft Research
Uday Shukla, IBM India
M. Vidyasagar, TCS

A new feature of this year's meeting is workshops organized by several volunteers. We would like to thank Srinivas Aluru, Rajendra Bera, S. Jagannathan,

Santosh Pande, Uday Shukla, and M. Vidyasagar, for their efforts in putting together the workshop programs.

We would like to thank IIIT, Hyderabad staff for their assistance with local arrangements: Rajeev Sangal who initiated IIIT's co-sponsorship, Kamal Karlapalem who acted as local arrangements chair, P. J. Narayanan who coordinated the cultural events, and Cdr. Raman for his assistance in using IIIT facilities. Continued sponsorship of the meeting by the IEEE Computer Society and ACM are much appreciated.

Finally, we would like to thank Henryk Chrostek, Sumit Mohanty, Bharani Thiruvengadam, and Bhaskar Srinivasan of USC, Swaroopa Rathna of Taj Hotels, and Siddhartha Tambat of IISc for their assistance over the past year.

September 2001   Viktor K. Prasanna
                                                       Sriram Vajapeyam

Conference Organization

General Co-chairs
Viktor K. Prasanna, University of Southern California
Sriram Vajapeyam, Consultant

Vice General Chair
Sajal K. Das, The University of Texas at Arlington

Program Chair
Burkhard Monien, Universität Paderborn, Germany

Program Vice-chairs
Algorithms
Bruce M. Maggs, Carnegie Mellon University

Applications
Horst D. Simon, NERSC, Lawrence Berkeley National Laboratory

Architecture
Allan Gottlieb, New York University and NEC Research

Communication Networks
Cauligi S. Raghavendra, University of Southern California

Systems Software
Guang R. Gao, University of Delaware

Keynote Chair
Vipin Kumar, AHPCRC, University of Minnesota

Poster / Presentation Chair
Sartaj Sahni, University of Florida

Tutorials Chair
Manavendra Misra, KBkids.com

Industry Liaison Chair
Sudheendra Hangal, Sun Microsystems

Scholarships Chair
Dheeraj Sanghi, Indian Institute of Technology, Kanpur

Awards Chair
Arvind, MIT

Finance Co-chairs
A. K. P. Nambiar, Software Technology Park, Bangalore
Ajay Gupta, Western Michigan University

Publicity Chair
Sandhya Dwarkadas, University of Rochester

Local Arrangements Chair
Kamal Karlapalem, Indian Institute of Information Technology, Hyderabad

Publications Chair
Nalini Venkatasubramanian, University of California at Irvine

Steering Chair
Viktor K. Prasanna, University of Southern California

Steering Committee
Jose Duato, Universidad Politecnica de Valencia
Viktor K. Prasanna, University of Southern California
N. Radhakrishnan, US Army Research Lab
Sartaj Sahni, University of Florida
Assaf Schuster, Technion-Israel Institute of Technology

PROGRAM COMMITTEE

Algorithms
Micah Adler, University of Massachusetts, Amherst
Richard Anderson, University of Washington
John Byers, Boston University
Ramesh Hariharan, Indian Institute of Science
Michael Mitzenmacher, Harvard University
S. Muthukrishnan, AT&T Research
Stephane Perennes, INRIA
Sushil K. Prasad, Georgia State University
Rajmohan Rajaraman, Northeastern University
Andrea Richa, Arizona State University
Christian Scheideler, Johns Hopkins University
Assaf Schuster, Technion-Israel Institute of Technology

Applications
Patrick Amestoy, ENSEEIHT-IRIT, Toulouse
Rupak Biswas, NASA Ames Research Center
Alvaro Coutinho, COPPE, Rio de Janeiro
Chris Ding, NERSC, Lawrence Berkeley National Laboratory
Jack Dongarra, Innovative Computing Lab, University of Tennessee

Robert Harrison, Pacific Northwest National Laboratory
Steve Jardin, Princeton Plasma Physics Laboratory
Ken Miura, Fujitsu, Japan
Franz-Josef Pfreundt, ITWM, Universität Kaiserslautern
Thomas Zacharia, Oak Ridge National Laboratory

Architecture

Alan Berenbaum, Agere Systems
Doug Burger, University of Texas at Austin
Vijay Karamcheti, New York University
Gyungho Lee, Iowa State University
Daniel Litaize, Institut de Recherche en Informatique de Toulouse
Vijaykrishnan Narayanan, Pennsylvania State University
Yale Patt, University of Texas at Austin
Shuichi Sakai, University of Tokyo
Jaswinder Pal Singh, Princeton University
Mateo Valero, Universidad Politecnica de Catalunya
Sudhakar Yalamanchili, Proceler Inc.

Communication Networks

Ian F. Akyildiz, Georgia Institute of Technology
Hari Balakrishnan, MIT
Joseph Bannister, University of Southern California/ISI
Deborah Estrin, University of California, Los Angeles/ISI
Ramesh Govindan, University of Southern California/ISI
Anurag Kumar, Indian Institute of Science
T. V. Lakshman, Lucent Technologies
Masayuki Murata, Osaka University
Marco Ajmone Marsan, Politecnico de Torino
Douglas Maughan, DARPA ITO
K. G. Shin, University of Michigan

Systems Software

Pradeep Dubey, IBM India Research Lab
R. Govindarajan, Indian Institute of Science
Wen-mei Hwu, University of Illinois at Urbana-Champaign
Laxmikant Kale, University of Illinois at Urbana-Champaign
Bernard Lecussan, SupAero, Toulouse
Yoichi Muraoka, Waseda University, Tokyo
Keshav Pingali, Cornell University
Martin Rinard, MIT Lab for Computer Science
Vivek Sarkar, IBM T. J. Watson Research Center
V. C. Sreedhar, IBM T. J. Watson Research Center
Theo Ungerer, Universität Karlsruhe

National Advisory Committee

R. K. Bagga, DRDL, Hyderabad
N. Balakrishnan, SERC, Indian Institute of Science
Ashok Desai, Silicon Graphics Systems (India) Pvt. Limited
Kiran Deshpande, Mahindra British Telecom Limited
H. K. Kaura, Bhabha Atomic Research Centre
Hans H. Krafka, Siemens Communication Software Limited
Ashish Mahadwar, Planetasia Limited
Susanta Misra, Motorola India Electronics Limited
Som Mittal, Digital Equipment (India) Limited
B. V. Naidu, Software Technology Park, Bangalore
N. R. Narayana Murthy, Infosys Technologies Limited
S. V. Raghavan, Indian Institute of Technology, Madras
V. Rajaraman, Jawaharlal Nehru Centre for Advanced Scientific Research
S. Ramadorai, Tata Consultancy Services, Mumbai
K. Ramani, Future Software Pvt. Limited
S. Ramani, Silverline Technologies Limited
Karthik Ramarao, Hewlett-Packard (India) Pvt. Limited
Kalyan Rao, Satyam Computers Limited
S. B. Rao, Indian Statistical Institute
H. Ravindra, Cirrus Logic
Uday S. Shukla, IBM Global Services India Pvt. Limited
U. N. Sinha, National Aerospace Laboratories

Workshop Organizers

Workshop on Embedded Systems

Co-chairs
S. Jagannathan, Infineon Technologies
Santosh Pande, Georgia Institute of Technology

Workshop on Cutting Edge Computing

Co-chairs
Uday S. Shukla, IBM Global Services India Pvt. Limited
Rajendra K. Bera, IBM Global Services India Pvt. Limited

Workshop on Bioinformatics and Computational Biology

Co-chairs
Srinivas Aluru, Iowa State University
M. Vidyasagar, Tata Consultancy Services

HiPC 2001 Reviewers

Micah Adler
Ian Akyildiz
Patrick Amestoy
Richard Anderson
Hari Balakrishnan
Joseph Bannister
Alan Berenbaum
Rupak Biswas
Doug Burger
John Byers
Alvaro L.G.A. Coutinho
Chris Ding
Jack Dongarra
Pradeep Dubey
Deborah Estrin
Ramesh Govindan
R. Govindarajan
Ramesh Hariharan
Robert Harrison
Wen-mei Hwu
Steve Jardin
Laxmikant Kale
Vijay Karamcheti
Anurag Kumar
T.V. Lakshman

Gyungho Lee
Bernard Leucussan
Daniel Litaize
Marco Ajmone Marsan
Douglas Maughan
Michael Mitzenmacher
Ken Miura
Yoichi Muraoka
Masayuki Murata
S. Muthukrishnan
Vijaykrishnan Narayanan
Yale Patt
Stephane Perennes
Franz-Josef Pfreundt
Keshav Pingali
Sushil K. Prasad
Rajmohan Rajaraman
Andrea Richa
Martin Rinard
Shuichi Sakai
Vivek Sarkar
Christian Scheideler
Assaf Schuster
K.G. Shin
Jaswinder Pal Singh
V.C. Sreedhar

TABLE OF CONTENTS

Keynote Address   149
High-Performance Scalable Java Virtual Machines
  Vivek Sarkar

Session III: Architecture
  Chair: Sriram Vajapeyam

Session IV: Systems Software
Chair: Guang R. Gao

Keynote Address

Session V: Communication Networks
Chair: Joseph Bannister

Stability Issues in Heterogeneous and FIFO Networks under the Adversarial Queueing Model

D.K. Koukopoulos, S.E. Nikoletseas, and Paul G. Spirakis*

Computer Technology Institute (CTI) and Patras University
Riga Fereou 61, P.O.Box 1122-26 110 Patras, Greece
koukopou,nikole,spirakis@cti.gr

Abstract. In this paper, we investigate and analyze the stability properties of heterogeneous networks, which use a combination of different universally stable queueing policies for packet routing, in the Adversarial Queueing model. We interestingly prove that the combination of SIS and LIS policies, LIS and NTS policies, and LIS and FTG policies leads to instability for specific networks and injection rates that are presented. It is also proved that the combination of SIS and FTG policies, SIS and NTS policies, and FTG and NTS policies is universally stable. Furthermore, we prove that FIFO is non-stable for any $r \geq 0.749$, improving significantly the previous best known bounds of [2,10], by using new techniques for adversary construction and tight analysis of the packet flow time evolution. We also show a graph for which FIFO is stable for any adversary with injection rate $r \leq 0.1428$, and, by generalizing, we present upper bounds for stability of any network under the FIFO protocol, answering partially an open question raised by Andrews et al. in [2]. The work presented here combines new and recent results of the authors.

1 Introduction, State-of-the-Art, and Our Results

In this paper, we study the behavior of packet-switched communication networks in which packets arrive dynamically at the nodes and are routed in discrete time steps across the edges. A crucial issue that arises in such a setting is that of *stability* - will the number of packets in the system remain bounded, as the system runs for an arbitrary long period of time?

The stability problem has been investigated under various models of packet routing and in a number of overlapping areas, see for example [6,7,4,5,8,1]. The *adversarial queueing model* of Borodin et al. [3], was developed as a robust model of queueing theory in network traffic, and replaces stochastic by worst case inputs. The underlying goal is to determine whether it is feasible to prove stability even when packets are injected by an adversary, rather than by an oblivious randomized process. Adversarial Queueing Theory considers the time evolution of a packet-routing network as a game between an adversary and a

---

* The work of all the authors was partially supported by the IST Programme of the EU under contract number IST-1999-14186 (ALCOM-FT)

B. Monien, V.K. Prasanna, S. Vajapeyam (Eds.): HiPC 2001, LNCS 2228, pp. 3–14, 2001.
© Springer-Verlag Berlin Heidelberg 2001

protocol. The adversary, at each time step, may inject a set of packets at some nodes. For each packet, the adversary specifies a simple network path that the packet must traverse and, when the packet arrives to its final destination, it is absorbed by the system. If more than one packet wish to cross an edge e in the current time step, then a contention resolution protocol is used to resolve the conflict. We use the equivalent term policy or service discipline for such a protocol.

A crucial parameter of the adversary is its *injection rate*. The rate of an adversary in this model, is specified by a pair (r, b) where $b \geq 1$ is a natural number and $0 < r < 1$. The adversary must obey the following rule: "Of the packets that the adversary injects in any interval I, at most $\lceil r|I| \rceil + b$ can have paths that contain any particular edge." Such a model allows for adversarial injection of packets that are "bursty".

In this paper, we consider only *greedy* (also known as *work-conserving*) protocols - those that advance a packet across an edge e whenever there is at least one packet waiting to use e. We. also, consider discrete time units $t = 0, 1, 2, \ldots$. In each time unit a queue can serve exactly one packet.

In particular, we study the stability properties of five greedy service disciplines. The *Shortest-in-System* (SIS) policy gives priority at every queue to the packet that was injected most recently in the system. On the contrary, the *Longest-in-System* (LIS) policy gives priority at every queue to the packet that has been in the system the longest. The *Farthest-to-Go* (FTG) policy gives priority to the packet that still has to traverse the largest number of queues, while the *Nearest-to-Source* (NTS) policy gives priority to the packet that has traversed the smallest number of edges. Finally, *First-In-First-Out* (FIFO) policy gives priority to the packet that has arrived first in the queue.

We emphasize the fact that until now the stability of these protocols has been studied in isolation i.e. in networks where all queues obey a single service policy. In this work, we study for the first time, *combinations* of these policies. The issue of the stability of combinations of policies studied in this paper is highly motivated by the fact that modern communication networks (and the Internet) are heterogeneous in the sense that more than one disciplines are used on the network servers.

We also focus on FIFO, since FIFO is one of the simplest queueing policies and has been used to provide best-effort services in packet-switched networks.

In [2,3], the authors highlight a very basic algorithmic question: when is a given contention-resolution protocol stable in a given network, against a given adversary? More specifically, these questions are based on the following definition.

Definition 1. *We say that a protocol P is stable on a network G against an adversary A if there is a constant C (which may depend on G and A) such that, starting from an empty configuration, the number of packets in the system at all times is bounded by C. We say that a graph G is universally stable if every greedy protocol is stable against every adversary of rate less than 1 on G. We* *say that a protocol P is universally stable if it is stable against every adversary of rate less than 1, on every network.*

The current state-of-the-art mostly focuses on FIFO and also other greedy protocols on networks whose queues use a single service discipline. In particular, in a fundamental work in the field of stability, Andrews et al. [2] have proved that LIS, SIS, NTS, and FTG policies are universally stable, while it has been shown that FIFO, Last-In-First-Out (LIFO), Nearest-to-Go (NTG) and Farthest-from-Source (FFS) policies can be unstable. FIFO is proved unstable for a particular network for $r \geq 0.85$. Later, Goel [9], in a very interesting work, presented an even simpler network of only 3 queues for which FIFO is unstable (see Figure 1).

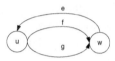

Fig. 1. Goel's network, N

In [10], the FIFO instability bound is slightly improved by our team by showing a network for which FIFO is unstable for $r \geq 0.8357$. Also, an open question raised by Andrews et al. [2] is *partially* answered in [10] by showing for the first time upper bounds on r for the stability of FIFO.

Summary of results: In this work,

1. We investigate stability properties of *combinations* of queueing disciplines.
2. We demonstrate instability of certain combinations of universally stable protocols in the same network.
3. We prove universal stability of certain other combinations of universally stable service policies in the same network.
4. We significantly improve the best previously known [2,10] FIFO instability lower bound by presenting a FIFO network where $r \geq 0.749$ leads to instability.
5. In the second result above, we interestingly show that for instability it suffices *to have only two queues* with a policy different from all other queues in the network.
6. We present here upper bounds on r for stability in FIFO networks. A first version of this result has appeared in [10].

Note that our results on combinations of queueing disciplines seem to suggest the need for an extended definition of the notion of universal stability of protocols, taking into account their sensitivity to changing the discipline of some (even a small number) of the queues in the network.

For lower bounds of the type we are interested in obtaining in this paper, it is advantageous to have an adversary that is as weak as possible. Thus, for these purposes, we say that an adversary A has rate r if for every $t \geq 1$, every interval I of t steps, and every edge e, it injects no more than $\lceil r|I| \rceil$ packets during I that require e at the time of their injection.

We will present our lower bounds for systems that start from a non-empty initial configuration. This implies instability results for systems with an empty initial configuration, by a lemma presented in Andrews et al. [2].

2   Non-stable Combinations of Universally Stable Policies

We split the time into periods. In each period we study the number of initial packets by considering corresponding time rounds. For each period, we inductively prove for instability that the number of packets in the system increases. This inductive argument can be applied repeatedly, thus showing instability. Our constructions use symmetric networks of two parts. The inductive argument has to be applied twice so that increased population appear in the same queues, since we use symmetric networks.

2.1   Mixing of LIS and FTG Is Unstable

Theorem 1. *Let* $r \geq 0.683$. *There is a network* G *that uses LIS and FTG as queueing disciplines and an adversary* A *of rate* r, *such that the* (G, A, LIS, FTG) *system is unstable, starting from a non-empty initial configuration.*

Proof. Let's consider the network in Figure 2.

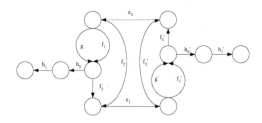

Fig. 2. A Network that uses LIS and FTG as queueing disciplines.

All the queues of this network use the LIS queueing discipline except from the queues that correspond to the edges g, g' that use the FTG queueing discipline.

Inductive Hypothesis: If at the beginning of phase j, there are s packets queued in the queues e_0, f_3' requiring to traverse edges e_0, g, f_3, then at the beginning of phase $j+1$ there will be more than s packets queued in the queues e_1, f_3 requiring to traverse edges e_1, g', f_3'.

From the inductive hypothesis, initially, there are s packets (called $S-flow$) in the queues e_0, f_3' requiring to traverse edges e_0, g, f_3.

Phase j consists of 3 rounds. The sequence of injections is as follows:

Round 1:

For s steps, the adversary injects in f_2' queue a set X of rs packets wanting to traverse edges $f_2', e_0, f_1, f_3, e_1, g', f_3'$. These packets are blocked by the $S - flow$ in queue e_0 because e_0 has LIS as queueing discipline and because for every arrival of an injected packet in e_0 there is at least one $S - flow$ packet there.

At the same time, the $S - flow$ is delayed by the adversary's injection $S_1 = rs$ packets in queue g that require to traverse queues g, h_0, h_1. This happens because g follows the FTG policy and S_1 has more edges to traverse than S. Thus, at the end of this round rs packets of S will remain in queue g because all the S_1 packets pass as they arrive in queue g and the size of S_1 is rs.

Round 2:

For the next rs steps, the adversary injects a set Y of r^2s packets in queue f_2' requiring to traverse edges $f_2', e_0, f_2, e_1, g', f_3'$. These packets are blocked by the set X in queue e_0 because e_0 uses LIS policy and the set X has arrived earlier in queue e_0.

At the same time, all X packets traverse e_0, f_1 but they are blocked in f_3 that uses LIS policy because of the remaining rs packets of $S - flow$ in g that want to traverse f_3 and are longer time in the system than X.

Round 3:

For the next $T = r^2s$ steps, the adversary injects a set Z of r^3s packets requiring to traverse edges e_1, g', f_3'. Moreover, the Y packets arrive in queue e_1 and $X_{e_1} = r^2s$ packets of X arrive in queue e_1, too. Also, $X_{f_3} = rs - r^2s$ X packets remain in queue f_3.

At the end of this round the number of packets queued in queues f_3, e_1 requiring to traverse edges e_1, g', f_3' is

$$s' = X_{e_1} + X_{f_3} + Y + Z - T = r^2s + rs - r^2s + r^2s + r^3s - r^2s = rs + r^3s$$

In order to have instability, we must have $s' > s$. Therefore, $rs + r^3s > s \Rightarrow r \geq 0.683$. □

2.2   Mixing of LIS and SIS, LIS and NTS Is Unstable

Theorem 2. *Let $r \geq 0.683$. There is a network G that uses LIS and SIS as queueing disciplines and an adversary A of rate r, such that the (G, A, LIS, SIS) system is unstable, starting from a non-empty initial configuration.*

Proof. See full paper in [11]. □

Theorem 3. *Let $r \geq 0.683$. There is a network G that uses LIS and NTS as queueing disciplines and an adversary A of rate r, such that the (G, A, LIS, NTS) system is unstable, starting from a non-empty initial configuration.*

Proof. See full paper in [11]. □

3   Universally Stable Combinations of Universally Stable Policies

Let $0 < \epsilon < 1$ be a real number. We assume that $r = 1 - \epsilon$, m is the number of network edges, and d is the length of the longest simple directed path in the network. Our techniques here are motivated by analogous techniques in Andrews et al. [2].

3.1   Mixing of FTG and NTS Is Universally Stable

Lemma 1. *Let p be a packet waiting in a queue e at time t and suppose there are currently $k - 1$ other packets in the system requiring e that have priority over p. Then p will cross e within the next $\frac{k+b}{\epsilon}$ steps if the queueing discipline in queue e is FTG or NTS.*

Proof. Let's assume that p does not cross e in the next $\frac{k+b}{\epsilon}$ steps from the moment it arrives in queue e. It will be proved that this assumption results in a contradiction. In order for packet p not to cross queue e in the next $\frac{k+b}{\epsilon}$ steps, other packets cross e in these steps (one distinct packet crosses queue e in each step because of the greedy nature of the protocol). These packets must either belong to the set of $k - 1$ packets existing in the system at time t requiring edge e that have priority over p or belong to the (at most) $(1 - \epsilon)\frac{k+b}{\epsilon} + b$ packets requiring queue e that can be injected in the system during the time period of $\frac{k+b}{\epsilon}$ steps.

Note that if FTG is used in the worst case new injections have more edges to traverse than packet p. Also, if NTS is used in the worst case new injections are nearest to their source than packet p. Therefore at most

$$ k - 1 + (1 - \epsilon)\frac{k + b}{\epsilon} + b < \frac{k + b}{\epsilon} $$

packets have priority over p during this time period. Hence, there is a contradiction. Thus, p will cross e within the next $\frac{k+b}{\epsilon}$ steps.

□

Let's define a sequence of numbers by the recurrence $k_j = \frac{mk_{j-1}+mb}{\epsilon}$, where $k_1 = \frac{mb}{\epsilon}$.

Lemma 2. *When a packet arrives at an edge e_{d-j+1} that has distance $d - j$ from the final edge on its path, there are at most $k_j - 1$ packets requiring to traverse e_j with priority over p if the used queueing policy is FTG or NTS.*

Proof. Induction will be used to prove the claim of this lemma. If queue e_{d-j+1} uses NTS then we will prove that the claim holds for $j = 1$. Let's define $X_i(t)$ the set of packets in the queue e_{d-j+1} that have crossed less than i edges at time t and let's assume that l_i is (at most at any time) the number of packets in the system that have crossed less than i edges ($i = j$). Let t be the current time and let t' be the most recent time in which $X_i(t')$ was empty. Any packet in $X_i(t)$ must either have had crossed less than $i-1$ edges at time t' or else it must have been injected after time t'. But, at every step t'' between times t' and t a packet from $X_i(t'')$ must have crossed edge e_{d-j+1}. Hence,

$$|X_i(t)| \leq l_{i-1} + (t-t')(1-\epsilon) + b - (t-t')$$
$$= l_{i-1} - (t-t')\epsilon + b$$

From the above inequality, we conclude that

$$t - t' \leq \frac{l_{i-1} + b}{\epsilon}$$

Hence, because there are m queues the total number of packets in the system that have crossed less than i edges is always at most $\frac{ml_{i-1} + mb}{\epsilon}$. We claim that, at most $\frac{ml_0 + mb}{\epsilon} - 1$ packets could have priority over p in queue $e_{d-j+1} = e_d$, which distance from the final edge on p's path is $d - 1$ when NTS is used, because in the queue e_d packet p hasn't crossed any edge. Therefore, no packet from the old ones have priority over p in queue e_d. So, $l_0 = 0$. Therefore, $\frac{b}{\epsilon} - 1 \leq k_1 - 1$ packets could have priority over p in queue e_d when NTS is used. Thus, the claim holds for $j = 1$ if NTS is used.

If queue e_{d-j+1} uses FTG then we will prove that the claim holds for $j = 1$. Let's define as $L_h(t)$ the set of packets in the queue e_{d-j+1} that still have to cross at least h edges at time t and let's assume that l_h is (at most) the number of packets in the system that still have to cross at least h edges ($h = d - j + 1$). Let t be the current time and let t' be the most recent time in which $L_h(t')$ was empty. Any packet in $L_h(t)$ must either have had at least $h + 1$ edges to cross at time t' or else it must have been injected after time t'. But, at every step t'' between times t' and t a packet from $L_h(t'')$ must have crossed edge e_{d-j+1}. Hence,

$$|L_h(t)| \leq l_{h+1} + (t-t')(1-\epsilon) + b - (t-t')$$
$$= l_{h+1} - (t-t')\epsilon + b$$

From the above inequality, we conclude that

$$t - t' \leq \frac{l_{h+1} + b}{\epsilon}$$

Hence, the number of packets in the system that have to cross h or more edges is always at most $\frac{ml_{h+1} + mb}{\epsilon}$. In the case of $j = 1$ and FTG as used policy, we claim that $k_{j=1} = l_d$ because in the queue e_d packet p will have to cross d edges. Therefore, only packets that have to cross $d + 1$ edges have priority over p in queue e_d. But, no packet have to cross $d + 1$ edges because d is the length of the longest path in the network. So, $l_{d+1} = 0$. Therefore, $\frac{0 + mb}{\epsilon} - 1 = k_1 - 1$ packets could have priority over p in queue e_1 when FTG is used. Thus, the claim holds for $j = 1$ if FTG is used.

Now suppose that the claim holds for some j. Then by lemma 5 p will arrive at the tail of e_{d-j} in at most another $\frac{k_j + b}{\epsilon}$ steps, during which at most $(1-\epsilon)\frac{k_j + b}{\epsilon} + b$

packets requiring edge e_{d-j} arrive with priority over p. Thus, when p arrives at the tail of e_{d-j} at most

$$k_j - 1 + (1 - \epsilon)\frac{k_j + b}{\epsilon} + b = \frac{k_j + b}{\epsilon} - 1 \leq \frac{mk_j + mb}{\epsilon} - 1 = k_{j+1} - 1$$

packets requiring queue e_{d-j} have priority over p and hence the claim holds.

□

Theorem 4. *The system (G, A, FTG, NTS) is stable, there are never more than k_d packets in the system and no queue contains more than $\frac{1}{\epsilon}(k_{d-1} + b)$ packets in the system where d is the length of the longest simple directed path in G.*

Proof. Assume there are $k_d + 1$ packets at some time all requiring the same edge. Then, the packet with the lowest priority of the $k_d + 1$ packets contradicts the claim of lemma 6. Combining both lemmas, a packet p takes at most $\frac{k_j + b}{\epsilon}$ steps to cross the e_{d-j+1} edge that has distance $d - j$ from the final edge on its path, once it is in the queue for this edge.   □

3.2   Mixing of SIS and FTG, SIS and NTS Is Universally Stable

Theorem 5. *The system (G, A, SIS, FTG) is stable, no queue ever contains more than k_d packets and no packet spends more than $\frac{1}{\epsilon}(db + \sum_{i=1}^{d} k_i)$ steps in the system, where d is the length of the longest simple directed path in G.*

Proof. See full paper ([11]).   □

Theorem 6. *The system (G, A, SIS, NTS) is stable, there are never more than k_d packets in the system and no queue contains more than $\frac{1}{\epsilon}(k_{d-1} + b)$ packets in the system where d is the length of the longest simple directed path in G.*

Proof. See full paper ([11]).   □

4   An Improved Lower Bound for Instability in FIFO Networks

We present a new adversary construction that significantly lowers the best previous known FIFO instability injection rate bounds to 0.749.

Theorem 7. *Let $r \geq 0.749$. There is a network G and an adversary A of rate r, such that the (G, A, FIFO) system is unstable, starting from a non-empty initial configuration.*

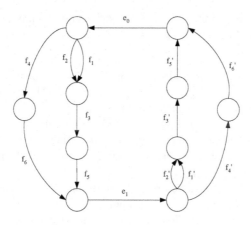

Fig. 3. Network G_1.

Proof. (Sketch, for details see [11]).

We consider the network G_1 in the above figure.

Inductive Hypothesis: If at the beginning of phase j, there are s packets queued in the queues $e_0, f'_3, f'_4, f'_5, f'_6$ requiring to traverse edges e_0, f_1, f_3, f_5, all these packets manage to depart their initial edges to the symmetric part of the network (f_1, f_3, f_5) as a continuous flow in s time steps, and the number of packets that are queued in queues f'_4, f'_6 is bigger than the number of packets queued in queues f'_3, f'_5 then at the beginning of phase $j+1$ there will be more than s packets (s' packets) queued in the queues f_3, f_5, f_4, f_6, e_1 requiring to traverse edges e_1, f'_1, f'_3, f'_5, all of which will be able to depart their initial edges to the symmetric part of the network (f'_1, f'_3, f'_5) in s' time steps as a continuous flow and the number of packets that are queued in queues f_4, f_6 is bigger than the number of packets queued in queues f_3, f_5.

From the inductive hypothesis, initially, there are s packets (called $S - flow$) in the queues $e_0, f'_3, f'_4, f'_5, f'_6$ requiring to traverse edges e_0, f_1, f_3, f_5.

Phase j consists of 3 rounds. The sequence of injections is as follows:

Round 1:

For s steps, the adversary injects in f'_4 queue a set X of rs packets wanting to traverse edges $f'_4, f'_6, e_0, f_2, f_3, f_5, e_1, f'_1, f'_3, f'_5$. These packets are blocked by the $S - flow$.

At the same time, the $S - flow$ is delayed by the adversary's single injections $S_1 = rs$ in queue f_1. The S_1 packets get mixed with the packets in the $S - flow$.

Notice that because of the FIFO policy, the packets of S, S_1 mix in consecutive blocks according to their initial proportion of their sizes (fair mixing property). Since $|S| = s$ and $|S_1| = rs$, these proportions are $\frac{1}{r+1}$ and $\frac{r}{r+1}$, respectively. Thus, during the s steps of round 1, the packets of S, S_1, which cross f_1 are, respectively, $s\frac{1}{r+1} = \frac{s}{r+1}$ and $s\frac{r}{r+1} = \frac{rs}{r+1}$.

Therefore, the remaining packets of each type in queue f_1 are for S_{rem}: $s - \frac{s}{r+1} = \frac{rs}{r+1}$ and for $S_{1,rem}$: $rs - \frac{rs}{r+1} = \frac{r^2 s}{r+1}$

Round 2:

For the next rs steps, the adversary injects a set Y of $r^2 s$ packets requiring edges $f_4', f_6', e_0, f_4, f_6, e_1, f_1', f_3', f_5'$. These packets are blocked by the set X. At the same time, the adversary pushes a set S_2 of single injections in the queue f_2, where $|S_2| = r^2 s$, a set S_3 of single injections in the queue f_3, where $|S_3| = r^2 s$ and a set S_4 of single injections in the queue f_5, where $|S_4| = r^2 s$.

By the proportions of the packet stream sizes, we can estimamte the remaining packets of each type, X_{rem,f_2} and S_{2,rem,f_2}.

Note that in queue f_1, there are the remaining $S - flow$ and the remaining $S_1 - flow$ packets. Since their total number is rs (which is equal to the duration of the round), the $S_{1,rem} - flow$ does not delay the $S_{rem} - flow$. Note also that, because the $S_{1,rem}$ packets are absorbed after they pass only f_1, only the S_{rem} packets require edge f_3. As a result the stream arriving from f_1 to f_3 contains empty spaces at the positions of the $S_{1,rem}$ packets. However, these empty spaces are uniformly spread in the system for the duration of the time period. Therefore, during round 2, three different flows of packets arrive to the f_3 queue: the $X_{pass,f_2} - flow$, which mixes with $S_{2,pass,f_2} - flow$, the $S_{rem} - flow$, and the S_3 single-injected packets.

The total number of packets in the three flows is $|T| = \frac{r^3 s + r^2 s + 2rs}{r+1}$. We can estimate the remaining packets in queue f_3 from each flow at the end of round 2. Note that during round 2 the stream arriving to f_5 contains 3 different flows of packets: the S_4 single-injected packets, the $S_{pass,f_3} - flow$ and the $X_{pass,f_3} - flow$. Note also that the S_3 packets that traverse f_3 are absorbed after they pass f_3.

The total number of packets in the three flows is $|T'| = \frac{r^4 s + r^3 s + 2r^2 s + 2rs}{r^2 + r + 2}$. The remaining packets in f_5 queue from each flow at the end of round 2 are: X_{pass,f_3}, S_{pass,f_3} and S_4.

Round 3:

For the next $|T_{round-3}| = r^2 s$ steps, the adversary injects a set S_5 of $r^3 s$ packets requiring to traverse edge f_4 (single injections). The set S_5 is mixed with the set Y. Call the packets Y, S_5 that remain in queue f_4 respectively, Y_{rem} and $S_{5,rem}$.

Furthermore, the adversary injects a set Z of $r^3 s$ packets requiring to traverse edges $f_6, e_1, f_1', f_3', f_5'$. The set Z is mixed with the set Y_{pass}. The remaining packets in f_6 queue are: Y_{rem,f_6} and Z_{rem,f_6}.

Note that the total number of packets in queue f_5 at the end of round 2 is $|T_1| = \frac{r^7 s + r^6 s + 3r^5 s + 3r^4 s + 2r^3 s + 2r^2 s}{(r^2 + r + 2)(r^3 + r^2 + 2r + 2)}$. However, it has been proved by MATLAB that $T_1 < r^2 s, \forall r > 0$. Therefore, the remaining time is $t_{rem} = r^2 s - T_1$. In t_{rem} steps the number of X_{rem,f_3} packets that traverses f_5, that is equal with the number of S_{rem,f_3} packets that absorbed when they pass f_5 is $|X_{pass,f_3,f_5}| = |S_{absorb,f_3,f_5}|$. Note that the total number of packets in queue f_3 at the beginning of round 3 is $|T_2| = \frac{r^5 s + r^4 s + 3r^3 s + r^2 s + 2rs}{(r+1)(r^2 + r + 2)}$. However, $r^2 s < |T_2|, \forall r < 1$ (it has been proved by MATLAB). Thus, a number of $X_{rem,f_3}, S_{rem,f_3}, S_{3,rem}$ packets remain in f_3. This number is $|T_3| = |T_2| - r^2 s$. From this number of packets, the number of packets that belong to $S_{3,rem}$ is $|S_{3,rem,f_3}| = |T_3| \frac{r^2+r}{r^2+r+2}$.

Also, the total number of packets that are in queue f_2 at the end of round 2 is $|T_3| = r^2 s$. Thus, all the $X - flow$ packets in queue f_2 are queued in queue f_3.

At the end of this round the number of packets that are in queues f_3, f_4, f_5, f_6, e_1 requiring to traverse the edges e_1, f_1', f_3', f_5' is

$$s' = r^3 s + r^2 s + \frac{r^2 s}{r+1} + \frac{r^3 s + rs}{(r+1)(r^2+r+2)} + \frac{r^4 s + r^2 s}{(r^2+r+2)(r^3+r^2+2r+2)} - r^2 s$$

For instability, we must have $s' > s$. Therefore, $r \geq 0.749$, by MATLAB.

In order to conclude the proof we should also show that $Q(f_3) + Q(f_5) \leq Q(f_4) + Q(f_6)$. But,

$$Q(f_3) + Q(f_5) = |X_{rem,f_2}| + (|X_{rem,f_3}| - |X_{pass,f_3,f_5}|) + (|S_{rem,f_3}| - |S_{absorb,f_3,f_5}|) + |S_{3,rem,f_3}| \text{ and}$$

$$Q(f_4) + Q(f_6) = (|Y_{rem}| + |S_{5,rem}|) + (|Y_{rem,f_6}| + |Z_{rem,f_6}|)$$

Using MATLAB, $Q(f_3) + Q(f_5) \leq Q(f_4) + Q(f_6) \implies r \geq 0.743$.

Notice that we have, till now, managed to reproduce the inductive hypothesis in queues f_3, f_5, f_4, f_6, e_1 but with some flows (in particular in queues f_4, f_3, f_5) having empty spaces (packets that don't want to traverse edges e_1, f_1', f_3', f_5'). In order for the induction step to work we must show that all these packets in these queues will manage to depart to the symmetric part of the network (f_1', f_3', f_5') in the s' time steps as a continuous flow. We now show this.

In order to show that, we should estimate the number of Y_{rem} packets that manage to traverse queue f_4 within the time steps the initial packets in queue f_6 need to traverse edges f_6, e_1.

The number of time steps needed for all the initial packets in queue f_6 to traverse edges f_6, e_1 is

$$|T_4| = |Z_{rem,f_6}| + |Y_{rem,f_6}| + (|Z_{pass,f_6}| + |Y_{pass,f_6}| + |X_{pass,f_3}| + |X_{pass,f_3,f_5}| - T_{round-3}$$

In T_4 time steps, the number of Y_{rem} packets that traverse f_4 is $A = |T_4| \frac{1}{r+1}$, and the number of $S_{5,rem}$ packets that traverse f_4 is $B = |T_4| \frac{r}{r+1}$, where $\frac{1}{r+1}, \frac{r}{r+1}$ are the mixing proportions of $Y_{rem}, S_{5,rem}$ correspondingly.

We should show that $A \geq |Y_{rem}| - A + |S_{5,rem}| - B$. It was proved by using MATLAB that this holds for $r \geq 0$.

Taking the maximum of the three values $\max\{0.749, 0.743, 0\} = 0.749$, the network G is unstable for $r \geq 0.749$. This concludes our proof.   □

5   An Upper Bound for the Stability in FIFO Networks

Let us consider the network N in Figure 1. We prove here, that given any initial packet configuration for N, and for any adversary A with injection rate $r \leq r_0$, for a given r_0 to be computed, the number of packets in the system remains bounded. So we can conclude that FIFO is stable for N, under any adversary with rate $r \leq r_0$.

Theorem 8. *For the network N, given any initial configuration, FIFO is stable against any adversary with injection rate $0 < r \leq 0.1428$.*

Proof. See full paper ([11]).   □

The previous argument, can be extended to work in full generality, for any network N with k edges, maximum in-degree α, and maximum directed path length β.

Theorem 9. *For any network N, given any initial configuration, there exist an $0 < r_N < 1$, such that FIFO is stable with respect any adversary with injection rate $r \leq r_N$, where r_N is independent of the initial configuration and it is a function of the maximum in-degree, the maximum directed path length and the number of edges.*

Proof. See full paper ([11]).   □

References

[1]   M. Andrews. Instability of FIFO in session-oriented networks. In 11th. ACM-SIAM Symposium on Discrete Algorithms (SODA 2000), pages 440-447, 2000.

[2]   M. Andrews, B. Awerbuch, A. Fernandez, J. Kleinberg, T. Leighton, and Z. Liu. Universal stability results for greedy contention-resolution protocols. Journal of the ACM, 48(1):39-69, January 2001.

[3]   A. Borodin, J. Kleinberg, P. Raghavan, M. Sudan, and D. Williamson. Adversarial queueing theory. Journal of the ACM, 48(1):13-38, January 2001.

[4]   M. Bramson. Instability of FIFO queuing networks. Annals of Applied Probability, 4:414-431, 1994.

[5]   M. Bramson. Instability of FIFO queueing networks with quick service times. Annals of Applied Probability, 4(3):693-718, 1994.

[6]   R. Cruz. A calculus for network delay. Part I: Network elements in isolation. IEEE Transactions on Information Theory, 37:114-131), 1991.

[7]   R. Cruz. A calculus for network delay. Part II: Network analysis. IEEE Transactions on Information Theory, 37:132-141, 1991.

[8]   D. Gamarnik. Stability of adversarial queues via models. In 39th. IEEE Symposium on Foundations of Computer Science (FOCS 1998), pages 60-76,1998.

[9]   A. Goel. Stability of networks and protocols in the adversarial queueing model for packet routing. In the 10th ACM-SIAM Symposium on Discrete Algorithms (SODA 1999) (short abstract), pages 315-324, 1999. Accepted for publication in Networks.

[10] J. Diaz, D. Koukopoulos, S. Nikoletseas, M. Serna, P. Spirakis and D. Thilikos. Stability and non-stability of the FIFO protocol. In Proc. of the 13th Annual ACM Symposium on Parallel Algorithms and Architectures (SPAA 2001), pages 48-52, Crete, Greece, July 2001.

[11] D. Koukopoulos, S. Nikoletseas and P. Spirakis. Stability Issues in Heterogeneous and FIFO Networks under the Adversarial Queueing Model (full paper), CTI Technical Report, July 2001, available at www.cti.gr/RD1/nikole/stability.ps

Mesh Algorithms for Multiplication and Division*
(Preliminary Version)

S. Rao Kosaraju

Department of Computer Science
The Johns Hopkins University
Baltimore, Maryland 21218

Abstract. We consider the implementation of multiplication and division operations on one and two dimensional mesh of processors. We develop an $O(\sqrt{n})$ step 2-dim mesh algorithm for multiplying two n-bit numbers. The algorithm is simple and does not rely upon discrete fourier transforms. We also develop an $O(n)$ step 1-dim mesh algorithm for dividing a $2n$-bit number by an n-bit number. This algorithm appears to be quite practical.

1 Introduction

We develop multiplication and division algorithms for multi-dimensional mesh architecture. In a d-dimensional mesh organization of n processors, the processors form an $n^{1/d} \times n^{1/d} \times \cdots \times n^{1/d}$ regular d-dimensional array. Each processor of the mesh can store $O(1)$ bits of memory. In a single step, each processor can receive information from each of its $2d$ neighbors and can perform one step of computation, changing the contents of its memory. This architecture has received considerable attention since algorithms developed for it can be easily implemented on standard hardware such as Field Programmable Gate Arrays (FPGA).

In section 2 we develop an $O(\sqrt{n})$ step algorithm for the multiplication of two n-bit numbers on a 2-dim mesh. In section 3 we develop an $O(n)$ step algorithm for the division of a $2n$-bit number by an n-bit number on a 1-dim mesh. This algorithm is a simple adaptation of the traditional method for division. In this implementation, subtractions are performed starting from the most significant position. The algorithm appears to be quite practical, and we plan to implement it on an FPGA. We have not been able to adapt our division algorithm as an $O(\sqrt{n})$ step algorithm for 2-dim meshes.

2 Multiplication

Given two n-bit numbers x and y, we develop an algorithm for computing their product $x \times y$ on a 2-dim $\sqrt{n} \times \sqrt{n}$ mesh in $O(\sqrt{n})$ steps. Initially each processor

* Supported by NSF Grant CCR-9821058 and ARO Grant DAAH04-96-1-0013

B. Monien, V.K. Prasanna, S. Vajapeyam (Eds.): HiPC 2001, LNCS 2228, pp. 17–23, 2001.
© Springer-Verlag Berlin Heidelberg 2001

stores two bits, one from each of x and y. We don't specify the exact mapping of these bits into the processors since our algorithm is described at a very high level, and, in addition, the ultimate speed of the algorithm is quite independent of any minor variations to the order of bit storage. The $2n$ bit result will be stored 2 bits per processor. During the computation phase, each processor makes use of $O(1)$ bits of additional storage.

2.1   Basic Approach

We first describe a simple procedure that fails. This procedure motivates the algorithm developed in the next subsection. Split each of x and y into two $n/2$-bit numbers:

$x \ = \ x_1 x_0$, and

$y \ = \ y_1 y_0$.

There are many known ways of expressing the product xy in terms of three $n/2$-bit multiplications. One scheme is given below. Let

$u \ = \ (x_1 + x_0)(y_1 + y_0)$,

$v \ = \ x_1 y_1$, and

$w \ = \ x_0 y_0$.

Then the product is given by

$xy \ = \ v2^n + (u - v - w)2^{n/2} + w$.

Implementation of this scheme on a 2-dim mesh runs into serious problems. Observe that two m-bit numbers can be added in $O(\sqrt{m})$ time if the numbers come appropriately stored within $O(\sqrt{m} \times \sqrt{m})$ mesh space (i.e. the corresponding processors). Thus on can try to implement the above scheme by initially computing $x_1 + x_0$ and $y_1 + y_0$ each in $O(\sqrt{n/2})$ time and then computing u, v, and w recursively and in parallel. Finally it appears that xy can be computed in $O(\sqrt{n})$ time. However, there is a fatal flaw in this argument. If we assume that an m-bit multiplication requires $S(m)$ mesh space, the recurrence relationship for the above implementation is as follows:

$S(n) \ = \ 3S(n/2)$, and

$S(1) \ = \ 1$.

Hence, $S(n) = O(n^{\log_2 3})$. Consequently, the resulting speed is well above $O(\sqrt{n})$. (The speed will be $O(n^{\log_2 1.5})$). Any alternative approach in which u, v, and w are not computed simultaneously will not lead to $O(\sqrt{n})$ speed.

We can easily observe that a simple extension of this approach to one that splits x and y into 3 or any fixed number of parts will also fail. Consequently, researchers had resorted to the discrete fourier transform approach [10].

Note that a $\sqrt{n} \times \sqrt{n}$ mesh operating for $\sqrt{n}$ steps performs a total of $n^{1.5}$ individual steps. Since the available number of steps is significantly more than $O(n \ polylog \ n)$, the steps needed to multiply two n-bit numbers by a sequential algorithm, it appears plausible that one should be able to design an $O(\sqrt{n})$ step multiplier without relying upon discrete fourier transforms. Such direct algorithms could ultimately lead to more practical hardware-based implementations. In the following we develop such a direct approach. However, our algorithm is not simple enough for hardware implementation. We believe that this algorithm opens up a new avenue of attack.

2.2   Our Scheme

Split each of x and y into 64 $n/64$-bit numbers:

$x = x_{63}x_{62}\cdots x_1 x_0$, and

$y = y_{63}y_{62}\cdots y_1 y_0$.

It is well known that the product xy can be expressed in terms of $2(64)-1=127$ $n/64$-bit products. The underlying algorithm can be given schematically as follows:

$u_i = i^{th}$ product, for $i = 0, \cdots, 126$, and

$xy = f_{126}()2^{126/64\ n} + f_{125}()2^{125/64\ n} + \cdots + f_1()2^{n/64} + f_0()$,

where each f_i is a weighted sum of a subset of u_i's in which the weights are constants. Note that after recursively computing all the u_i's, each f_i can be computed in $O(\sqrt{n})$ steps. Now we present the key details of the algorithm.

We partition the $\sqrt{n} \times \sqrt{n}$ space into 64 subsquares, each occupying $\sqrt{n}/8 \times \sqrt{n}/8$ area. We then compute the u_i's in 3 phases as follows.

In the first phase, we generate the multiplicand and the multiplier for each of $u_0, u_1, \cdots, u_{42}$ and then move them into 43 subsquares (the multiplicand and the multiplier for any u_i go to the same subsquare) while storing the original inputs x and y in the remaining 21 subsquares. This can be accomplished in $O(\sqrt{n})$ time. Note that the implementation will not need more than a constant number of bits of storage per location. Assume that c_1 bits are needed for this phase.

We recursively and in parallel compute each of $u_0, u_1, \cdots, u_{42}$ in its subsquare. We then move these products into the 21 subsquares occupied by x and y. This requires an additional constant number of bits, say c_2, per location.

In the second phase, we similarly compute the next 43 products $u_{43}, \cdots, u_{85}$ and store the products in the 21 subsquares occupied by x and y. Let this require an additional c_3 bits per location.

In the third phase, we compute the remaining 41 products $u_{86}, \cdots, u_{126}$.

Finally, the product xy is computed in an additional $O(\sqrt{n})$ steps by making use of all the u_i's. Observe that the mesh space used for performing the recursion doesn't store any information being carried forward. (This is the reason for storing the two input numbers in a separate area.) Hence the recursive call requires $O(1)$ bits of storage per location since $c_1 + c_2 + c_3$ is $O(1)$. (Note that if we carry forward even 1 bit of storage in the space where the recursions are performed, the storage requirement jumps to $\Omega(\log n)$ bits per location.) Consequently, the complete algorithm makes use of only $O(1)$ bits of storage per location.

If we denote the time needed to multiply two n-bit numbers by $T(n)$, the recurrence relationship for the above implementation is given by

$T(n) = 3T(n/64) + O(\sqrt{n})$, and

$T(1) = 1$.

Hence $T(n) = O(\sqrt{n})$.

3   Division

In this section we outline a method for implementing the division of a $2n$-bit number x by an n-bit number y on a 1-dim mesh of n processors. The result z is rounded to n bits. Each processor initially stores 2 bits of x and 1 bit of y, and each processor will store 1 bit of the result z at the end. As before, during the computation, each processor can make use of $O(1)$ bits of storage. The algorithm runs in $O(n)$ steps. Multiplying two n-bit numbers on a 1-dim mesh in $O(n)$ steps is fairly simple [17]; but division is significantly more complex.

First we develop a redundant notation for representing the intermediate results. Other redundant notations such as the Booth's scheme [1] are well known. The redundant representation permits us to start the division from the most significant bit without performing a complete subtraction. We specify the division algorithm based on the redundant representation. However, the resulting pattern of communication between different locations of the mesh is complex. In the final version, we transform the resulting algorithm into an $O(n)$ time implementation on 1-dim mesh. Finally, in $O(n)$ steps, the result in redundant notation can be easily transformed into the standard binary notation.

In the next three subsections we specify a simple redundant positional number representation, and show that arithmetic subtraction, addition and division can be performed starting from the most significant position. (Multiplication can also be performed starting from the most significant position, but we don't need it here.) We assume that, in each case, the second operand is in standard binary notation. This simplifies the presentation and is sufficient for our purposes. We also assume specific initialization conditions.

3.1   Redundant Notation

The positional weights are as in the standard binary representation, and each position permits values 2, 1, 0, -1, and -2 instead of the standard 0,1 values. For example, $200(-1)1$ represents $2(2^4) + 0(2^3) + 0(2^2) - 1(2^1) + 1(2^0)$. With a slight abuse of notation, we call the value in a position a *bit*. Note that $200(-1)1$ has 5 bits.

3.2   Subtraction

Assume that two m-bit numbers $A = a_{m-1}a_{m-2}\cdots a_1a_0$, and $B = b_{m-1}b_{m-2}\cdots b_1b_0$, with A in the redundant representation and B in the standard binary representation are given. In addition, assume that $b_{m-1} = 1$, $a_{m-1} \geq 1$, and $a_{m-2} - b_{m-2} \geq -1$. We compute $A - B = r_{m-1}r_{m-2}\cdots r_1r_0$ in the redundant notation starting from the most significant position. During the subtraction, the resulting bit r_i in position i depends upon the bits in positions i and $i-1$ and the "carry/borrow" adjusted bit a_i' from position $i+1$. The value of a_i' is the adjusted value of a_i taking into consideration the "borrowing" or "carrying" needed due to the subtraction performed upto position $i+1$. In addition, position i generates a carry/borrow adjusted bit a_{i-1}', and transmits it to position $i-1$.

Initially we set $a'_{m-1} = a_{m-1}$, and after computing $r_{m-1}, r_{m-2}, \cdots, r_{i+1}$, we maintain a value a'_i such that $a'_i - b_i \geq -1$. Note that this inequality holds initially since $a'_{m-1} - b_{m-1} \in \{0, 1\}$. We compute the values of r_i and a'_{i-1} from a'_i, b_i, a_{i-1}, and b_{i-1} as follows:

Case 1: $|a_{i-1} - b_{i-1}| \leq 1$

Then let $r_i = a'_i - b_i$, and $a'_{i-1} = a_{i-1}$. Note that the value of r_i requires no borrowing from or carrying to the $i - 1$st position. Hence no adjustment to a_{i-1} is needed.

Case 2: $a_{i-1} - b_{i-1} \leq -2$

This happens when $a_{i-1} = -2$ and $b_{i-1} \in \{0, 1\}$, or $a_{i-1} = -1$ and $b_{i-1} = 1$. Then let $r_i = a'_i - b_i - 1$, and $a'_{i-1} = a_{i-1} + 2$. Here a carry of 2 is transmitted to the $i - 1$st position.

Case 3: $a_{i-1} - b_{i-1} = 2$

This happens when $a_{i-1} = 2$ and $b_{i-1} = 0$. Then let $r_i = a'_i - b_i$, and $a'_{i-1} = a_{i-1}$. In this case, no carry is transmitted to the $i - 1$st position.

It can be easily verified that in every one of the cases the inequality $a'_{i-1} - b_{i-1} \geq -1$ holds.

If the above algorithm is applied for computing 110(-1)121 - 1011001, the result will be 01(-2)0120.

3.3   Addition

The addition of two m-bit numbers A and B in which A is in the redundant notation and B is in the standard binary notation is very similar to the above algorithm. Here we assume that $b_{m-1} = 1$, $a_{m-1} \leq -1$, and $a_{m-2} - b_{m-2} \leq 1$. During the addition process, we maintain the inequality $a'_i - b_i \leq 1$. The various cases are given below.

Case 1: $|a_{i-1} + b_{i-1}| \leq 1$

Then let $r_i = a'_i + b_i$, and $a'_{i-1} = a_{i-1}$.

Case 2: $a_{i-1} + b_{i-1} \geq 2$

This happens when $a_{i-1} = 2$ and $b_{i-1} \in \{0, 1\}$, or $a_{i-1} = 1$ and $b_{i-1} = 1$. Then let $r_i = a'_i + b_i + 1$, and $a'_{i-1} = a_{i-1} - 2$.

Case 3: $a_{i-1} + b_{i-1} = -2$

This happens when $a_{i-1} = -2$ and $b_{i-1} = 0$. Then let $r_i = a'_i + b_i$, and $a'_{i-1} = a_{i-1}$.

In the next subsection we outline our overall division algorithm. In the algorithm, we make use of the following adjustment operation at some position, say at position k.

Adjustment

If $x_k \geq 1$ and $x_{k-1} = -2$, then set $x_k = x_k - 1$ and $x_{k-1} = 0$. If $x_k \leq -1$ and $x_{k-1} = 2$, then set $x_k = x_k + 1$ and $x_{k-1} = 0$.

Note that this adjustment doesn't change the value of x, but insures that if $x_k \geq 1$ then $x_{k-1} \geq -1$, and if $x_k \leq -1$ then $x_{k-1} \leq 1$.

3.4   Overall Division Algorithm

In this subsection we outline a method for dividing a $2n$-bit number $x = x_{2n-1}$ $x_{2n-2} \cdots x_1 x_0$ by an n-bit number $y = y_{n-1} y_{n-2} \cdots y_1 y_0$ - both represented in standard binary notation. The result will be in the redundant notation. We first specify the algorithm without any implementation considerations. Without any loss of generality, we assume that the most significant bits of x and y are $1's$. Let the result of the division be $z = z_{n-1} z_{n-2} \cdots z_1 z_0$. The n bits of z are computed in n iterations. (In fact an additional iteration is needed for proper rounding of the value of z.)

In the first iteration, we let $z_{n-1} = 1$, and subtract y from $x_{2n-1} \cdots x_{n+1} x_n$; the resulting n-bit redundant number replaces the previous values of $x_{2n-1} \cdots$ $x_{n+1} x_n$. This results in $x_{2n-1} = 0$. We then advance to the next iteration, iteration 2.

In the ith iteration, we first initialize z_{n-i} to 0. We then perform Adjustment to position $2n - i$ and then apply one of the following cases.

Case 1: $x_{2n-i} = 0$

This completes the iteration, resulting in a final value of z_{n-i}.

Case 2: $x_{2n-i} > 0$

In this case $x_{2n-i} \in \{1, 2\}$. We subtract y from $x_{2n-i} x_{2n-i-1} \cdots x_{n-i+2} x_{n-i+1}$, and add 1 to z_{n-i}. Once again the resulting bits replace the old values of the corresponding x bits. Then the ith iteration is repeated.

Case 3: $x_{2n-i} < 0$

In this case $x_{2n-i} \in \{-1, -2\}$. We add y to $x_{2n-i} \cdots x_{n-i+1}$, and subtract 1 from z_{n-i}. Once again the resulting bits replace the old x bits. Then the ith iteration is repeated.

We can show that within 2 repetitions of the above cases, x_{2n-i} becomes 0, resulting in the completion of the iteration. If case 2, respectively case 3, is repeated q times, $q \in \{1, 2\}$, then z_{n-i} will have a final value of q, respectively $-q$.

It is somewhat nontrivial to implement the above algorithm as an $O(n)$ step algorithm on a 1-dim mesh. The complete details will be presented in the final version.

In addition, we will generalize our approach and derive general simulation results for implementing certain broadcasting capabilities of meshes. In this generalization, the regular mesh of processors has a movable *head* which at any instant accesses a location of the mesh. In a step the head can broadcast the information it accesses to all the processors in the array, and then can move to any adjacent location. We will establish that such a generalized mesh of processors can be simulated without any slowdown by the standard mesh of processors. This result generalizes the corresponding result of Seiferas [14] for a fixed head mesh.

References

1. B. Parhami, *Computer Arithmetic: Algorithms and Hardware Design*, Oxford University Press, New York, 2000.
2. M.D. Ercegovac, L. Imbert, D.W. Matula, J.-M. Muller, and G. Wei, *Improving Goldschmidt division, square root, and square root reciprocal*, IEEE Trans. on Computers, July 2000, pp. 759-762.
3. E. Savas, and C.K. Koc, *The Montogomery modular inverse - revisited*, IEEE Trans. on Computers, July 2000, pp. 763-766.
4. J.H. Reif (ed), *Synthesis of parallel algorithms*, Morgan Kaufmann Publishers, San Mateo, CA, 1993.
5. C.D. Walter, *Systolic modular multiplications*, IEEE Trans. on Computers, March 1993, pp. 376-378.
6. S.E. Eldridge, and C.D. Walter, *Hardware implementation of Montgomery's modular multiplication algorithm*, IEEE Trans. on Computers, July 1993, pp. 693-699.
7. A. Aggarwal, J.L. Carter, and S.R. Kosaraju, *Optimal tradeoffs for addition on systolic arrays*, Algorithmica, 1991, pp. 49-71.
8. N. Shankar, and V. Ramachandran, *Efficient parallel circuits and algorithms for division*, Inf. Processing Letters, 1988, pp. 307-313.
9. H.J. Sips, *Bit-sequential arithmetic for parallel processors*, IEEE Trans. on Computers, Jan 1984, pp. 7-20.
10. K. Mehlhorn, and F.P. Preparata, *Area-time optimal VLSI integer multiplier with minimum computation time*, Information and Control, 1983, pp. 137-156.
11. H.T. Kung, and C. Leiserson, *Algorithms for VLSI Processor Arrays*, Proc. Symp. on Sparse Matrix Computations, Knoxville, TN, 1981.
12. C.E. Leiserson, and J.B. Saxe, *Optimizing synchronous systems*, Proc. of IEEE FOCS, 1981, pp. 23-36.
13. D.P. Agrawal, *High-speed arithmetic arrays*, IEEE Trans. on Computers, March 1979, pp. 215-224.
14. J. Seiferas, *Iterative arrays with direct central control*, Acta Informatica, 1977, pp. 177-192.
15. K.S. Trivedi, and M.D. Ercegovac, *On-line algorithms for division and multiplication*, IEEE Trans. on Computers, July 1977, pp. 681-687.
16. S.R. Kosaraju, *Speed of recognition of context-free languages by array automata*, SIAM J. Comput., Sept. 1975, pp. 331-340.
17. A. Atrubin, *One-dimensional real time iterative multiplier*, IEEE Trans. on Computers, 1965, pp. 394-399.

Compact Routing in Directed Networks with Stretch Factor of Two

Punit Chandra and Ajay D. Kshemkalyani

Dept. of Computer Science
University of Illinois at Chicago, Chicago, IL 60607-7053, USA.
pchandra@eecs.uic.edu, ajayk@cs.uic.edu

Abstract. This paper presents a compact routing algorithm with stretch less than 3 for directed networks. Although for stretch less than 3, the lower bound for the total routing information in the network is $\Omega(n^2)$ bits, it is still worth examining to determine the best possible saving in space. The routing algorithm uses header size of $4 \log n$ and provides round-trip stretch factor of 2, while bounding the local space by $[n - \sqrt{n}(\sqrt{1 - 7/4n}) - 1/2] \log n$. These results for stretch less than 3 for directed networks match those for undirected networks.

1 Introduction

1.1 Background

As high-speed networking gains popularity, the routing bottleneck shifts from the propagation delay to the route decision function. Therefore, simple routing schemes implemented in hardware will be preferred in practice. The decision function is bounded by the size of the routing table. The bigger the routing table, the more time it takes to determine the next hop. Also, it is desirable that the routing table be kept in fast memory such as cache. Furthermore, it is not desirable for the memory requirements to grow fast with the size of the network, since it means adding more hardware to all the routers in the network.

Compact routing addresses this problem by decreasing the table size and hence the decision time. The trade-off involved here is the communication cost of passing messages between a pair of nodes.

1.2 Existing Work

Early work on compact routing focused on routing schemes for special networks such as rings, trees [14], and grids [9,10]. Peleg and Upfal [11,12] were the first to construct an universal compact routing scheme. An universal routing scheme is an algorithm which generates a routing scheme for every given network, that is, for all sorts of topologies. A trivial version of an universal routing strategy stores at each node a routing table which specifies an output port for each destination. Although this routing strategy can guarantee routing through the shortest path,

B. Monien, V.K. Prasanna, S. Vajapeyam (Eds.): HiPC 2001, LNCS 2228, pp. 24–35, 2001.

each router has to store locally $O(n \log d)$ bits of memory, where d is the degree of the node and n is the total number of nodes in the network. As the network grows in size, it becomes necessary to reduce the amount of memory kept at each router. But there is a trade-off involved here between the memory and the stretch factor, denoted here by s and defined as the maximum ratio between the length of the path traversed by a message and that of the shortest path between its source and destination. The lower bounds for routing information in the network are:

For stretch factor	$s \geq 1$	$\Omega(n^{1+1/(2s+4)})$ bits	[13]
For stretch factor	$s < 3$	$\Omega(n^2)$ bits	[4,6]
For stretch factor	$s = 1$	$\Theta(n^2 \log n)$ bits (optimal)	[5]

There are various algorithms for compact routing. The algorithm in [1] achives a stretch of 3 while using $O(n^{3/2} \log n)$ bits in total, but some individual nodes use $O(n \log n)$ bits. A recent algorithm of [7] uses interval routing for undirected graphs to achieve a stretch of 5 while using local routing table size of $O(n^{1/2} \log n)$ bits. More recently, [2] presented an hierarchical scheme which uses a local routing table size of $O(n^{2/3} \log^{4/3} n)$ and guarantees a stretch of 3.

Most of the recent work done in routing has been for undirected graphs. Directed networks are much harder than the undirected case because a deviation from the exact shortest route has serious consequences. That is, if we take a walk in the wrong direction for a few steps from u, in an undirected graph it is easier to retrace our steps, but in a directed graph it might be extremely hard to get back to u. So to overcome this difficulty, [3] used round-trip distance. The round-trip distance is a measure of how easy it is to get back from an exploratory trip. They achieved a roundtrip stretch of $2^{l+1} - 1$ using $O(l \log n)$ size addresses and a $O(ln^{1/(l+1)})$ sized routing table on the average on each node, where l is an integer greater than 0 and denotes the level of landmark.

Although designing routing algorithms for stretch of less than 3 does not seem so attractive because of the lower bound of $\Omega(n^2)$ as shown in [4,6], it is still worth examining to determine the best possible saving in space. Recently, [8] addressed the problem of compact routing for stretch factor of less than 3 in undirected networks. It presented an algorithm whose local table size is $(n - \sqrt{n} + 2) \log n$ for a stretch factor of 2. In this paper, we propose a simple routing scheme for directed networks. We use a scheme similar to [8], coupled with the concept of round-trip distance which makes it a harder problem. The algorithm uses node names of size $4 \log n$ and has a round-trip stretch factor of 2, while bounding the local space by $[n - \sqrt{n}(\sqrt{1 - 7/4n}) - 1/2] \log n$.

2   Some Definitions Used in Compact Routing

Definition 1. *Routing scheme (R)*

Routing scheme (R) is a distributed algorithm that consists of distributed data structures in the network and a delivery protocol.

Definition 2. *Round-trip distance*

Round-trip distance between any two nodes, say x and y, in a network is the shortest distance from x to y and back. Formally,

$$RT(x,y) = d(x,y) + d(y,x)$$

where $d(x,y)$ represents the shortest distance between x and y.

Definition 3. *Round-trip distance for a routing scheme (R)*

Round-trip distance for a routing scheme (R) between any two nodes, say x and y, is the distance of the path taken by a packet going from x to y and back, in accordance with the routing algorithm (R). Mathematically,

$$RT(R,x,y) = d(R,x,y) + d(R,y,x)$$

where $d(R,x,y)$ represents the distance of the path taken by a packet going from x to y in accordance with the algorithm.

Definition 4. *Round-trip stretch of a routing scheme (R)*

Round-trip stretch of a routing scheme (R) is defined as follows

$$\text{Round-trip stretch (R)} = max\frac{d(R,x,y) + d(R,y,x)}{d(x,y) + d(y,x)}$$

3   Overview of the Algorithm

The initial step of the algorithm is to construct two sets U and W from the set of the network nodes V such that U and W satisfy the following conditions: $V = U \cup W$, $U \cap W = \phi$ and $|U| = |W| = n/2$. The sets are constructed in an incremental fashion. At each step i, u_i and w_i are added to U and W respectively, where $1 \leq i \leq n/2$ and $u_i, w_i \in V - \{U + W\}$. The pair (u_i, w_i) is chosen such that $RT(u_i, w_i)$, the round-trip distance between u_i and w_i, is minimum among all the pairs in $V - \{U + W\}$.

Now suppose $u_i, u_j \in U$, $w_i \in W$, and $i < j$. Also let $RT(u_i, u_j) = z$, $RT(u_i, w_i) = y$, and $RT(u_j, w_i) = x$. See figure 1.

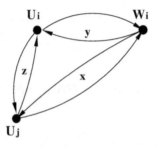

Fig. 1. Round-trip distance for u_i, w_i and u_j

Observation 1. *Given the above construction, $RT(w_i, u_i) \leq RT(u_j, w_i)$ and $RT(w_i, u_i) \leq RT(u_j, u_i)$, i.e., $y \leq x$ and $y \leq z$.*
Proof. The proof follows from the fact that if $RT(u_j, w_i)$ was less than $RT(u_i, w_i)$, then the pair (u_j, w_i) would have been selected before (u_i, w_i). Similar proof applies for (u_j, u_i). □

Note the way the above construction scheme for U and W differs from that used in [8]. The above scheme guarantees $RT(w_i, u_i) \leq RT(u_j, w_i)$ and $RT(w_i, u_i) \leq RT(u_j, u_i)$ while that in [8] guarantees only $d(w_i, u_i) \leq d(u_j, w_i)$. The second condition i.e., $RT(w_i, u_i) \leq RT(u_j, u_i)$, is crucial for the algorithm to work in the case of directed networks.

Consider the case when $z \leq x$. Using Observation 1, we can conclude that $z + y \leq 2x$. In other words, a packet going from u_j to w_i can be routed via u_i if we are ready to tolerate a round-trip stretch of 2. Thus we can omit the entry for w_i in the routing table of u_j. Let us now consider the other case when $x \leq z$. Again by Observation 1, we can conclude that $x + y \leq 2z$. Thus packets from u_j to u_i can be routed via w_i, making it possible to omit u_i in u_j's routing table.

By the above discussion, it is clear that we can always omit $j - 1$ entries from u_j and w_j. The drawback of this scheme is that there is no lower bound on the local space saved. For example at u_1 and w_1, no space is saved. To overcome this, we use the approach used by Cowen [2] which stores the routing information in the header of the packets. For the node u_1, we choose k nodes from the set $V - \{u_1 + w_1\}$, and use headers of the form $H(x_i) = (x_i, y_i, (u_1, P_{(x_i, u_1)}))$, where k is the minimum number of entries we want to save at each node. Here, $H(x_i)$ represents the header of the packet with destination x_i (one of the k nodes), x_i and y_i are the nodes added to U and W respectively in the ith step , and $P_{(x_i, u_1)}$ is the port on u_1 which connects to the first node on the shortest path to x_i. Thus, we need not store an entry for x_i or any of the k nodes at u_1. In a similar manner, we choose k nodes for w_1 from the set $V - \{u_1 + w_1\}$.

In general for u_i, $k - i + 1$ nodes are selected from the set $V - \{u_1 + u_2 + ...u_i + w_1 + w_2 + ...w_i\}$. If x_i is one of the $k - i + 1$ nodes, we set the header of the packet with destination x_i as $H(x_i) = (x_i, y_i, (u_i, P_{(x_i, u_i)}))$. This way we can omit a total of $(i - 1 + k - i + 1) = k$ entries at every node.

Note that for u_i, $k - i + 1$ nodes are selected from the set $V - \{u_1 + u_2 + ...u_i + w_1 + w_2 + ...w_i\}$, which is a dynamic set, i.e., it varies with i, while in [8] the set from which $k - i + 1$ nodes were selected is $V - \{u_1 + u_2 + ...u_k + w_1 + w_2 + ...w_k\}$, which is static. This leads to a more optimized bound on k in case of the above algorithm.

4   The Algorithm

The routing algorithm consists of four parts: an initial construction where nodes are divided and paired according to some set criterion, a header construction where the header is constructed for each destination, a table construction which describes the routing table stored at each node, and the delivery protocol which describes how to identify the output port on which the packet gets sent, based on the header and local routing table.

4.1   Initial Construction

Assume $G = (V, E)$ is a connected directed network with n nodes. Here we construct two equal sets U and W such that $U \cup W = V$, $U \cap W = \phi$ and $|U| = |W| = n/2$.

- *Initiate*: $U = W = \phi$
  For $i = 1$ to $n/2$
      Find (u_i, w_i) such that $RT(u_i, w_i)$ is minimum in $V - \{U + W\}$
      Add u_i to U and w_i to W

Note u_1, w_1 will have the minimum round-trip distance among all the nodes.

4.2   Header Construction

The header for any packet consists of a 3-tuple. The first entry for the header of a packet with destination u_i is u_i itself, the second consists of w_i. The third entry consists of a pair: a node and a port on that node which leads to the shortest path to u_i.

Here we construct a mapping which helps in selecting the third entry in the header. Map u_k to $u_{n/2}$ and u_{k-1} to $\{u_{n/2-1}, u_{n/2-2}\}$ where $k < n$. In general, u_{k-i} is mapped to $\{u_{n/2-\sum_{j=1}^{i} j}, \cdots, u_{n/2-(\sum_{j=1}^{i+1} j)+1}\}$, denoted by $M(u_{k-i})$. Note $|M(u_{k-i})| = |M(w_{k-i})| = i + 1$. For the nodes $M(u_{k-i})$, the third entry in the header is selected as $(u_{k-i}, P_{(x,u_{k-i})})$, where $x \in M(u_{k-i})$. Formally, the $i + 1$ headers for packets whose destination is a node in $M(u_{k-i})$ are

$$H(u_{n/2-\sum_{j=1}^{i} j}) = (u_{n/2-\sum_{j=1}^{i} j}, w_{n/2-\sum_{j=1}^{i} j}, (u_{k-i}, P_{(u_{n/2-\sum_{j=1}^{i} j}, u_{k-i})}))$$

$$\cdot$$

$$\cdot$$

$$H(u_{n/2-(\sum_{j=1}^{i+1} j)+1}) =$$

$$(u_{n/2-(\sum_{j=1}^{i+1} j)+1}, w_{n/2-(\sum_{j=1}^{i+1} j)+1}, (u_{k-i}, P_{(u_{n/2-(\sum_{j=1}^{i+1} j)+1}, u_{k-i})}))$$

The headers for destinations in W are constructed similarly.

4.3   Constructing Tables

The following procedure is used to construct the routing table at node u_j. See figure 2.

- A node, say u_i, is not included in the routing table of u_j if any of the following conditions are satisfied.

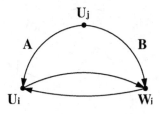

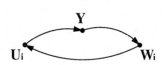

Fig. 2. Table construction for u_j **Fig. 3.** The route from u_i to w_i

- The node $u_i \in M(u_j)$, i.e., the third element in the header of u_i contains the field u_j.
- $A \geq B$ and $RT(u_i, w_i) < RT(u_j, u_i)$.
- $A \geq B$, $RT(u_i, w_i) = RT(u_j, u_i)$, and either of the two conditions is satisfied:
  * u_j does not lie on the round-trip path from u_i to w_i.
  * u_j lies on the round-trip path from u_i to w_i and $j > i$.
- Else, u_i is included in the routing table of u_j.

Similar procedure holds when deciding about w_i, except that w_i is not omitted from $u'_j s$ routing table in the case when $j > i$ and u_i has been omitted from $u'_j s$ routing table.

4.4 Delivery Protocol

A packet with header $(u, w, (v, P_{(u,v)})$ at node x is routed according to the following procedure.

- If $x = v$, then route along the port $P_{(u,v)}$.
- Else if u is in $x's$ routing table, then route along the port $P_{(u,x)}$.
- Else if w is in $x's$ routing table, then route along the port $P_{(w,x)}$.
- Else route fails.

5 Correctness Proof

Theorem 1. *The routing algorithm is correct.*

Proof. There are four cases to consider.

Case 5.1. *A packet has to be routed from x ($= u_i$) to w_i.*
From the table construction scheme, it can be concluded that u_i will have an entry for w_i. Let y ($= u_j$) be any node between u_i and w_i (Fig. 3). Now we need to show that y has an entry for w_i. This gives five subcases.

1. $d(y, u_i) > d(y, w_i)$: y will have an entry for w_i because this subcase falls in the else part of the table construction scheme.
2. $d(y, u_i) \leq d(y, w_i)$ and $RT(y, w_i) < RT(u_i, w_i)$: again y will include w_i because of the same reason as above.

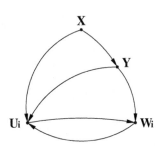

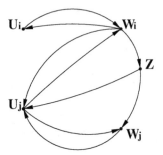

Fig. 4. The route from x to w_i **Fig. 5.** The route from u_j to w_i and back

3. $d(y, u_i) \leq d(y, w_i)$, $RT(y, w_i) = RT(u_i, w_i)$, and $j < i$: this subcase also falls in the else part of the table construction scheme. Hence, y will have an entry for w_i.
4. $d(y, u_i) \leq d(y, w_i)$, $RT(y, w_i) = RT(u_i, w_i)$, and $j > i$: By using Observation 1, $RT(u_i, w_i) \leq RT(y, u_i)$. Also $RT(u_i, w_i) \geq RT(y, u_i)$, as y is on the round-trip path from u_i to w_i. This gives $RT(u_i, w_i) = RT(y, u_i)$. Expanding the above expression gives

$$d(u_i, y) + d(y, w_i) + d(w_i, u_i) = d(u_i, y) + d(y, u_i)$$

This implies

$$d(y, w_i) + d(w_i, u_i) = d(y, u_i)$$

As $d(w_i, u_i) > 0$, we get $d(y, w_i) < d(y, u_i)$ which contradicts the condition assumed. Hence this subcase cannot exist.

Note that if a packet has to be routed from y to u_i and $j > i$, then the path taken by the packet is via w_i. Also this path is equal to the shortest path from y to u_i as $d(y, w_i) + d(w_i, u_i) = d(y, u_i)$.
5. $d(y, u_i) \leq d(y, w_i)$ and $RT(y, w_i) > RT(u_i, w_i)$: As y is an intermediate node on a round-trip from u_i to w_i, hence $RT(y, w_i) \leq RT(u_i, w_i)$. This is again a contradiction. Thus this subcase cannot exist.

Hence the packet can always be routed from u_i to w_i.

Case 5.2. *A packet has to be routed from x ($\neq u_i$) to w_i and $d(x, u_i) > d(x, w_i)$ (Fig. 4).*
As $d(x, u_i) > d(x, w_i)$, x will have an entry for w_i in its routing table. Let y be any node on the shortest path from x to w_i. Using the triangle inequality, we have $d(x, y) + d(y, u_i) > d(x, u_i)$. As $d(x, u_i) > d(x, y) + d(y, w_i)$, we get

$$d(x, y) + d(y, u_i) > d(x, u_i) > d(x, y) + d(y, w_i)$$

Thus, $d(y, u_i) > d(y, w_i)$. This implies that y will also contain an entry for w_i and the packet will eventually reach w_i.

Case 5.3. *A packet has to be routed from x ($\neq u_i$) to w_i with the following conditions:*

$$d(x, u_i) \leq d(x, w_i) \text{ and } RT(x, w_i) < RT(u_i, w_i)$$

The table construction scheme and the latter condition imply that x will contain an entry for w_i. Now suppose y is any node on the shortest path from x to w_i (Fig 4). Using the fact that $RT(y, w_i) \leq RT(x, w_i)$ and $RT(x, w_i) < RT(u_i, w_i)$, it can be concluded that

$$RT(y, w_i) < RT(u_i, w_i)$$

Hence, y will have an entry for w_i.

Case 5.4. *A packet has to be routed from* x ($\neq u_i$) *to* w_i *with the following conditions:*

$$d(x, u_i) \leq d(x, w_i) \text{ and } RT(u_i, w_i) \leq RT(x, w_i)$$

If x is on the round-trip path from u_i to w_i, this reduces to Case 5.1. Otherwise the entry for w_i is omitted from x. Hence the packet moves towards u_i. Let z ($= u_j$) be a node on the shortest path from x to u_i. There are two subcases.

- z does not have an entry for w_i. The packet is routed towards u_i. Once it reaches u_i, this reduces to Case 5.1.
- z has an entry for w_i. This is possible under three scenarios.
  - $d(z, w_i) < d(z, u_i)$
  - $d(z, w_i) \geq d(z, u_i)$ and $RT(z, w_i) < RT(u_i, w_i)$
  - z lies on the round-trip path from u_i to w_i, $j < i$, and $RT(z, w_i) = RT(u_i, w_i)$.

  The first two scenarios represent Case 5.2 and Case 5.3, respectively. The third scenario reduces to Case 5.1. Hence the packet will be routed to w_i via z.

Similar proof holds for destination u_i.   □

6   Complexity Analysis

In this section, we find the bounds on maximum space and round-trip stretch.

6.1   Round-Trip Stretch

Theorem 2. *The upper bound on round-trip stretch for the algorithm presented in Section 4 is two.*

Proof. There are four cases to be considered. See figure 5.

Case 1. *A packet has to be routed from* x ($= u_i$) *to* w_i.

This corresponds to the first case in the previous section. Remember that any node between u_i and w_i will have an entry for w_i and a node between w_i and u_i will have entry for u_i. Thus the round-trip stretch for (u_i, w_i) is 1.

For the rest of the cases, we will consider the round-trip from x ($\neq u_i$) to w_i as the route for the packet.

Case 2. *A packet has to be routed on the round-trip path from x (where $x \neq u_i$ and $x = u_j$) to w_i with the condition that x has an entry for w_i and w_i has an entry for u_j.*

As u_j has an entry for w_i, so any node between u_j and w_i will also have an entry for w_i (Case 5.2.). Similar argument holds for any node between w_i and u_j. Hence the packet will travel through the shortest round-trip path between u_j and w_i. This again gives a round-trip stretch of 1.

Case 3. *A packet has to be routed on the round-trip path from x (where $x \neq u_i$ and $x = u_j$) to w_i and x has an entry for w_i while w_i does not have an entry for u_j.*

The packet will travel from u_j to w_i along the shortest path. From w_i, it will be routed towards w_j. There are two subcases to be considered.

- None of the nodes between w_i and w_j has an entry for u_j. In this subcase, the packet reaches w_j and from there it goes to u_j. This gives

$$RT(R, u_j, w_i) = d(u_j, w_i) + d(w_i, w_j) + d(w_j, u_j)$$

  Note, $d(w_i, w_j) \leq d(w_i, u_j)$ and $RT(u_j, w_j) \leq RT(u_j, w_i)$ because w_i does not have an entry for u_j. Thus,

$$\begin{aligned} RT(R, u_j, w_i) &\leq d(u_j, w_i) + d(w_i, u_j) + d(w_j, u_j) \\ &\leq RT(u_j, w_i) + RT(u_j, w_j) \\ &\leq 2RT(u_j, w_i) \end{aligned}$$

  Hence we obtain a round-trip stretch of 2.
- There is a node between w_i and w_j which has an entry for u_j. Let the first node between w_i and w_j which has entry for u_j be z. The path taken by the packet is from w_i to z and then z to u_j. As $d(w_i, z) + d(z, u_j) \leq d(w_i, w_j) + d(w_j, u_j)$ and $d(w_i, w_j) \leq d(w_i, u_j)$, we get

$$\begin{aligned} RT(R, u_j, w_i) &= d(u_j, w_i) + d(w_i, z) + d(z, u_j) \\ &\leq d(u_j, w_i) + d(w_i, u_j) + d(w_j, u_j) \\ &\leq RT(u_j, w_i) + RT(u_j, w_j) \\ &\leq 2RT(u_j, w_i) \end{aligned}$$

  This also gives a round-trip stretch of 2.

Case 4. *A packet has to be routed on the round-trip path from x (where $x \neq u_i$ and $x = u_j$) to w_i and w_i has an entry for u_j while u_j does not have an entry for w_i.*

Using a similar argument as in the previous case, it can be proved that a round-trip stretch of 2 results.

Note that we cannot have a case where neither u_j has an entry for w_i nor w_i has an entry for u_j. Thus, the round-trip stretch for the algorithm is bounded by 2.   □

6.2   Maximum Local Space

Lemma 1. *Node u_i has an entry for either u_j or w_j, where u_i, u_j, $w_j \in V$ and $i > j$.*

Proof.   By using Observation 1, it can be concluded that $RT(u_j, w_j) \leq RT(u_j, u_i)$ and $RT(u_j, w_j) \leq RT(w_j, u_i)$. There are 2 cases to be considered.

Case 6.1. *The round-trip path from u_j to w_j contains u_i.*

As u_i is an intermediate node on the round-trip path from u_j to w_j, hence $RT(u_i, w_j) \leq RT(u_j, w_j)$. Thus we can conclude that $RT(u_i, w_j) = RT(u_j, w_j)$. It can be easily verified that for the above condition, either w_j or u_j is included in $u_i's$ routing table.

Case 6.2. *u_i does not lie on the round-trip path from u_j to w_j.*

There are three subcases to be considered.

- $d(u_i, u_j) < d(u_i, w_j)$. From the table construction scheme, it can be concluded that only u_j is included in $u_i's$ routing table.
- $d(u_i, u_j) = d(u_i, w_j)$. Only w_j will be included in $u_i's$ routing table as $RT(u_i, u_j) \geq RT(u_j, w_j)$.
- $d(u_i, u_j) > d(u_i, w_j)$. Again, only w_j will be included in $u_i's$ routing table.

   □

Theorem 3. *The maximum local space at any node for the algorithm in Section 4 is $[n - \sqrt{n}(\sqrt{1 - 7/4n}) - 1/2] \log n$.*

Proof.   To get the upper bound for maximum local space, consider a node u_i, where $i < k$. Using Lemma 1 and the header construction scheme, we can omit $i - 1 + |M(u_i)| = i - 1 + k - i + 1 = k$ entries at u_i. To bound k, note that

$$\sum_{i=1}^{k} |M(u_i)| = \sum_{j=1}^{k} j \leq n/2 - 1$$

which gives $k(k+1) \leq n - 2$. It can be easily verified the $k = (-1 + \sqrt{4n - 7})/2$ satisfies the inequality. Thus the maximum local space at any node is $(n - 1 - k) \log n = [n - \sqrt{n}(\sqrt{1 - 7/4n}) - 1/2] \log n$.   □

6.3 Maximum Global Space

From the algorithm, it can be seen that for any node u_i or w_i, where $i \leq k$, k entries are saved. This means that the total space occupied by the routing table on the first k nodes in U and W is $2(n - 1 - k)k \log n$. For other nodes u_i and w_i, $i - 1$ entries are saved on each node. This gives a total table space of $2 \sum_{i=n/2}^{n-1-k} i \log n$. Thus the total space for all the nodes is $2((n - k - 1)k + \sum_{i=n/2}^{n-1-k} i) \log n$. Substituting the value of k in the above expression gives the total space as $(3n^2/4 - 3n/2 + 2) \log n$.

Note that this scheme saves on an additional $n - 2$ entries by using the technique used by Cowen [2] to store the routing information in the packet's header.

7 Conclusion

We presented a compact routing algorithm with stretch less than 3 for directed networks. Although for stretch less then 3, the lower bound for the total routing information on the network is $\Omega(n^2)$ bits, nevertheless we examined it for directed networks to determine the best possible saving in space. The minimum local memory needed for stretch between 2 and 3 is $(n - 2\sqrt{n}) \log n$ for undirected networks [8], while this algorithm requires a maximum local space of $[n - \sqrt{n}(\sqrt{1 - 7/4n}) - 1/2] \log n$ with a stretch of 2 for directed networks.

Acknowledgements. This work was supported by the U.S. National Science Foundation grant CCR-9875617.

References

1. Awerbuch, B., Bar-Noy, A., Linial, N., Peleg, D.: Improved routing strategies with succinct tables. Journal of Algorithms 11 (1990) 307-341
2. Cowen, L. J.: Compact routing with minimum stretch. In Proceedings of the 10th Annual ACM-SIAM Symposium on Discrete Algorithms (1999) 255-260
3. Cowen, L. J., Wagner, C. G.: Compact roundtrip routing in directed networks. In Proceedings of the 19th Annual ACM Symposium on Principles of Distributed Computing (2000) 51-59
4. Fraigniaud, P., Gavoille, C.: Memory requirement for universal routing schemes. In Proceedings of the 14th Annual ACM Symposium on Principles of Distributed Computing (1995) 223-230
5. Gavoille, C., Perennes, S.: Memory requirement for routing in distributed networks. In Proceedings of the 15th Annual ACM Symposium on Principles of Distributed Computing (1996) 125-133
6. Gavoille, C., Gengler, M.: Space-efficiency of routing schemes of stretch factor three. In Proceedings of the 4th International Colloquium on Structural Information and Communication Complexity (1997)
7. Gavoille, C., Peleg, D.: Compact routing scheme with low stretch factor. In Proceedings of the 17th Annual ACM Symposium on Principles of Distributed Computing (1998) 11-20

8. Iwama, I., Kawachi, A.: Compact routing with stretch factor of less than three. In Proceedings of the 19th Annual ACM Symposium on Principles of Distributed Computing (2000) 337

9. Leeuwen, J. van, Tan, R.: Routing with compact routing tables. In G. Roznenberg and A. Salomaa, editors, The Book of L. Springer-Verlag, New York, New York (1986) 256-273

10. Leeuwen, J. van, Tan, R.: Interval routing. The Computer Journal 30 (1987) 259-273

11. Peleg, D., Upfal, E.: A tradeoff between size and efficiency for routing tables. In Proceedings of the 20th Annual ACM Symposium on Theory of Computing (1988) 43-52

12. Peleg, D., Upfal, E.: A tradeoff between size and efficiency for routing tables. Journal of the ACM 36 (1989) 510-530

13. Peleg, D.: An overview of locality-sensitive distributed computing. Unpublished Monograph, The Weizmann Institute, Rehovot, Israel, 1997

14. Santoro, N., Khatib, R.: Implicit routing in networks. The Computer Science Journal 28 (1985) 5-8

Parametric Scheduling – Algorithms and Complexity

K. Subramani

Department of Computer Science and Electrical Engineering,
West Virginia University,
Morgantown, WV
ksmani@csee.wvu.edu

Abstract. Parametric Scheduling is a scheduling technique used to partially address the inflexibility of static scheduling in real-time systems. Real-Time scheduling confronts two issues not addressed by traditional scheduling models viz. parameter variability and the existence of complex relationships constraining the execution of tasks. Accordingly, modeling becomes crucial in the specification of scheduling problems. In this paper, we study the complexity of parametric scheduling as modeled within the E-T-C scheduling framework [Sub00]. We establish that (a) The parametric schedulability query is in PSPACE for generalized constraints, and (b) Parametric dispatching is not harder than deciding the schedulability query. To the best of our knowledge, our results are the first of their kind for this problem.

1 Introduction

An important feature in Real-time systems is *parameter impreciseness*, i.e. the inability to accurately determine certain parameter values. The most common such parameter is *task execution time*. A second feature is the presence of complex relationships between tasks that constrain their execution. Traditional models do not accomodate either feature completely: (a) Variable execution times are modeled through a fixed value (*worst-case*), and (b) Relationships are limited to those that can be represented by precedence graphs. We present a task model that effectively captures *variable task execution time*, while simultaneously permitting arbitrary linear relationships between tasks. Real-Time systems (and the associated scheduling problems) can be classified as *static, co-static or parametric* depending upon the information available at dispatching [Sub00]. In this paper we focus on *parametric real-time systems* wherein the only information available before a job is to be dispatched is the *actual execution time of every job sequenced before it.* The primary scheduling goal is to provide an offline guarantee that the input constraints will be met at run-time. In a Parametric System, the dispatch time of a job will, in general, be a parameterized function of the start and execution times of jobs sequenced before it.

The parametric scheduling problem is concerned with the following two issues:

B. Monien, V.K. Prasanna, S. Vajapeyam (Eds.): HiPC 2001, LNCS 2228, pp. 36–46, 2001.

1. Deciding the schedulability predicate for a parametrically specified system (Section §2), and
2. Determining the dispatch time of a job, given the start and execution times of all jobs sequenced before it.

2   Statement of Problem

2.1   Job Model

Assume an infinite time-axis divided into windows of length L, starting at time $t = 0$. These windows are called *periods* or *scheduling windows*. There is a set of non-preemptive, ordered jobs, $\mathcal{J} = \{J_1, J_2, \ldots, J_n\}$ that execute in each scheduling window.

2.2   Constraint Model

The constraints on the jobs are described by System (1):

$$\mathbf{A}.[\vec{s}\ \vec{e}]^{\mathbf{T}} \leq \vec{b}, \quad \vec{e} \in \mathbf{E}, \tag{1}$$

where,

- $\mathbf{A}$ is an $m \times 2.n$ rational matrix,
- $\mathbf{E}$ is an axis-parallel rectangle aph represented by:

$$\mathbf{\Upsilon} = [l_1, u_1] \times [l_2, u_2] \times \ldots [l_n, u_n] \tag{2}$$

The aph $\mathbf{\Upsilon}$ models the fact that the execution time of job J_i can assume any value in the range $[l_i, u_i]$ i.e. it is not constant.
- $\vec{s} = [s_1, s_2, \ldots, s_n]$ is the start time vector of the jobs, and
- $\vec{e} = [e_1, e_2, \ldots, e_n] \in \mathbf{E}$ is the execution time vector of the jobs

There are 3 types of constraints that are very relevant to the real-time scheduling domain, viz.

1. Standard Constraints - These are strict relative constraints that capture the distance requirements between two jobs, e.g. $s_1 + e_1 + 3 \leq s_2$ captures the requirement that J_2 should wait at least 3 units after J_1 finishes before starting;
2. Network Constraints - These are constraints that can be expressed in the form $a.s_i + b.s_j \leq c.e_i + d.e_j + k$, where a, b, c, d, k are arbitrary rational numbers. Network constraints serve to approximate performance metrics such as *Sum of Completion times*;
3. Generalized Constraints - These are constraints of the form $\sum_{i=1}^{n} w_i.s_i \leq k$, $w_i, i = 1, 2, \ldots, n$ and k are arbitrary rationals - These constraints are used to specify performance metrics such as *Weighted Sum of Completion Times*.

2.3   Query Model

Suppose that job J_a has to be dispatched. We assume that the dispatcher has access to the start times $\{s_1, s_2, \ldots, s_{a-1}\}$ and execution times $\{e_1, e_2, \ldots, e_{a-1}\}$ of the jobs $\{J_1, J_2, \ldots, J_{a-1}\}$.

Definition 1. *A parametric schedule of an ordered set of jobs, in a scheduling window, is a vector $\vec{s} = [s_1, s_2, \ldots, s_n]$, where s_1 is a rational number and each $s_i, i \neq 1$ is a function of the start time and execution time variables of jobs sequenced prior to job J_i, i.e. $\{s_1, e_1, s_2, e_2, \ldots, s_{i-1}, e_{i-1}\}$. Further, this vector should satisfy the constraint system (1) for all execution time vectors $\vec{e} \in \Upsilon$.*

The combination of the Job model, Constraint model and the Query model constitutes a scheduling problem specification within the E-T-C scheduling framework [Sub00].

The Parametric Scheduling problem is concerned with the following two issues:

1. Determining whether the given job-set has a parametric schedule, i.e. a schedule as defined in Definition (1);
2. Computing the start-time of a job in each scheduling window, assuming that
   a) The parametric schedulability query has been decided affirmatively, and
   b) The start and execution times of all jobs sequenced before it are provided. *This corresponds to the online dispatching phase.*

The discussion above directs us to the following formulation of the parametric schedulability query:

$$\exists s_1 \, \forall e_1 \in [l_1, u_1] \, \exists s_2 \, \forall e_2 \in [l_2.u_2], \ldots \exists s_n \, \forall e_n \in [l_n, u_n] \quad \mathbf{A}.[\vec{s} \ \vec{e}]^{\mathbf{T}} \leq \vec{b} \quad ? \ (3)$$

3   Motivation

Parametric scheduling provides a means of addressing the lack of flexibility in static scheduling.

Example 1. Consider the two job system $J = \{J_1, J_2\}$, with start times $\{s_1, s_2\}$, execution times in the set $\{(e_1 \in)[2, 4] \times (e_2 \in)[4, 5]\}$ and the following set of constraints:

− Job J_1 must finish before job J_2 commences; i.e. $s_1 + e_1 \leq s_2$;
− Job J_2 must commence within 1 unit of J_1 finishing; i.e. $s_2 \leq s_1 + e_1 + 1$;

Clearly a static approach would declare the constraint system to be infeasible, i.e. there do not exist rational $\{s_1, s_2\}$ which can satisfy the constraint set for all execution time vectors. Now consider the following start time dispatch vector:

$$\vec{s} = \begin{bmatrix} s_1 \\ s_2 \end{bmatrix} = \begin{bmatrix} 0 \\ s_1 + e_1 \end{bmatrix} \tag{4}$$

This assignment satisfies the input set of constraints and is hence a valid dispatch schedule. The key feature of the schedule provided by Equation (4) is that the start time of job J_2 is no longer an absolute time, but a (parameterized) function of the start and execution times of job J_1. This phenomenon where a static scheduler declares a schedulable system infeasible is called *Loss of Schedulability*. Thus, the parametric schedulability query provides the flexibility to partially address the loss of schedulability phenomenon.

Parametric schedulability is particularly useful in real-time Operating Systems such as Maruti [LTCA89,MAT90] and MARS [DRSK89], wherein program specifications can be efficiently modeled through constraint matrices, and interactions between processes are permitted through linear relationships between their start and execution times. The flexibility of parametric schedulability greatly enhances the schedulability of a job-set, at the expense of computing dispatch vectors online.

4   Complexity of Parametric Schedulability

Prior to analyzing the complexity of the query, let us discuss a simple algorithm to decide the query. Quantifier elimination procedures help us to work through query (3), by eliminating one quantified variable at a time.

Function PARAMETRIC-SCHEDULER $(\mathbf{\Upsilon}, \mathbf{A}, \vec{\mathbf{b}})$

```
 1: for  (i = n down to 2)  do
 2:     ELIM-UNIV-VARIABLE(e_i)
 3:     ELIM-EXIST-VARIABLE(s_i)
 4:     if  (CHECK-INCONSISTENCY()) then
 5:        return(false)
 6:     end if
 7: end for
 8: ELIM-UNIV-VARIABLE (e_1)
 9: if  (a ≤ s_1 ≤ b,   a, b ≥ 0) then
10:     Valid Parametric Schedule Exists
11:     return
12: else
13:     No Parametric Schedule Exists
14:     return
15: end if
```

Algorithm 4.1:   A Quantifier Elimination Algorithm for determining Parametric Schedulability

---

Algorithm (4.2) describes the procedure for eliminating the universally quantified execution variable $e_i \in [l_i, u_i]$. We implement the ELIM-EXIST-VARIABLE() procedure using the Fourier-Motzkin elimination technique [DE73].

The correctness of Algorithm (4.2) has been argued in [Sch87], while the correctness of the Fourier-Motzkin procedure is discussed in [DE73] and [Sch87]

Function ELIM-UNIV-VARIABLE $(\mathbf{A}, \vec{b})$

  1: Substitute $e_i = l_i$ in each constraint that can be written in the form $e_i \geq ()$
  2: Substitute $e_i = u_i$ in each constraint that can be written in the form $e_i \leq ()$

Algorithm 4.2: Eliminating Universally Quantified variable $e_i \in [l_i, u_i]$

Observation 1. *Eliminating a universally quantified execution time variable does not increase the number of constraints.*

Observation 2. *Eliminating an existentially quantified variable s_i in general, leads to a quadratic increase in the number of constraints, i.e. if there are m constraints, prior to the elimination, there could be $O(m^2)$ constraints after the elimination (See [DE73]). Thus, the elimination of k existential quantifiers could increase the size of the constraint set to $O(m^{2^k})$ [VR99]. Clearly, the exponential size blow-up, makes the Algorithm (4.1) impractical for general constraint sets.*

Lemma 1. *Algorithm (4.1) correctly decides the parametric schedulabilty query (3).*

Proof: *Follows from the discussion above.* □

Lemma 2. *Query (3) is locally convex over the execution time variables, i.e. if the query is true for $e_i = a_1$ and $e_i = b_1$, then it is true for $e_i = \lambda.a_1 + (1 - \lambda).b_1, 0 \leq \lambda \leq 1$.*

  Proof: *Follows from the correctness of Algorithm (4.2).* □
  Lemma (2) allows us to restate the parametric schedulability query as:

$$\exists s_1 \, \forall e_1 \in \{l_1, u_1\} \, \exists s_2 \, \forall e_2 \in \{l_2.u_2\}, \ldots \exists s_n \, \forall e_n \in \{l_n, u_n\} \quad \mathbf{A}.[\vec{s} \ \vec{e}]^{\mathbf{T}} \leq \vec{b} \quad ? \tag{5}$$

  We shall refer to query (5) as the parametric schedulability query in Subsection §4.1. Note that query (3) is equivalent to query (5) in that query (3) is true if and only if query (5) is true. The advantage of the discrete form is that it simplifies the complexity analysis of the query.

4.1   Parametric Schedulability as a 2−Person Game

We show that query (5) can be modeled as a 2−person game; the modeling helps us to establish that the query is in PSPACE.

Lemma 3. *The parametric schedulability query can be decided in* ATIME(m^2).

  Proof: *Algorithm (4.3) describes the Alternating Turing machine that decides query (5) in polynomial time. The e_i guesses are of size polynomial in the size of the input, since l_i and u_i are provided as part of the input. What needs to be shown is that the s_i guesses are of length polynomial in the size of the input too. Observe that for each guessed vector $\vec{e} = [e_1, e_2, \ldots, e_n]$, the system $\mathbf{A}.[\vec{s} \ \vec{e}]^{\mathbf{T}} \leq \vec{b}$ is transformed into a system $\mathbf{G}.\vec{s} \leq \vec{b'}$, which is a rational polyhedron in the* *start-time variables and hence has basic solutions of size polynomial in the size of the input [Sch87]. It follows that the* ATM *can confine itself to s_i guesses of size polynomial in the size of the input. The only non-trivial computation is at the leaf of the tree and this computation involves the multiplication of a $m \times 2n$ matrix with a $2.n \times 1$ vector, giving a running time of* ATIME$(m.n)$, *which can be expressed as* ATIME(m^2), *since $m > n$ except in trivial cases.* □

Function PARAM-DECIDE$(\mathbf{A}, \vec{b}, \Upsilon)$
1: **for** $(i = 1$ **to** $n)$ **do**
2:     Guess a value for $s_i \in [0, L]$ in state K_{OR}, of the computation tree
3:     Guess a value for $e_i \in \{l_i, u_i\}$ in state K_{AND}, of the computation tree
4: **end for**{ We now have both $\vec{s} = [s_1, s_2, \ldots, s_n]$ and $\vec{e} = [e_1, e_2, \ldots, e_n]$. }
5: **if** $(\mathbf{A}.[\vec{s} \ \vec{e}]^{\mathbf{T}} \leq \vec{b})$ **then**
6:     There exists a parametric schedule
7: **else**
8:     There does not exist a parametric schedule
9: **end if**

Algorithm 4.3: An Alternating Turing Machine to decide Parametric Schedulability

---

Corollary 1. *Parametric schedulability is in* PSPACE.

Proof: *From [Pap94], we know that,* PSPACE $= \cup_{i=0}^{\infty}$ATIME(m^i). □

5   Special Case Analysis

We now analyze the parametric schedulability problem when the constraints are standard. This restriction is referred to as `<aph|stan|param>` in the E-T-C Scheduling framework. We provide a dual interpretation of the quantifier elimination algorithm, discussed in [Sak94].

Given an instance of `<aph|stan|param>`, we construct the dual graph as discussed in [Sub00]. Algorithm (5.1) takes the dual graph as input and decides the schedulability query over the dual. The principal difference between this algorithm and Algorithm (4.1) lies in the implementation of the existential variable elimination procedure. When the constraints are standard, existential variable elimination can be performed through vertex contraction, in the manner discussed in Algorithm (5.2). The exponential increase in the number of constraints that is a feature of the Algorithm (4.1) is prevented in this case; Algorithm (5.3) deletes redundant edges (constraints) as new edges (constraints) are created as a result of vertex contraction, thereby bounding the total number of constraints at any time in the computation.

Observation 3. *For deleting redundant edges between a vertex s_a and another vertex s_b, where $a > b$, we use a separate function $\geq_{back} (e_{old}, e_{new})$, which retains the edge with the smaller (positive) weight.*

Function PARAM-SCHEDULE-STANDARD $(G =< V, E >)$

1: **for** $(i = n$ **down to** $2)$ **do**
2:     Substitute $e_i = u_i$ on all edges where e_i is prefixed with a negative sign
3:     Substitute $e_i = l_i$ on all other edges { We have now eliminated e_i in
        $\forall e_i \in [l_i, u_i]$ }
4:     $G' =< V', E' >=$VERTEX-CONTRACT(s_i)
5: **end for**
6: Substitute $e_1 = u_1$ on all edges, where e_1 is prefixed with a negative sign
7: Substitute $e_1 = l_1$ on all other edges { At this point, the only edges in the
    graph are between s_1 and s_0 and the weights on these edges are rational
    numbers.}
8: Find the edge $s_1 \rightsquigarrow s_0$ with the smallest weight, say u. { $u \geq 0$ }
9: Find the edge $s_0 \rightsquigarrow s_1$ with the largest weight in magnitude, say l. { $l \leq 0$ }
10: **if** $(-l \leq u)$ **then**
11:     **return** $[-l, u]$ as the range in which J_1 can begin execution
12: **else**
13:     There is no Parametric Schedule
14:     **return**
15: **end if**

Algorithm 5.1: Implementing quantifier elimination over the dual graph

---

Function VERTEX-CONTRACT $(G =< V, E >, s_i)$

1: **for** each edge $s_j \rightsquigarrow s_i$, with weight w_{ji} **do**
2:     **for** each edge $s_i \rightsquigarrow s_k$, with weight w_{ik} **do**
3:         Add an edge (say e_{new}) $s_j \rightsquigarrow s_k$ with weight $w_{ji} + w_{ik}$
4:         **if** $j = k$ **then**
5:             **if** $(w_{ji} + w_{ik} < 0)$ **then**
6:                 No Parametric Schedule Exists { See Observation (5)}
7:                 **return**(false)
8:             **end if**{ Eliminating self-loops }
9:         **else**
10:             Discard e_{new}
11:             **continue** {We do not add self-loops to the edge set}
12:         **end if**
13:         $E' = E \cup e_{new}$
14:         REMOVE-REDUNDANT$(G =< V, E' >, s_j, s_k, e_{new})$
15:     **end for**
16:     $E' = E' - (s_j \rightsquigarrow s_i)$
17: **end for**
18: **for** each edge $s_i \rightsquigarrow s_k$, with weight w_{ik} **do**
19:     $E' = E' - (s_i \rightsquigarrow s_k)$
20: **end for**
21: $V' = V - \{s_i\}$ { We have now eliminated s_i in $\exists s_i$ }

Algorithm 5.2: Vertex Contraction

Observation 4. *The number of edges from any vertex s_a to any other vertex s_b (say $a < b$) does not exceed four at any time; in fact, after the elimination of e_b, the number of edges does not exceed two.*

Function REMOVE-REDUNDANT($G = < V, E >, s_j, s_k, e_{new}$)

1: { W.l.o.g. assume that $j < k$.}
2: { e_{new} is the new edge that is added from s_j to s_k. }
3: Let t_o = type of e_{new}
4: **if** (there does not exist an edge from s_j to s_k having type t_o) **then**
5:    Retain e_{new} {The addition of e_{new} does not cause the number of edges from s_j to s_k to increase beyond four.}
6: **else**
7:    Let e_{old} be the existing edge with the same type as e_{new}
8:    **if** ($\geq_{for} (e_{old}, e_{new})$) **then**
9:       Discard e_{new} { See [Sub01] for the implementation of $\geq_{for}$}
10:    **else**
11:       Discard e_{old}
12:    **end if**
13: **end if**

Algorithm 5.3: Redundant edge elimination

Observation 5. *The class of standard constraints is closed under execution time variable elimination, i.e. the elimination of the execution time variables does not alter the network structure of the graph.*

Observation 6. *The class of standard constraints is closed under vertex contraction.*

Observation 7. *The only manner in which infeasibility is detected is through the occurrence of a negative weight self-loop on any vertex (Step (5) of Algorithm (5.2) and Steps (7-13) of Algorithm (5.1).) These loops could occur in two ways:*

1. *The contraction of a vertex results in a self-loop with a negative rational number on another vertex; e.g. in the constraint set $s_1 + 8 \leq s_2, s_2 \leq s_1 + 7$, contraction of s_2 results in a self-loop at s_1 of weight -1;*
2. *The contraction of a vertex results in a self-loop of the following form: $e_a - 7$, on vertex s_a. In this case, either $l_a \geq 7$, in which case the edge can be discarded (redundant), or the system is infeasible.*

5.1   Correctness and Complexity

The correctness of the algorithm follows from the correctness of the quantifier elimination algorithm Algorithm (4.1).

The elimination of an universally quantified execution time variable e_i takes time proportional to the degree of vertex s_i, since e_i occurs only on those edges that represent constraints involving s_i. Hence eliminating e_i takes time $O(n)$ in the worst case. The total time taken for execution time variable elimination over all n vertices is thus $O(n^2)$. The contraction of a single vertex takes time $O(n^2)$ in the worst-case, since every pair of incoming and outgoing edges has to be combined. In fact $O(n^2)$ is a lower-bound on the contraction technique. However, the total number of edges in the graph is always bounded by $O(n^2)$; the total time spent in vertex contraction is therefore $O(n^3)$.

Thus the complexity of `<aph|stan|param>` is $O(n^3)$.

6   Complexity of Online Dispatching

In [Sak94], query (3) was addressed by providing a parametrized list of linear functions for the start time of each job.During actual execution, s_1 can take on any value in the range $[a, b]$. Upon termination of job J_1, we know e_1 which along with s_1 can be plugged into $f_1()$ and $f_1'()$, thereby providing a range $[a', b']$ for s_2 and so on, till job J_n is scheduled and completes execution.

The principal problem with the creation of the function lists is that we cannot *a priori* bound the length of these lists. In the case of standard constraints, it can be argued that the length of these lists is at most $O(n)$. However, there appears to be no easy way to bound the length of the function lists, when the constraint matrix is an arbitrary totally unimodular matrix. We argue here that explicit construction of the parameterized function list is unnecessary. Determination of feasibility is sufficient, thereby eliminating the need for storing the parameterized function list. Observe that at any point in the scheduling window, the first job that has not yet been scheduled has a start time that is independent of the start and execution times of all other jobs. Once this job is executed, we can determine a rational range, e.g.$[a'', b'']$ for the succeeding job and the same argument applies to this job. In essence, all that is required to be determined is *the start time of the first unexecuted job in the sequence.*

Let us assume the existence of an oracle $\mathbf{\Omega}$ that decides query (3) in time $T(\mathbf{\Omega})$. Algorithm (6) can then be used to determine the start time of the first unexecuted job (say J_ρ [1]) in the schedule :

The end of the period L is the deadline for all jobs in the job-set. We must have $0 \leq s_\rho \leq L$. The goal is to determine the exact value that can be safely assigned to s_ρ without violating the current set. Let

$$\mathbf{G}^\rho.\vec{s^\rho} + \mathbf{H}^\rho.\vec{e^\rho} \leq \vec{b^\rho} \tag{6}$$

denote the current constraint set, where

- $\mathbf{G}^\rho$ is obtained from $\mathbf{G}$, by dropping the first $(\rho-1)$ columns; $\mathbf{G}^{1-\rho}$ represents the first $(\rho - 1)$ columns of $\mathbf{G}$,
- $\mathbf{H}^\rho$ is obtained from $\mathbf{H}$, by dropping the first $(\rho-1)$ columns; $\mathbf{H}^{1-\rho}$ represents the first $(\rho - 1)$ columns of $\mathbf{H}$,
- $\vec{s^\rho} = [s_\rho, s_{\rho+1}, \ldots, s_n]; \ \vec{s^{1-\rho}} = [s_1, s_2, \ldots s_{\rho-1}];$
- $\vec{e^\rho} = [e_\rho, e_{\rho+1}, \ldots, e_n]; \ \vec{e^{1-\rho}} = [e_1, e_2, \ldots e_{\rho-1}],$ and
- $\vec{b^\rho} = \vec{b} - (\mathbf{G}^{1-\rho}.\vec{s^{1-\rho}} + \mathbf{H}^{1-\rho}.\vec{e^{1-\rho}}).$

We exploit the local convexity of s_ρ, i.e. if $s_\rho \geq a$ is valid and $s_\rho \leq b$ is valid, then any point $s_\rho = \lambda.a + (1 - \lambda).b, 0 \leq \lambda \leq 1$ is valid. The cost of this strategy is $O(\log L)$ calls to the oracle $\mathbf{\Omega}$, i.e. $O(T(\mathbf{\Omega}).\log L)$. We have thus established that the principal complexity of the parametric scheduling problem is in deciding query (3). *This result is significant because it decouples dispatching complexity from decidability. In many complexity analyses involving* **PSPACE** *problems, the*

---

[1] At commencement, $\rho = 1$

Function DETERMINE-START-TIME $(\mathbf{G}^\rho, \mathbf{H}^\rho, \vec{\mathbf{b}^\rho}, [a_l, a_h])$

1: { Initially $[a_l, a_h] = [0, L]$; the interval is reduced to half its original length at
   each level of the recursion}
2: Let $m' = \frac{a_h - a_l}{2}$
3: **if** $(\Omega(\mathbf{G}^\rho, \mathbf{H}^\rho, \vec{\mathbf{b}^\rho}, s_\rho \geq m'))$ **then**
4:   {We now know that there is a valid assignment for s_ρ in the interval
       $[m', a_h]$; the exact point in time needs to be determined}
5:   **if** $(\Omega(\mathbf{G}^\rho, \mathbf{H}^\rho, \vec{\mathbf{b}^\rho}, s_\rho = m'))$ **then**
6:     $s_\rho = m'$
7:     **return**
8:   **else**
9:     {m' is not a valid point; however we can still recurse on the smaller
         interval}
10:     DETERMINE-START-TIME $(\mathbf{G}^\rho, \mathbf{H}^\rho, \vec{\mathbf{b}^\rho}, [m', a_h])$
11:   **end if**
12: **else**
13:   { We know that the valid assignment for s_ρ must lie in the interval $[a_l, m']$ }
14:   DETERMINE-START-TIME $(\mathbf{G}^\rho, \mathbf{H}^\rho, \vec{\mathbf{b}^\rho}, m')$
15: **end if**

Algorithm 6.1: Parametric Dispatcher to determine s_ρ

*size of the output is used to provide a lower bound for the running time of the
problem, e.g. if we can provide a problem instance where the start time of a job
must have exponentially many dependencies, due to the nature of the constraints,
then we have an exponential lower bound for the dispatching scheme in [GPS95].
Algorithm (6.1) assures us that efficient dispatching is contingent only upon
efficient decidability.*

7 Summary

In this paper, we studied the computational complexity of the parametric schedulability query. A naive implementation of the quantifier elimination algorithm,
as suggested in [Sak94], requires exponential space and time to decide the query.
Our work establishes that the problem is in PSPACE for arbitrary constraint
classes.

References

[DE73]   G. B. Dantzig and B. C. Eaves. Fourier-Motzkin Elimination and its Dual.
          Journal of Combinatorial Theory (A), 14:288–297, 1973.
[DRSK89] A. Damm, J. Reisinger, W. Schwabl, and H. Kopetz. The Real-Time Oper-
          ating System of MARS. *ACM Special Interest Group on Operating Systems*,
          23(3):141–157, July 1989.
[GPS95]   R. Gerber, W. Pugh, and M. Saksena. Parametric Dispatching of Hard
          Real-Time Tasks. *IEEE Transactions on Computers*, 1995.

[LTCA89] S. T. Levi, S. K. Tripathi, S. D. Carson, and A. K. Agrawala. The Maruti Hard Real-Time Operating System. *ACM Special Interest Group on Operating Systems*, 23(3):90–106, July 1989.

[MAT90]   D. Mosse, Ashok K. Agrawala, and Satish K. Tripathi. Maruti a hard real-time operating system. In *Second IEEE Workshop on Experimental Distributed Systems*, pages 29–34. IEEE, 1990.

[Pap94]   Christos H. Papadimitriou. *Computational Complexity*. Addison-Wesley, New York, 1994.

[Sak94]   Manas Saksena. *Parametric Scheduling in Hard Real-Time Systems*. PhD thesis, University of Maryland, College Park, June 1994.

[Sch87]   Alexander Schrijver. *Theory of Linear and Integer Programming*. John Wiley and Sons, New York, 1987.

[Sub00]   K. Subramani. *Duality in the Parametric Polytope and its Applications to a Scheduling Problem*. PhD thesis, University of Maryland, College Park, July 2000.

[Sub01]   K. Subramani. An analysis of partially clairvoyant scheduling, 2001. Submitted to Journal of Discrete Algorithms.

[VR99]   V.Chandru and M.R. Rao. Linear programming. In *Algorithms and Theory of Computation Handbook, CRC Press, 1999*. CRC Press, 1999.

An Efficient Algorithm for Computing Lower Bounds on Time and Processors for Scheduling Precedence Graphs on Multicomputer Systems[*]

B.S. Panda[1] and Sajal K. Das[2]

[1] Department of Computer and Information Sciences
University of Hyderabad, Hyderabad 500 046, INDIA
bspcs@uohyd.ernet.in

[2] Department of Computer Science and Engineering
The University of Texas at Arlington
Arlington, TX 76019-0015, USA
das@cse.uta.edu

Abstract. In this paper, we propose new lower bounds on minimum number of processors and minimum time to execute a given program on a multicomputer system, where the program is represented by a directed acyclic task graph having arbitrary execution time and arbitrary communication delays. Additionally, we propose an $O(n^2 + m \log n)$ time algorithm to compute these bounds for a task graph with n nodes and m arcs. The key ideas of our approach include: (i) identification of certain points called *event points* and proving that the intervals having event points as both ends are enough to compute the desired bounds; and (ii) the use of a *sweeping technique*. Our bounds are shown to be as sharp as the current best known bounds due to Jain and Rajaraman [7]. However, their approach requires $O(n^2 + m \log n + nW_{erl}^2)$ time, where W_{erl} is the earliest execution time of the task graph when arbitrary number of processors are available. Thus, in general, our algorithm performs as good as their algorithm, and exhibits better time complexity for task graphs having $W_{erl} > O(\sqrt{n})$.

1   Introduction

A multicomputer is a multiple instruction stream, multiple data stream (MIMD), distributed memory parallel computer. It consists of several processors each of which has its own memory. There is no global memory and processors communicate via message passing. The objective of multiprocessing is to minimize the overall computation time of a problem (and hence gain speedup compared to a sequential algorithm) by solving it on a multicomputer. However, it is a challenge

---

[*] Part of this work was done while the first author was visiting the Department of Computer Science and Engineering, the University of Texas at Arlington. This work was supported by NASA Ames Research Center under Cooperative Agreement Number NCC 2-5395.

B. Monien, V.K. Prasanna, S. Vajapeyam (Eds.): HiPC 2001, LNCS 2228, pp. 47–57, 2001.

to develop efficient software that takes full advantage of hardware parallelism offered by a multicomputer. A parallel program is a collection of tasks that may run serially or in parallel. These tasks must be optimally placed on the processors with the shortest possible execution time. This is known as the *scheduling problem*, which is NP-Hard in its general form [13].

A program on a multicomputer can be modeled as a directed acyclic graph $G(\Gamma, \rightarrow, \mu, \eta)$, where $\Gamma = \{T_1, T_2, \ldots, T_n\}$ represents the set of tasks constituting the program, $\rightarrow$ represents the temporal dependencies between the tasks, $\mu(T_i)$ denotes the execution time for task T_i, and each arc $T_i \rightarrow T_j$ is associated with a positive integer $\eta(T_i, T_j)$ representing the number of messages sent from the task T_i to its immediate successor T_j. To chacterize the underlying system, a processor graph $P(\tau)$ is also introduced such that $\tau(i, j)$ represents the communication delay (i.e., the time to transfer a message) between processors i and j. This model is called Enhanced Directed Acyclic Graph (EDAG) [6].

Given a parallel program or equivalently an EDAG, the problem is to assign the tasks to different identical processors of a multicomputer so that the overall execution time or makespan is minimum. As this problem is NP-Hard [13], it is unlikely to design a polynomial time optimal algorithm for scheduling EDAG or precedence graphs on multicomputer systems. Therefore, many heuristic algorithms have been proposed in recent years [3,7,9-12].

This paper deals with two basic problems concerning the theory of scheduling. Given a precedence graph $G = (\Gamma, \rightarrow, \mu, \eta)$,

1. Find the minimum number of identical processors that are needed in order to execute the program represented by the precedence graph in time not exceeding that taken when arbitrary number of processors are available .
2. Determine the minimum time to process the given precedence graph when a fixed number of processors are available.

These problems being NP-Hard [13], researchers have proposed lower bounds for (1), i.e., minimum number of processors (LBMP), and lower bounds for (2), i.e., execution time (LBMT) for EDAG model [2,7,8].

Al-Mouhammed [2] was the first to suggest LBMT and LBMP. His algorithm runs in $O(m \log n + n^2)$ time, where m is the number of edges and n is the number of tasks in the precedence graph. Jain and Rajaraman [7] have proposed an algorithm which gives tighter LBMP and LBMT than that in [2] but takes $O(nW_{erl}^2 + m \log n)$ time where W_{erl} is the earliest completion time of the task graph when arbitrary number of processors are available. Using a graph partitioning technique, the authors [7] also suggested another algorithm which gives the current best LBMP and LBMT in $O(m \log n + n^2 + nW_{erl}^2)$ time.

In this paper, we propose new LBMP and LBMT needed to execute a given program on a multicomputer system, where the program is represented by an EDAG. We, then, show that these bounds are as sharp as the current known best bounds due to Jain and Rajaraman [7]. We present an algorithm to find these bounds in $O(m \log n + \min(n^2, W_{erl}^2))$ time. So, in general, our algorithm is as good as the algorithm due to Jain and Rajaraman [7] and exhibits better time complexity for task graphs having $W_{erl} > O(\sqrt{n})$. The key ideas of our approach include: (i) identification of certain points called *event points* and proving that the interval having event points as both ends are enough to compute the desired bounds; and (ii) the use of a *sweeping* technique.

2   Preliminaries

Let $G(\Gamma, \rightarrow, \mu, \eta)$ be a precedence graph having n nodes. The earliest time at which a task T_i can be started on any processor, called *earliest starting time*, is denoted by est_i. The earliest time at which a task can be completed on any processor, called *earliest completion time*, is denoted by ect_i. Let W_{erl} be the *earliest completion time* of the task graph. Then $W_{erl} = Max_{1 \leq i \leq n}\{ect_i\}$. Let $d(T_i)$ denote the *maximum time* (or delay) by which a task T_i can be postponed without increasing W_{erl} of the task graph. Let lst_i, the *latest starting time* of task T_i, be the latest time at which a task T_i can be started without increasing W_{erl}. So, $lst_i = est_i + d(T_i)$. Similarly, the *latest completion time* of a task T_i is given by $lct_i = ect_i + d(T_i)$.

Let Δt_m be the lower bound on the minimum increase in time to execute the precedence graph G over W_{erl} when m processors are used, and m_{min} be the lower bound on the number of processors required to complete G in W_{erl} time.

An $O(m \log n)$-time algorithm for finding est_i and $d(T_i)$ for a precedence graph is given in [2].

The **load density function** [4], $f(T_i, t_{s_i}, t)$, associated with a task T_i represents its activity which starts at time t_{s_i} and ends at time $t_{s_i} + \mu(T_i)$. For $t_{s_i} \in [est_i, lst_i]$, this is defined as:

$$f(t_i, t_{s_i}, t) = \begin{cases} 1, \text{ if } t \in [t_{s_i}, t_{s_i} + \mu(T_i)] \\ 0, \text{ otherwise} \end{cases}$$

The **load density function** $F(t_s, t)$ of a graph G is the sum of all the task activities at time t and is given by

$$F(t_s, t) = \sum_{i=1}^{n} f(T_i, t_{s_i}, t).$$

Let $\phi(t_1, t_2, t)$ represent the load density function within interval $[t_1, t_2] \subset [0, W_{erl}]$, after all tasks have been shifted to yield minimum overlap in $[t_1, t_2]$. Thus $\phi(t_1, t_2, t)$ can be viewed as the essential activity in time $[t_1, t_2]$ which should be executed within this time interval. Let $(m_{min})_{[a,b]} = \lceil \frac{1}{b-a} \int_a^b \phi(a, b, t)dt \rceil$, and $(\Delta t_m)_{[a,b]} = \lceil \frac{1}{m} \int_a^b \phi(a, b, t)dt - (b - a) \rceil$.

The LBMP and LBMT due to Al-Mouhammed [2] can be stated as follows:

$$m_{min} = Max_{1 \leq i \leq n}\{(m_{min})_{[est_i, lct_i]}\} \tag{1}$$

$$\Delta t_m = Max_{1 \leq i \leq n}\{(\Delta t_m)_{[est_i, lct_i]}\} \tag{2}$$

In other words, the author in [2] considered only the n intervals of the type $[est_i, lct_i]$, $1 \leq i \leq n$, to find the bounds. The time complexity to calculate Eqs. (1) and (2) is $O(m \log n + n^2)$.

Jain and Rajaraman [7] showed that sharper bounds can be obtained if all integer intervals in $[0, W_{erl}]$ are considered to compute LBMP and LBMT. They proposed the following sharper bounds:

$$m_{min} = Max_{0 \le t_1 < t_2 \le W_{erl}}\{(m_{min})_{[t_1, t_2]}\} \qquad (3)$$

$$\Delta t_m = Max_{0 \le t_1 < t_2 \le W_{erl}}\{(\Delta t_m)_{[t_1, t_2]}\} \qquad (4)$$

where t_1 and t_2 are integers.

However, $O(nW_{erl}^2 + m \log n)$ time is required to compute Eqs. (3) and (4). To get tighter bounds, the authors [7] have suggested an $O(n^2)$ time graph partitioning algorithm. They have shown that under certain conditions, the task graph can be partitioned into two or more parts in such a way that the LBMT (the LBMP) for the original task graph is the sum (minimum) of the LBMT (LBMP) of the different parts of the partition. So $O(m \log n + n^2 + nW_{erl}^2)$ time is needed to compute the current best known LBMT and LBMT due to Jain and Rajaraman [7].

3   Proposed Bounds and Algorithm

The heart of the algorithm due to [2] to compute the lower bounds is the calculation of $\phi(t_1, t_2, t)$ within $[t_1, t_2] \subset [0, W_{erl}]$. This is calculated by summing up the contribution of all the tasks to this interval. The contribution of the task T_i is computed by shifting the tasks to yield minimum overlap within this interval. We denote $\mu(T_i)$ by μ_i if no confusion arises.

A task T_i can start at any time $t_i \in [est_i, lst_i]$, without increasing W_{erl}. However, we show below that a task gives minimum overlap within an interval if it is scheduled to start at either est_i or lst_i.

Lemma 3.1: Let $[c, d] \subset [0, W_{erl}]$. A task T_i yields minimum overlap within $[c, d]$ if it is scheduled to start either at est_i or at lst_i.

Proof: Let s_i be the time when T_i is started. So, $est_i \le s_i \le lst_i$. We consider two cases.

Case I: $s_i \le c$.

If $s_i + \mu_i \le c$, then overlap of T_i in [c, d]$=0=$ overlap of T_i in [c, d] when T_i is scheduled at est_i. If $s_i + \mu_i > c$, then Overlap of T_i in [c, d] is $[c, d] \cup [s_i, s_i + \mu_i]$,
$= \min\{s_i + \mu_i, d\} - c$
$\ge \min\{est_i + \mu_i, d\} - c$, since $est_i \le s_i$
$=$ overlap when T_i is scheduled at est_i.
So, overlap of T_i in [c, d] is minimum if T_i is scheduled at est_i.

Case II: $s_i > c$.

If $s_i > d$, then overlap of T_i in [c, d]$=0=$ overlap of T_i in [c, d] when T_i is scheduled at lst_i. If $s_i \le d$, then overlap of T_i in $[c, d] = \min\{s_i + \mu_i, d\} - s_i$
$\ge \min\{lst_i + \mu_i, d\} - lst_i$, since $lst_i \le s_i$
$=$ overlap when T_i is scheduled at lst_i.

Thus, a smaller overlap is achieved when a task T_i is scheduled either at est_i or at lst_i than any other point. □

The heart of our lower bounds is the identification of certain points, called *event points* . Jain and Rajaraman [7] consider all interger subintervals of $[0, W_{erl}]$ for calculation of lower bounds. We will show that it is enough to consider all subintervals of $[0, w_{erl}]$ having both the ends as event points.

Let us now sort $est_i, lst_i, ect_i,$ and lct_i for all $i = 1, 2, ..., n$ in non-decreasing order and eliminate duplicate points to obtain distinct points $e_1, e_2, ..., e_p$, where $p \leq 4n$. Each point, so obtained, is called an **event point**.

Let $I = [c, d]$ be the interval for which minimum overlap is to be calculated. Let $[a, b] \subseteq I$ and $[a, b] \subseteq [e_i, e_{i+1}i]$, for some i, $1 \leq i \leq p - 1$.

The **activity**, $P_{[a,b]}(I)$, of $[a, b]$ with respect to I, is defined as the number of tasks which have a nonzero overlap with $[a, b]$ after being shifted to yield minimum overlap within I .

Lemma 3.2 Let $[a, b] \subset [e_i, e_{i+1}]$ for some i, $1 \leq i \leq p - 1$, and let $[e_i, e_{i+1}] \subset [x, y] \subseteq [0, W_{erl}]$. Then $P_{[a,b]}([x, y]) = P_{[e_i, e_{i+1}]}([x, y])$.

Proof: Follows directly from Lemma 3.1. $\qquad \square$

3.1 New LBMP and LBMT

In the following we propose a new LBMP and LBMT and we prove that they are as sharp as that given by Eqs. (2.3) and (2.4), respectively.

$$m_{min} = Max_{1 \leq i < j \leq p}\{(m_{min})_{[e_i, e_j]}\} \qquad (5)$$

$$\Delta t_m = Max_{1 \leq i < j \leq p}\{(\Delta t_m)_{[e_i, e_j]}\} \qquad (6)$$

Lemma 3.3: The bounds obtained by Eqs. (5) and (6) are as sharp as those obtained by Eqs. (3) and (4), respectively.

Proof: Let $I = [c, d]$ be any integer subinterval of $[0, W_{erl}]$.

Case I: $I \subseteq [e_i, e_{i+1}] = I'$, for some i, $1 < i < P$.

Let $L = e_{i+1} - e_i$, and $d - c = l$. Now by Lemma 3.2, $P_I(I'') = P_{I'}(I'') = Q$, say, where $I' \subset I'' \subset [0, W_{erl}]$.

Now, $(m_{min})_{[a,b]} = \lceil \frac{Q*l}{l} \rceil$, and $(m_{min})_{I'} = \lceil \frac{Q*L}{L} \rceil$. So, $(m_{min})_I = (m_{min})_{I'}$.

Also, $(\Delta t_m)_I = \lceil \frac{Q*l}{l} - l \rceil \leq \lceil L\frac{(Q-m)}{m} \rceil = \lceil \frac{Q*L}{m} - L \rceil = (\Delta t_m)_{I'}$.

Hence $(\Delta t_m)_I \leq (\Delta t_m)_{I'}$. In other words, it is better to consider the interval $[e_i, e_{i+1}]$ rather than any of its subintervals for both bounds.

Case II: There exists k, $2 \leq k \leq p - 1$, such that $e_k < d \leq e_{k+1}$.

Let i be the largest integer for which $e_i \leq c$, and let j be the smallest integer for which $d \leq e_j$. So, $e_i \leq c < d \leq e_j$ and $j \geq i + 2$. Hence, $c < e_{i+1} < d$.

Claim (i) $Max\{ (m_{min})_{[c,e_{j-1}]}, (m_{min})_{[c,e_j]}\} \geq (m_{min})_{[c,d]}$,

(ii) $Max\{ (\Delta t_m)_{[c,e_{j-1}]}, (\Delta t_m)_{[c,e_j]}\} \geq (\Delta t_m)_{[c,d]}$.

Proof of the claim:

Let $X = P_{[e_{j-1}, e_j]}(I)$, and $A = \int_c^{e_{j-1}} \phi(c, d, t)dt$. Now, $(m_{min})_{[c,d]} = \lceil \frac{A+X(d-e_{j-1})}{d-c} \rceil$, $(m_{min})_{[c,e_j]} = \lceil \frac{A+X(e_j-e_{j-1})}{e_j-c} \rceil$, and $(m_{min})_{[c,e_{j-1}]} = \lceil \frac{A}{e_{j-1}-c} \rceil$.

If $(m_{min})_{[c,d]} > (m_{min})_{[c,e_{j-1}]}$, then $\frac{A+X(e_j-e_{j-1})}{d-c} > \frac{A}{e_{j-1}-c}$. So,

$$A < X(e_{j-1} - c) \qquad (7)$$

If $(m_{min})_{[c,d]} > (m_{min})_{[c,e_j]}$, then $\lceil\frac{A+X(d-e_{j-1})}{d-c}\rceil > \lceil\frac{A+X(e_j-e_{j-1})}{e_j-c}\rceil$. So,

$$A > X(e_{j-1}-c) \qquad (8)$$

Now (7) and (8) cannot hold simultaneously. So, either $(m_{min})_{[c,e_{j-1}]} \geq (m_{min})_{[c,d]}$, or $(m_{min})_{[c,e_j]} \geq (m_{min})_{[c,d]}$
So claim (i) is true.

Now, $(\Delta t_m)_{[c,d]} = \lceil\frac{A+X(d-e_{j-1})}{m} - (d - c)\rceil$,
$(\Delta t_m)_{[c,e_{j-1}]}= \lceil\frac{A}{m} - (e_{j-1} - c)\rceil$, and $(\Delta t_m)_{[c,e_j]} = \lceil\frac{A+X(e_j-e_{j-1})}{m} - (e_j - c)\rceil$.
If, $(\Delta t_m)_{[c,d]} > (\Delta t_m)_{[c,e_{j-1}]}$, then

$$m < P \qquad (9)$$

Again, if $(\Delta t_m)_{[c,d]} > (\Delta t_m)_{[c,e_j]}$, then

$$m > P \qquad (10)$$

Now (9) and (10) cannot hold simultaneously. So, either, $(\Delta t_m)_{[c,e_{j-1}]} \geq (\Delta t_m)_{[c,d]}$ or $(\Delta t_m)_{[c,e_j]} \geq (\Delta t_m)_{[c,d]}$.
Hence claim (ii) is true.

So, it is sufficient to consider all intervals $[c,d]$, where d is an event point, to find LBMP and LBMT. Using a similar argument it can be proved that it is sufficient to consider all intervals $[c,d]$ where c is an event point, to find LBMP and LBMT.

Hence, the lemma is proved.   □

3.2   Computation of Minimum Overlap in an Interval

Let $S_1^c = \{T_i|est_i > c\}$, $S_2^c = \{T_i|est_i \leq c \leq lst_i$ and $ect_i > c\}$, and $S_3^c = \{T_i|lst_i < c$ and $ect_i > c\}$, where $c \in [0, W_{erl}]$. Note that tasks T_i with $ect_i \leq c$ do not belong to any of the above sets and gives zero overlap within $[c,d]$, $c < d \leq W_{erl}$.

We define b_i and l_i for each task T_i as follows
Rule 1: If $T_i \in S_1^c$, then $b_i = lst_i$ and $l_i = lct_i$.
Rule 2: If $T_i \in S_2^c$, then $b_i = lst_i$ and $l_i = ect_i + lst_i - c$.
Rule 3: If $T_i \in S_3^c$, then $b_i = c$ and $l_i = ect_i$.
Rule 4: If $T_i \notin (S_1^c \cup S_2^c \cup S_3^c)$, then $b_i = 0$ and $l_i = 0$.

The numbers b_i and l_i are the starting and ending time when the task T_i begins and ends its contribution for minimum overlap within the interval $[e_i, e_j]$ where $e_i = c$ and $i < j \leq p$. If $b_i \geq d$, then clearly the minimum overlap of the task T_i within $[c,d]$ is 0.

Let minOv denote the minimum overlap in a given interval. The following lemma gives a formula for computing the minimum overlap of a task T_i within $[c,d]$ in terms of b_i and l_i.
Lemma 3.4 Let $d > b_i$ for a task T_i. Then the minimum overlap of a task T_i within $[c,d]$ is $\min\{l_i, d\} - b_i$.
Proof Let ov_1 and ov_2 be the overlap of task T_i within $[c,d]$ when scheduled at est_i and lst_i, respectively. We consider the following cases corresponding to the above four rules.
Case I: $T_i \in S_1^c$.

So, $c < est_i$. Now, $ov_1 = \min\{d, ect_i\} - est_i = \min\{d - est_i, \mu(T_i)\}$
and $ov_2 = \min\{d, lct_i\} - lst_i = \min\{d - lst_i, \mu(T_i)\}$.

So, $\min\{ov_1, ov_2\} = ov_2 = \min\{d - lst_i, \mu(T_i)\} = \min\{l_i, d\} - b_i$.

Case II: $T_i \in S_2^c$.

So, $est_i \leq c \leq lst_i$. Considering the subcases (i) $d \leq ect_i$, (ii) $ect_i < d < ect_i + lst_i - c$, (iii) $d = ect_i + lst_i - c$, (iv) $d > ect_i + lst_i - c$, and the computation of ov_1 and ov_2 as in **case I**, it is a routine job to show that min $\{ ov_1, ov_2\} = \min\{l_i, d\} - b_i$ in this case.

Case III: $T_i \in S_3^c$.

So, $c > lst_i$. Now, $ov_1 = \min\{d, est_i\} - c$ and $ov_2 = \min\{d, lct_i\} - c$. So, $ov_1 \leq ov_2$. Hence, $minOv = \min\{d, est_i\} - c = \min\{d, l_i\} - b_i$.

Case IV: $T_i \notin (S_1^c \cup S_2^c \cup S_3^c)$.

So, $l_i = b_i = 0$. So, $minOv = \min\{l_i, d\} - b_i = l_i - b_i = 0$.

The Lemma is thus proved. $\qquad\square$

3.3 Algorithm Pseudo-Code

Below, we give an algorithm that computes a matrix $OV_{p \times p}[]$, where $OV[i, j] = \int_{e_i}^{e_j} \phi(e_i, e_j, t)dt$ for $1 \leq i < j \leq p$.

Lemma 3.4 tells that the minimum overlap in an interval can be computed by adding the l's and substracting the b's of the tasks. Again, if we know the minimum overlap in an interval [c,d], then minimum overlap in any other interval [c,e], where $e > d$ can be computed by making use of the minimum overlap in [c,d] and by adding certain l's and subtracting certain b's of tasks. Making use of this fact and a *sweeping* technique, we compute the minimum overlap. We find the overlap of the intervals in the following sequence:

$[e_1, e_2], [e_1, e_3], \ldots, [e_1, e_p], \qquad [e_2, e_3], \ldots, [e_2, e_p], \ldots, [e_{p-2}, e_{p-1}], [e_{p-2}, e_p],$
$[e_{p-1}, e_p]$

i.e, we fix the left boundary and vary the right boundary till we reach to e_p, then we increment the left boundary and continue the process till the left boudary becomes e_p. So, at the end of the ith stage we have the overlap of all subintervals $[e_j, e_i]$ for $j = 1, 2, ..., i-1$. Algorithm Minoverlap, which is given below is a clever use of the Lemma 3.4.

Algorithm MinOverlap
Input: A task graph $G(\Gamma, \rightarrow, \mu, \eta)$ and a processor graph $P(\tau)$.
Output: A matrix $[OV]_{p \times p}$ such that $OV[i, j] = \int_{e_i}^{e_j} \phi(e_i, e_j, t)dt$, $1 \leq i < j \leq p$ gives minimum overlap within the interval $[e_i, e_j]$.
$\{$
1. Find $est_i, lst_i, ect_i, lct_i, 1 \leq i \leq n$;
2. Sort $est_i, lst_i, ect_i, lct_i$, $1 \leq i \leq n$ to get distinct event points $e_1, e_2, \ldots, e_p, p \leq 4n$;
3. Insert all tasks in S_1;
   Find corresponding b_i and l_i according to Rule (1);
   Sort all b_i in non-decreasing order to get sorted list B_1;
   Sort all l_i in non-decreasing order to get sorted list L_1;

$S_2 = S_3 = B_2 = B_3 = L_2 = L_3 = \phi$;
OV[i,j] = 0, $1 \le i < j \le p$;
$e_0 = e_1$;
4. for $i = 1$ to $p - 1$ do
/* Here $c = e_i$ */
{ Call Update(i);
Merge sorted lists $B_1, B_2, B_3, L_1, L_2, L_3$ to get sorted list $R = r_1, r_2, \ldots, r_q$
(all need not be distinct);
for $j = 1$ to q do
if ($r_j \in B_k$ for some $k = 1, 2, 3$) then type(r_j) = b; else type(r_j) = l;
count = 0; pos = 1; $r_0 = r_1$; $B = 0$; $L = 0$;
for $j = i + 1$ to p do /* Here $d = e_j$ */
{ while ($e_j > r_{pos}$) do {
if (type(r_{pos}) = 'b') then
count = count + 1; $B = B + r_{pos}$;
else count = count $-$ 1; $L = L + r_{pos}$; pos = pos + 1; }
OV[i,j] = L-B+count * e_j; }
}
}

As c varies from e_1 to e_{p-1}, S_1^c, S_2^c and S_3^c also change. The following procedure obtains $S_1^{e_{i+1}}, S_2^{e_{i+1}}$, and $S_3^{e_{i+1}}$ from $S_1^{e_i}, S_2^{e_i}$, and $S_3^{e_i}$.

Procedure Update(i) {
for all tasks $T_j \in S_1$ do
if ($est_j \ge e_i$) then Move_and_Adjust(T_j, S_1, S_2);
for all tasks $T_j \in S_2$ do
if ($ect_j \le e_i$) then
{ Delete T_j from S_2;
Delete b_j and l_j from B_2 and L_2, respectively; }
else if ($lst_j > e_i$) then
Move_and_Adjust(T_j, S_2, S_3);
for all tasks $T_j \in S_2$ do
$l_j = l_j - (e_i - e_{i-1})$;
for each task $T_j \in S_3$ do
if ($ect_j \le e_i$) then { Delete T_j from S_3;
Delete b_j and l_j from B_3 and L_3, respectively; }
for each task $T_j \in S_3$ do
$b_j = e_i$; }

Move_and_Adjust(T_i, S_j, S_k) {
Move T_i from S_j to S_k;
Delete b_i and l_i from B_j and L_j, respectively;
Find new b_i and l_i according to Rule k, and
insert in B_k and L_k, respectively; }

We are now in a position to give an algorithm to compute the lower bounds.

Algorithm L_Bound
Input: A task graph $G(\Gamma, \rightarrow, \mu, \eta)$ and processor graph $P(\tau)$ and number of available processors m.
Output: $m_{\min}$, the LBMP and Δt_m, the LBMT.
{
1. Compute matrix OV[i,j] using algorithm MinOverlap.
2. Compute the following using the matrix OV.

$$m_{min} = Max_{1 \leq i < j \leq p}\{\frac{OV[i,j]}{(j-i)}\}$$

$$\Delta t_m = Max_{1 \leq i < j \leq p}\{\frac{OV[i,j]}{m} - (j-i)\}$$

}

4   Proof of Correctness and Complexity

The proof of correctness of Algorithm L_Bound follows from the proof of correctness of MinOverlap, which we shall prove below. Also, we show that it can be implemented in $O((m+n)\log n + Min\{nW_{erl}, n^2\})$ time.

Let $[c,d] \subseteq [0, W_{erl}]$. If $T_i \in S_1^c \cup S_2^c \cup S_3^c$, then clearly $b_i \geq c$. So, if $l_i \leq d$, then $b_i \in [c,d]$. Let $X_1 = \{T_i | l_i < d\}$ and $X_2 = \{T_i | l_i \geq d \text{ and } b_i < d\}$.

So, Minimum overlap within an interval $[c,d] = \sum_{T_i \in X_1} (l_i - b_i) + \sum_{T_i \in X_2} (d - b_i) = (\sum_{T_i \in X_1} l_i - \sum_{T_i \in X_1 \cup X_2} b_i) + d.r$, where $r = |X_2|$.

In the MinOverlap algorithm, $L = \sum_{l_i < d} l_i$, and $B = \sum_{b_i < d} b_i$, and *count* is the difference between the number of b's and number of l's in $[c,d] = [e_i, e_j]$. Hence we have the following results.

Theorem 4.1 Algorithm MinOverlap correctly computes the matrix OV[i, j], $1 \leq i < j \leq p$, such that

$$OV[i,j] = \int_{e_i}^{e_j} \phi(e_i, e_j, t)dt.$$

Theorem 4.2 Algorithm L_Bound requires $O((n+m)\log n + Min\{nW_{erl}, n^2\})$ time.
Proof: The complexity of the algorithm L_Bound is dominated by the complexity of algorithm MinOverlap. So, it is enough to show that algorithm MinOverlap runs in $O((m+n)\log n + Min\{nW_{erl}, n^2\})$ time. Now, we analyze algorithm MinOverlap. Step 1 can be implemented in $O(m \log n)$ time [2]. Step 2 can be implemented in $O(n \log n)$ time by using merge Sort. In Step 3, the procedure Update is invoked p times. In every call of Update, each task $T_i \in (S_1^{e_{i-1}} \cup S_2^{e_{i-1}} \cup S_3^{e_{i-1}})$ is scanned once. Each scan takes $O(n)$ time. Again each task is deleted from $S_j^{e_{i-1}}$ and inserted in $S_{j+1}^{e_i}$, $1 \leq j \leq 2$, for some i.

The lists B_i and L_i can be implemented as height balanced trees [1]. Hence each insertion and deletion takes $O(\log n)$ time. Throughout the p calls of Update, each task is inserted and deleted a maximum of three times. This is because a task from $S_1^{e_{i-1}}$ may be moved to $S_2^{e_i}$ and that from $S_2^{e_{i-1}}$ may be moved to $S_3^{e_i}$ but not vice-versa. Thus $O(n)$ insertions and deletions take place. So Step 3 takes $O(n \log n + n.p)$ time. Since $p \leq 4n$ and $p \leq W_{erl}$, Step 3 takes $O(n \log n + Min\{nW_{erl}, n^2\})$ time. Now for each i, Step 4 takes $O(p - i)$ time. So Step 4 takes an overall time $O(p^2 = O(Min\{n^2, W_{erl}^2\})$. Hence, the algorithm L_Bound takes $O((n+m) \log n + Min\{nW_{erl}\}, n^2) + Min(n^2, W_{erl}^2))$ time which can be reduced to $O((n + m) \log n + Min\{n^2, nW_{erl}\})$ time.   □

Jain and Rajaraman [7] have introduced the concept of partitioning technique. They have shown that under certain conditions, the task graph can be decomposed into two or more parts so that the lower bounds for the original task graph is sum of the lower bounds of the parts. If we incorporate the partitioning technique of Jain and Rajaraman [7], we can compute the current best LBMP and LBMT in $O(m \log n + n^2)$ time as partitioning the precedence graph takes $O(n^2)$ time [7].

5   Conclusion

In this paper, we have proposed new lower bounds for time and processors to execute a task graph and have also shown that our bounds are as sharp as those obtained by Jain and Rajaraman [7]. We then proposed an algorithm to compute LBMP and LBMT. Our algorithm is at least as good as the one in [7], but performs better if $W_{erl} > O(\sqrt{n})$, where W_{erl} is the earliest execution time of the task graph having n nodes when arbitrary number of processors are available. More precisely, in this case, our algorithm has an improvement of $O(n)$ over that due to [7].

References

1. Aho, A. V., Hopcroft, J. E., and Ullman, J. D. *The Design and Analysis of Computer Algorithms*, 1974, Addison-Wesley Publishing Company.
2. Al-Mouhammed, M. A. "Lower bound on the number of processors and time for scheduling precedence graphs with communication costs". *IEEE Trans. Soft. Engg.* 12 (1990) 1390-1401.
3. El-Rewini, H. and Lewis, T. "Scheduling parallel program tasks onto arbitrary target machines". *J. of Parallel and Dist. Comput.* 9, 2 (1990) 138-153.
4. Fernandez, E. V. and Bussell, "B. Bounds on the number of processors and time for multiprocessors optimal schedules". *IEEE Trans Comput* 22 (1973) 745-751.
5. Graham, R. "Bounds on certain multiprocessing timing anomalies", *SIAM J. Appl. Math.* 17, 2 (1969) 416-429.
6. Hwang, J. J., Chow, Y.C., Anger, F., and Lee, C. Y. "Scheduling Precedence graphs in systems with inter-processor communication times". *SIAM J. of Comput.* 18, 2 (1989), 244-257.

7. Jain, K. K. and Rajaraman, V. "Improved lower bounds on time and processors for scheduling precedence graphs on multicomputer systems." *J. of Parallel and Dist Comput.* 28 (1995), 101-108.

8. Jung, H., Kirousis, L. M., and Spirakis, P. "Lower bounds and efficient algorithms for multiprocessor scheduling of directed acyclic graphs with communication delays". *Inform. and Comput.* 105 (1993), 94-104.

9. Kwok Yu-Kwong and Ahmad I. "Static scheduling algorithms for allocating directed task graphs to multiprocessors", *ACM Computing Surveys* , 31, 4 (1999) 406-471.

10. Kwok Yu-Kwong and Ahmad I. "Benchmarking and comparison of the task graphs scheduling algorithms", *J. of Parallel and Dist Comput* 59, 103 (1999) 381-422.

11. Lee C.Y., Hwang J.J. , Chow Y.C. and Anger, F. "Multiprocessor scheduling with interprocessor communication delays". *Operations Res. Letters* 7 (1988), 141-145.

12. Manoharan, S. and Thanisch, P. "Assigning dependency graphs onto processor networks". *Parallel Computing* 17 (1991), 63-73.

13. Ullman, J. D. "NP-complete scheduling problems". *J. Comput System Sci.* 10 (1975), 384-393.

On Job Scheduling for HPC-Clusters and the dynP Scheduler

Achim Streit

PC^2– Paderborn Center for Parallel Computing,
33102 Paderborn, Germany,
streit@upb.de
http://www.upb.de/pc2

Abstract. Efficient job-scheduling strategies are important to improve the performance and usability of HPC-clusters. In this paper we evaluate job-scheduling strategies (FCFS, SJF, and LJF) used in the resource management system CCS (Computing Center Software). As input for our simulations we use two job sets that are generated from trace files of CCS.

Based on the evaluation we introduce the dynP scheduler which combines the three scheduling strategies and dynamically changes between them online. The average estimated runtime of jobs in the waiting queue and two bounds are used as the criterion whether or not dynP switches to a new policy. Obviously the performance of dynP depends on the setting of the two bounds. Diverse parameter pairs were evaluated, and with the proper setting dynP achieves an equal or even better (+9%) performance.

1   Introduction

In recent years clusters of workstations became more and more popular which is surely based on the good price-performance ratio compared to common super-computers (SP2, T3E, SX-5, Origin, or VPP). Nowadays two types of clusters are found: HTC-clusters (High Throughput Computing) are very popular as the nodes hardware consists of low-cost "components of the shelf" from the computer shop around the corner connected with one Fast Ethernet network. This network limits the scalability of the system, so that tightly coupled applications are less suitable for HTC-clusters.

In contrast HPC-clusters (High Performance Computing) combine a fast (high bandwidth and low latency) interconnect and powerful computing nodes. The 192-processor cluster at the PC^2 is such a HPC cluster. It was installed in 1998 as a prototype of the new Fujitsu-Siemens *hpcLine* product line. The cluster consists of 96 double processor Intel Pentium III 850 MHz nodes. The SCI interconnect combines high bandwidth (85 MByte/s) with low latency (5.1 μs for ping-pong/2). We use our own resource management system called CCS (Computing Center Software) [9,10] for accessing and managing the cluster. The cluster is operated in space-sharing mode as users often need the complete compute power of the nodes and the exclusive access to the network interface.

B. Monien, V.K. Prasanna, S. Vajapeyam (Eds.): HiPC 2001, LNCS 2228, pp. 58–67, 2001.

Currently three space-sharing scheduling strategies are implemented in CCS: First Come First Serve (FCFS), Shortest Job First (SJF), and Longest Job First (LJF) each combined with conservative backfilling. As the scheduler has a strong influence on the system performance, it is worth while to improve the performance of the scheduling strategy. In the past preliminary evaluations based on simulations were done by Feitelson et al. [1,14] for the IBM SP2, or by Franke, Moreira et al. [7,12] for the ASCI Blue-Pacific.

The remainder of this paper is organized as follows. Section 2 describes the scheduling strategies and backfilling in detail. The two job sets for the evaluation are described in Section 3. After the performance metrics are described in Section 4, the concept of the new dynP scheduler is presented in Section 5. Finally Section 6 compares the performance of dynP to SJF, FCFS, and LJF.

2   Scheduling Strategies

Today's resource management systems generally operate after a queuing system approach: jobs are submitted to the system, they might get delayed for some time, resources are assigned to them, and they leave the queueing system. A classification of queuing systems was done by Feitelson and Rudolph in [4]. Here we look at an open, on-line queueing model. In a simulation based evaluation of job scheduling strategies the two important phases of a scheduling strategy are: 1) putting jobs in the queue (at submit time) and 2) taking jobs out of the queue at start time.

Different alternatives are known for the two phases (details in [2]). Here we sort the waiting queue by increasing submission time (FCFS) and increasing (SJF) or decreasing (LJF) estimated runtime. Furthermore we always process the head of the queue. With that the scheduling process looks as follows:

```
WHILE (waiting queue is not empty) {
    get head of queue (job j);
    IF (enough resources are free to start j) {
        start j;
        remove j from queue;
    } ELSE {
        backfill other jobs from the queue;
    }
}
```

Backfilling is a well known technique to improve the performance of space-sharing schedulers. A backfilling scheduler searches for free slots in the schedule and utilizes them with suitable jobs without delaying other already scheduled jobs. For that, users must provide an estimated runtime for their jobs, so that the scheduler can predict when jobs are finished and others can be started. Jobs are cancelled, if they run longer than estimated. Two backfilling variants are known: *conservative* [3] and *aggressive* [11]. The aggressive variant is also known as EASY (Extensible Argonne Scheduling sYstem [13]) backfilling.

As the name implies aggressive backfilling might delay the start of jobs in the queue. The drawbacks are 1) an unbounded delay for job start, and 2) an unpredictability of the schedule [1]. Primarily due to the unbounded delay (especially for interactive jobs), aggressive backfilling is not used in CCS.

3   Workload Model

The workload used in this evaluation is based on trace data generated by CCS. In these traces CCS logs the decisions (start and end time of a job) of the scheduler. Since CCS did not log the submit time of a job in the past, the waiting time (from submit to start) could not be derived from the here used original trace.

Using this trace again as input for a scheduler (and for the evaluation) would result in meaningless performance numbers. Each job will be started immediately with zero waiting time, regardless of the scheduling strategy. To achieve useful results the trace needs to be modified to generate a higher workload. When the workload is increased the scheduler is no longer able to schedule each newly submitted job immediately. This leads to a filled waiting queue (i.e. backlog) and waiting times greater zero. As the backlog grows, a backfilling scheduler has more opportunities to find suitable jobs to backfill.

Methods for modifying the trace are:

1. A smaller average interarrival time: Here the duration between each two job submits is reduced. Hence it is possible to simulate more users on the cluster. By reducing the duration down to 0, an offline scheduling is simulated.
2. Extension of job runtime: As the estimated runtime is also given, both values (actual and estimated runtime) have to be modified. This would remove common characteristics from the job set, like e.g. specific bounds for the estimated runtime (i.e. 2 hours). Furthermore the used area of jobs would be changed.

In this case we use the first method and introduce the *shrinking factor*. The submit time of each job is multiplied with this factor. We use values ranging from 0.75 down to 0.25 in steps of 0.05. This enables us to evaluate the performance of different scheduling strategies over a wide range of workloads.

As trace data from two years is available, we extracted two significant job sets from the data pool. These two job sets (Tab. 1) were chosen, because of their different characteristics and because these job-sets quite well represent our current users. In set1 shorter jobs are submitted at a high rate compared to set2. In set2 the jobs are more than twice as long, but are submitted with a greater lag.

As can be seen in Fig. 1 both job sets have a lot of narrow (small number of requested nodes) jobs and only a moderate number of wide (highly parallel) jobs. Furthermore peaks are noticeable at powers of 2. Computing the product of width (requested nodes) and length (actual runtime) leads to Fig. 2. Nodedays more likely represent the resource usage of jobs. A more detailed look on the estimated runtime of jobs can be found in [2].

Table 1. Job set characteristics

job set	number of jobs	average requested nodes	average actual runtime [sec.]	average estimated runtime [sec.]	average inter-arrival time [sec.]
set1	8469	13.49	2835.83	6697.88	628.19
set2	8166	10.11	6697.88	21634.00	1277.34

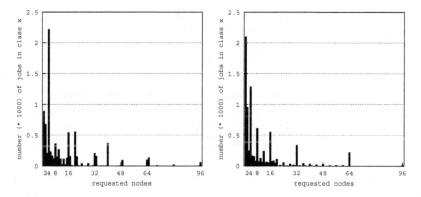

Fig. 1. Distribution of job width: number of jobs for each job width (requested nodes). set1 left, set2 right.

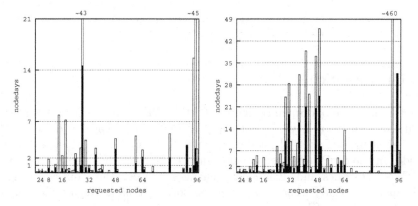

Fig. 2. Nodedays for actual (black boxes) and estimated (white boxes) runtimes. set1 left, set2 right.

4   Performance Metrics

For evaluating the simulated scheduling strategies and comparing them with each other, we basically focus on two metrics: the *average response time* and the *utilization* of the whole cluster. Herein the average response time is user centric whereas the utilization is owner centric. Note, that both numbers are contrary.

With these numbers we plot diagrams showing the utilization on the x-axis and the average response time on the y-axis. Combined with the shrinking factor we get several data points for each strategy and plot a curve. As the objective is to achieve high utilization combined with low average response times, curves have to converge to the x-axis.

Using

$t_j.submitT$	for the submit time of job j
$t_j.startT$	for the start time of job j
$t_j.endT$	for the end time of job j
n_j	for the occupied number of nodes of job j
J	for the total number of jobs
N	for the total number of available nodes
$firstSubmitT$	for the time the first submit event occurred $(\min_J t_j.submitT)$
$lastEndT$	for the time the last end event occurred $(\max_J t_j.endT)$

We define the average response time:

$$ART = \frac{1}{J} * \sum_J (t_j.endT - t_j.submitT)$$

and the utilization of the cluster:

$$UTIL = \frac{\sum_J (n_j * (t_j.endT - t_j.startT))}{N * (lastEndT - firstSubmitT)}$$

5   The dynP Scheduler

A first look at the performance figures of SJF, FCFS, and LJF shows that each scheduling strategy achieves an optimal performance for a specific job set and/or range of utilization (see [2] for details, or Fig. 3 and Fig. 4). That lead us to the idea of the dynP scheduler (for **dyn**amic **P**olicy), a scheduler that combines the three strategies and dynamically changes between them online. Now the problem is to decide when to change the sorting policy. We use the average estimated runtime of jobs currently in the waiting queue as criterion, because: 1) the performance of each strategy strongly depends on the characteristics of the job set (i.e. jobs in the waiting queue), and 2) SJF and LJF sort the queue according to the estimated runtime.

A similar approach was introduced in the IVS (Implicit Voting System) scheduler [6,8]. The IVS Scheduler uses the utilization of the machine and the ratio of batch and interactive jobs to decide which policy to take. The decider of IVS works as follows:

```
CASE
    (MPP is not utilized) --> switch to FCFS;
```

```
    (# interactive jobs > (# batch jobs + delta)) --> switch to SJF;
    (# batch jobs > (# interactive jobs + delta)) --> switch to LJF;
END;
```

We use two bounds to decide when the policy should be changed. If the average estimated runtime in the queue is smaller than the *lower_bound* the policy is changed to SJF. LJF is used when the average estimated runtime is greater than the *upper_bound*. A change of policy also necessitates a reordering of the waiting queue, as always the head of the queue is processed. If the reordering is omitted, certainly the wrong job according to the new policy is scheduled. The decider of the dynP scheduler works as follows:

```
IF (jobs in waiting queue >= 5)
    AERT = average estimated runtime of all \
           jobs currently in the waiting queue;
    IF (0 < AERT <= lower_bound) {
            switch to SJF;
    } ELSE IF (lower_bound < AERT <= upper_bound) {
            switch to FCFS;
    } ELSE IF (upper_bound < AERT) {
            switch to LJF;
    }
    reorder waiting queue according to new policy;
}
```

We use a threshold of 5 to decide whether or not a check for policy change is done. With that we prevent the scheduler from unnecessary policy changes.

6   Performance of the dynP Scheduler

Obviously the performance of the dynP scheduler strongly depends on the parameter settings. We simulated various combinations of lower and upper bound settings [2]. After a first evaluation we restricted the parameters for a second run. We found out that the combination of 7200 s (2h) for the lower bound and 9000 s (2h:30min) for the upper bound is superior to all other combinations.

Table 2. Lower and upper bounds used in a first (top) and second (bottom) run.

lower	600, 1200, 1800, 3600
upper	3600, 7200, 14400, 21600, 43200

lower	6600, 7200, 7800
upper	8400, 9000, 9600

Note, if lower and upper bound are equal, FCFS is not used at all. Then the decider switches only between SJF and LJF. In a real environment the parameters for lower and upper bound have to be set by the system administrator based on a profound knowledge. This fine-tuning is necessary to get the best performance out of the dynP scheduler and its environment (machine configuration, user behavior).

As the dynP scheduler has been developed to be superior to FCFS, SJF, and LJF independent from the job set, we have to compare the performance to these strategies.

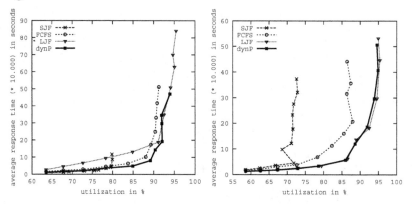

Fig. 3. Performance (average response time) focusing on maximal utilization in saturated state for the dynP scheduler with lower bound 7200 s and upper bound 9000 s. set1 left, set2 right.

When looking at the maximum utilization achieved in the saturated state dynP achieves similar results as LJF (~92% for set1 and ~94% for set2). Especially in the diagram for set1 the dynamic character of dynP is obvious. At low utilizations the performance (i.e. average response time) is as good as SJF. With increasing utilization the dynP curve follows FCFS and is even slightly better, even though LJF is rarely selected.

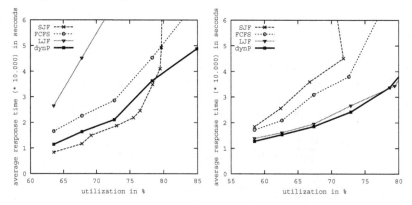

Fig. 4. Performance (average response time) focusing on medium utilizations for the dynP scheduler with lower bound 7200 s and upper bound 9000 s. set1 left, set2 right.

Focusing again on medium utilizations (Fig. 4) the performance of dynP is almost as good as SJF for set1 and LJF for set2. In set2 the performance can even be increased by about 8% (see Tab. 4).

Table 3. Average response times for SJF, FCFS, and LJF at medium utilizations. FCFS is used as reference for computing the percentages. set1 left, set2 right.

	SJF	FCFS	LJF
~63%	8335 s	16587 s	26459 s
	(+49%)	(0%)	(-60%)
~78%	34840 s	45292 s	93641 s
	(+23%)	(0%)	(-106%)

	SJF	FCFS	LJF
~58%	18388 s	17182 s	13883 s
	(-7%)	(0%)	(+19%)
~73%	45080 s	37998 s	26673 s
	(-19%)	(0%)	(+30%)

Table 4. Comparing average response times for dynP with a lower bound of 7200 s and an upper bound 9000 s at medium utilizations. As reference for computing the percentages SJF for set1 and LJF for set2 is used (best in each case).

set1	SJF	dynP_7200_9000
~63%	8335 s	11454 s
	(0%)	(-37%)
~78%	34840 s	36324 s
	(0%)	(-4%)

set2	LJF	dynP_7200_9000
~58%	13883 s	12769 s
	(0%)	(+8%)
~73%	26673 s	24214 s
	(0%)	(+9%)

As the dynP scheduler dynamically changes its policy, Fig. 5 shows how long each policy is used. The utilization is again plotted on the x-axis. The time each policy is used (in percent) is shown on the y-axis. Note, that the percentages are accumulated. Therefore the percentage for SJF is found between the x-axis and the first curve, the gap between the first and second curve represents the percentage of FCFS, and finally the gap between the second curve and the 100% percentage line represents LJF. For set1 the dynP scheduler utilizes the LJF policy only for 15% and less, as the pure LJF scheduler achieves only poor results (cf. Fig. 3). On the other hand FCFS and SJF are used more often. FCFS starts at about 30% for low utilizations and ends up at ~80%. At the same time the usage of SJF drops from ~56% to 19%. Obviously a different

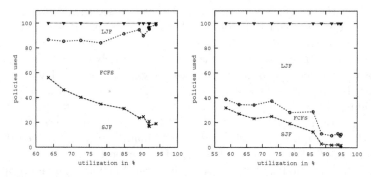

Fig. 5. Accumulated percentages of policies used by the dynP scheduler with lower bound 7200 s and upper bound 9000 s. set1 left, set2 right.

characteristic can be seen for set2. LJF dominates from the beginning, starting at 61% and ending at slightly over 90%. FCFS and SJF are less used with increasing utilization (FCFS: between 8 and 16%, SJF: from 31% down to 1%).

Based on its adaptability the dynP scheduler achieves a good performance regardless of the used job sets, i.e. the jobs in the waiting queue in a real environment. Of course the performance strongly depends on the parameter settings for the lower and upper bound. The complete set of simulated combinations (cf. 1st run in Tab. 2) shows that if awkward pairs of lower and upper bounds are chosen, the performance can drop significantly.

7   Conclusion

In this paper we compared space-sharing strategies for job-scheduling on HPC-clusters with a scheduling simulator. Therein we focused on FCFS, SJF, and LJF which are used in CCS (Computing Center Software). Each strategy was combined with conservative backfilling to improve its performance. To compare the results we used the average response time as an user centric, and the utilization of the whole cluster as an owner centric criterion.

As job input for the simulations we used two job sets that were extracted from trace data of the *hpcLine* cluster. To increase the workload for the scheduler we used a shrinking factor (0.75 down to 0.25 in steps of 0.05) to reduce the average interarrival time. With that a wide range of utilization (55 - 95%) was achieved. Each scheduling strategy reached a saturated state, in which a further increase of workload does not lead to a higher utilization, but only to higher average response times. LJF achieved the highest utilization for both job sets, with a peak at 95%. The performance at medium utilizations up to approx. 80% showed an unbalanced behavior. None of the three strategies was superior for both job sets. SJF performs best for set1 with a performance gain of $\tilde{2}3\%$ at 78% utilization compared to FCFS. For the other job set at an equal utilization SJF was $\tilde{1}9\%$ worse (cf. Tab. 3). FCFS proved to be a good average. These anticipated results rest upon the different characteristics of the job sets combined with the basic ideas of the scheduling strategies.

Based on these results we presented the dynP scheduler, which changes its policies (FCFS, SJF, and LJF) online. As criterion when to change the policy, dynP uses the average estimated runtime of jobs currently in the queue. A lower and upper bound specifies when to change the policy. Obviously the performance of the dynP scheduler strongly depends on the chosen bounds. The best setting found for the two jobs sets were 7200 s (2h) for the lower and 9000 s (2h30m) for the upper bound. Compared to SJF for set1 and LJF for set2 the dynP scheduler performed only slightly worse (-4%) for set1 and even better (+9%) with set2 at medium utilizations. The maximum utilization achieved in the saturated state (~95%) is as good as with LJF for both job sets.

As already mentioned the performance of dynP depends on the proper settings of the bounds. To ease the usability of the dynP scheduler we are currently working on an adaptive version of dynP with does not require anymore startup parameters. Self-tuning systems [5] might be used for this. In the future we will also add the dynP scheduler to CCS and use it for everyday work. This will show, if the simulated results from this paper can also be achieved in a real environment.

References

1. A. Mu'alem and D. Feitelson. Utilization, Predictability, Workloads, and User Runtime Estimates in Scheduling the IBM SP2 with Backfilling. In *IEEE Trans. Parallel & Distributed Systems 12(6)*, pages 529–543, June 2001.
2. A. Streit. On Job Scheduling for HPC-Clusters and the dynP Scheduler. TR-001-01, PC2 - Paderborn Center for Parallel Computing, Paderborn University, July 2001.
3. D. Feitelson and A. Weil. Utilization and Predictability in Scheduling the IBM SP2 with Backfilling. In *Proceedings of the 1st Merged International Parallel Processing Symposium and Symposium on Parallel and Distributed Processing (IPPS/SPDP-98)*, pages 542–547, Los Alamitos, March 1998. IEEE Computer Society.
4. D. G. Feitelson and L. Rudolph. Metrics and Benchmarking for Parallel Job Scheduling. *Lecture Notes in Computer Science*, 1459:1–24, 1998.
5. D.G. Feitelson and M. Naaman. Self-Tuning Systems. In *IEEE Software 16(2)*, pages 52–60, April/Mai 1999.
6. F. Ramme and T. Romke and K. Kremer. A Distributed Computing Center Software for the Efficient Use of Parallel Computer Systems. *Lecture Notes in Computer Science*, 797:129–136, 1994.
7. H. Franke and J. Jann and J. Moreira and P. Pattnaik and M. Jette. An Evaluation of Parallel Job Scheduling for ASCI Blue-Pacific. In *Proceedings of SC'99, Portland, Oregon*, pages 11–18. ACM Press and IEEE Computer Society Press, 1999.
8. J. Gehring and F. Ramme. Architecture-Independent Request-Scheduling with Tight Waiting-Time Estimations. *Lecture Notes in Computer Science*, 1162:65–80, 1996.
9. A. Keller, M. Brune, and A. Reinefeld. Resource Management for High-Performance PC Clusters. *Lecture Notes in Computer Science*, 1593:270–281, 1999.
10. A. Keller and A. Reinefeld. CCS Resource Management in Networked HPC Systems. In *Proc. of Heterogenous Computing Workshop HCW'98 at IPPS, Orlando, 1998; IEEE Computer Society Press*, pages 44–56, 1998.
11. D. A. Lifka. The ANL/IBM SP Scheduling System. *Lecture Notes in Computer Science*, 949:295–303, 1995.
12. J.E. Moreira, H. Franke, W. Chan, and L. L. Fong. A Gang-Scheduling System for ASCI Blue-Pacific. *Lecture Notes in Computer Science*, 1593, 1999.
13. J. Skovira, W. Chan, H. Zhou, and D. Lifka. The EASY – LoadLeveler API Project. *Lecture Notes in Computer Science*, 1162:41–47, 1996.
14. D. Talby and D. G. Feitelson. Supporting Priorities and Improving Utilization of the IBM SP2 Scheduler Using Slack-Based Backfilling. TR 98-13, Hebrew University, Jerusalem, April 1999.

An Adaptive Scheme for Fault-Tolerant Scheduling of Soft Real-Time Tasks in Multiprocessor Systems*

R. Al-Omari, Arun K. Somani, and G. Manimaran

Dependable Computing & Networking Laboratory
Dept. of Electrical and Computer Engineering
Iowa State University, Ames, IA 50011
{romari,arun,gmani}@iastate.edu

Abstract. The scheduling of real-time tasks with primary-backup based fault-tolerant requirements has been an important problem for several years. Most of the known scheduling schemes are non-adaptive in nature meaning that they do not adapt to the dynamics of faults and task's parameters in the system. In this paper, we propose an adaptive fault-tolerant scheduling scheme that has a mechanism to control the overlap interval between the primary and backup versions of tasks such that the overall performance of the system is improved. The overlap interval is determined based on the estimated primary fault probability and task's soft laxity. We also propose a new performance index that integrates schedulability (S) and reliability (R) into a single metric, called SR index. To evaluate the proposed scheme, we have conducted analytical and simulation studies under different fault and deadline scenarios, and found that the proposed adaptive scheme adapts to system dynamics and offers better SR index than that of the non-adaptive schemes.

1 Introduction

Real-time systems are defined as those systems in which the correctness of the system depends not only on the logical result of computation, but also on the time at which the results are produced [1]. Multiprocessors and multicomputer systems are emerging as a powerful computing means for real-time applications due to their capability for high performance and reliability. In most real-time applications, there is a need for reliable execution of tasks even in the presence of faults. Reliable (fault-tolerant) execution of a task is usually achieved by scheduling multiple versions of the task [2]-[9].

Primary-Backup (PB) scheduling is an important fault-tolerant approach in which two versions of a task are scheduled on two different processors and an *acceptance test* is used to check the correctness of the result [3]-[8]. The three variants [8] of this approach are: primary-backup exclusive (PB-EXCL) [5,7],

---

* This research was supported in part by the NSF under grant CCR-0098354.

B. Monien, V.K. Prasanna, S. Vajapeyam (Eds.): HiPC 2001, LNCS 2228, pp. 68–78, 2001.

primary-backup concurrent (PB-CONCUR) [8,9], and primary-backup overlap (PB-OVER) [8].

In PB-EXCL approach, the primary and backup versions of a task are excluded in time as well as in space (i.e., scheduled on two different processors in an non-overlapping manner). The backup version is executed only if the output of the primary fails the acceptance test, otherwise it (backup) is deallocated from the schedule. This approach uses less resources if faults rarely occur, because backups are mostly deallocated. Nevertheless, it requires the *execution interval* (defined as the interval between start time of primary and finish time of backup) of a task to be at least twice that of its computation time. In soft real-time systems, the execution interval of a task is critical in determining the utility of the task's output. The lower the execution interval, the higher the utility.

In PB-CONCUR approach, the primary and the backup versions of a task are executed concurrently. In this approach, the *consumption time* (defined as the total processor time used by a task for its execution) of a task is always twice that of its computation time irrespective of the fault rate in the system. Nevertheless, the execution interval for the task is always equal to its computation time. The consumption time is inversely proportional to the processor utilization. The lower the consumption time, the better the schedulability.

It is evident that the PB approaches offer a trade-off between the number of tasks scheduled to the total utility of their output to the system. It can be noted that the reduction in both consumption time and execution interval will contribute to better performance. In summary, PB-EXCL offers better performance when fault rate is low or tasks have high soft laxities (Laxity of a task is equal to its deadline minus current time minus its remaining execution time; Informally, the higher the laxity, the lesser the urgency), whereas PB-CONCUR offers better performance when the fault rate is high and tasks have low laxities. The PB-OVER is a flexible approach wherein the primary and the backup versions of a task are scheduled to overlap in execution. This approach has the potential to exploit the advantages of the other two approaches if the overlap interval is suitably adapted.

In this paper, we propose an adaptive PB-OVER based scheduling scheme that makes use of an estimate of the primary fault probability and task's laxity to control the degree of overlap between its (task) versions. There exists some adaptive fault-tolerant scheduling algorithms (e.g., [6]) wherein the task specifies to the scheduler the adaptiveness strategy. Clearly, such an algorithm does not enjoy the benefits of feedback-based adaptiveness wherein the scheduler adapts to the dynamics of the system.

2   System Model

Task Model

1. The system has soft real-time aperiodic tasks. Each task T_i has the attributes arrival time (a_i), ready time (r_i), worst case computation time (c_i), a relative soft deadline (d_i^s), and a relative firm deadline (d_i^f). Tasks are non-preemptable. **2.**

Each task T_i has two versions, namely primary version (Pr_i) and backup version (Bk_i). The versions have identical attributes. **3.** $d_i^s = k_1 \times c_i$ and $d_i^f = k_2 \times c_i$, where $k_2 \geq k_1$ and $k_2 \geq 2$.

Fault Model

Assumption 1: We assume that each processor, except the scheduler, may fail due to hardware fault which results in task's failure. The faults can be transient or permanent and are independent of each other.

Assumption 2: We assume that $(d_i^f - r_i) \ \forall \ T_i$ are much smaller than the mean time to failure $(MTTF)$ of the system. This assumption is used to enhance the probability of successfully executing the backup of a task, if its primary fails.

Assumption 3: We assume that there exists fault-detection mechanisms such as fail-signal and acceptance test to detect processor and task failures, respectively. The scheduler will not schedule tasks to a known faulty processor.

3   Proposed Adaptive Scheduling Scheme

In a dynamic scheduling, tasks arrive at the scheduler, from where they are distributed to other processors for execution. The communication between the scheduler and the processors is through dispatch queues (Figure 1). The scheduler runs in parallel with the processors and periodically updates the schedules in the dispatch queues. In our work, we employ a fault monitor to periodically observe the fault rate in the system using a feedback as shown in Figure 1. The fault rate is then used to estimate the primary (task) fault probability in the system. The fault probability together with tasks' soft laxity is used to control the overlap interval between the primary and backup versions of each task.

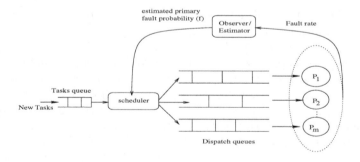

Fig. 1. An adaptive scheduler model

In this paper, we propose an adaptive PB-OVER scheduling scheme that makes use of an estimate of the primary fault probability and task's laxity to control the degree of overlap between the versions of a task. Two variations of the adaptive scheme are proposed by varying the adaptation mechanism. The adaptation can be done in a continuous manner which leads to an approach called

PB-OVER continuous (PB-OVER-CONT), or it can be done in a discrete manner which leads to an approach called PB-OVER switch (PB-OVER-SWITCH). In PB-OVER-CONT, the overlap interval varies from no overlap to full overlap in a continuous manner as the fault probability varies from 0 to 1. In PB-OVER-SWITCH, the scheduler uses a threshold value of fault probability to switch from PB-EXCL to PB-CONCUR, i.e., if the probability is less than the threshold, the adaptive scheduler behaves like PB-EXCL, otherwise it behaves like PB-CONCUR. In PB-OVER-SWITCH, the threshold value is adapted with task's laxity.

3.1   Fault Monitoring System

A fault monitoring system periodically (with period p) monitors the completion of tasks in the system. For each period, the monitoring system counts the number of faulty primaries $(n^f(t))$ and the total number of primaries completed $(n(t))$ during that period.

According to the *frequency interpretation* probability concepts, the probability of an event (primary fail) is the proportion of the time that events of the same kind will occur in the long run. Hence, we define the primary fault probability (f) as the ratio of the number of faulty primaries to the total number of primaries executed in a given interval $[0, t]$: $f(t) = \frac{n^f(t)}{n(t)}$. This probability is calculated every p time units, where p is the sample period for the monitor.

3.2   Performance Index

The performance index that we use to compare the fault-tolerant approaches capture the trade-off between the processor utilization and task's value. This performance index is called the *schedulability-reliability* (SR) index.

For a task T_i, we define the processor utilization of the task $(UTIL_i)$ as the ratio of worst case execution time (c_i) to the expected consumption time $(\overline{PT_i})$. $UTIL_i = c_i/\overline{PT_i}$. We also define a value function (VAL_i) that is used to credit the successful output from a task depending on its expected execution interval $(\overline{ET_i})$. In the value function (Equation 1), the task contributes a value of one if it finishes before its soft deadline, a monotonically decreasing value if it finishes between its soft and firm deadlines, a value of zero if it misses its firm deadline. The monotonically decreasing task value is inversely proportional to its expected execution interval.

$$VAL_i = \begin{cases} 1 & \text{for } \overline{ET_i} \le d_i^s \\ \frac{d_i^s}{\overline{ET_i}} & \text{for } d_i^s \le \overline{ET_i} \le d_i^f \\ 0 & \text{for } \overline{ET_i} > d_i^f \end{cases} \tag{1}$$

The performance index of a soft real-time system is usually measured as the sum of the values contributed by the admitted tasks. For a given system capacity, there are two options: (i) admit a few tasks with a higher value for each task or

(ii) admit a large number of tasks with a lower value for each task. Therefore, the SR_i index for a task (T_i) is:

$$SR_i = UTIL_i \times VAL_i. \tag{2}$$

The effect on the number of admitted tasks is inherently captured by the processor utilization of each task. The SR index for a set of n tasks that are admitted into the system is computed as $SR = \frac{\sum_{i=1}^{n} SR_i}{n}$.

4   Analysis of PB-Based Fault-Tolerant Approaches

To analyze and compare the fault-tolerant approaches, we make the following assumptions: (i) The primary fault probability (f) is estimated using the fault monitoring mechanism. (ii) The worst case execution time for the tasks is fixed and is equal to c. (iii) The relative soft deadline (d^s) for a task is c, and its relative firm deadline (d^f) is $2c$.

Primary-Backup EXCLusive (PB-EXCL): In this approach, the primary and the backup versions of a task are excluded in space as well as in time. The expected execution interval for a task T_i using this approach is: $\overline{ET_i} = 2cf + (1-f)c = c(1+f)$. The expected consumption time for the task is given by $\overline{PT_i} = 2cf + (1-f)c = c(1+f)$. The processor utilization is given by: $UTIL_i = \frac{c}{(1+f)c} = \frac{1}{(1+f)}$. Since $d_i^s = c$ and $d_i^f = 2c$, the value function for this approach is $VAL_i = \frac{c}{(1+f)c} = \frac{1}{1+f}$. The SR index for this approach is then $SR_i = \frac{1}{(1+f)^2}$ (refer Figure 2a for the plot).

Primary-Backup CONCURrent (PB-CONCUR): In this approach, the primary and the backup versions of tasks are executed concurrently. In this approach, $\overline{ET_i}$ and $\overline{PT_i}$ are equal to c and $2c$, respectively. The processor utilization for this approach is given by $UTIL_i = \frac{c}{2c} = 0.5$. The value function for this approach is $VAL_i = \frac{c}{c} = 1$. The SR index for this approach is then $SR_i = 0.5$ (refer Figure 2a).

Primary-Backup OVERlap (PB-OVER): In this approach, the primary and the backup versions of a task can overlap in execution by an amount equal to γc. The $\overline{ET_i}$ and $\overline{PT_i}$ for the task using this approach are given by

$$\begin{aligned} \overline{ET_i} &= f(c + (1-\gamma)c) + (1-f)c = cf(1-\gamma) + c \\ \overline{PT_i} &= 2cf + (1-f)(c + \gamma c) = (1-\gamma)cf + (1+\gamma)c \end{aligned} \tag{3}$$

The processor utilization and the value function for this approach are respectively given by

$$UTIL_i = \frac{1}{1 + f + \gamma(1-f)} \quad and \quad VAL_i = \frac{1}{1 + f(1-\gamma)} \tag{4}$$

The SR index for this approach is then

$$SR_i = \frac{1}{(1 + f + \gamma(1 - f))(1 + f(1 - \gamma))} \tag{5}$$

The most important property that is expected from PB-OVER is to combine the advantages of PB-EXCL and PB-CONCUR approaches. That is, when $f = 0$, the desirable values of $\overline{ET_i}$ and $\overline{PT_i}$ is c. When $f = 1$, the desirable values of $\overline{ET_i}$ and $\overline{PT_i}$ is c and $2c$, respectively.

5   Adaptive PB-OVER Fault-Tolerant Approaches

To achieve the adaptive property, we propose two adaptive PB-OVER approaches by varying the adaptation mechanism: PB-OVER continuous (PB-OVER-CONT) and PB-OVER switch (PB-OVER-SWITCH).

5.1   Primary-Backup OVERlap CONTinuous (PB-OVER-CONT)

In PB-OVER-CONT, the overlap interval varies from no overlap to full overlap in a continuous manner as the fault probability varies from 0 to 1. From Equation (3), we found that this can be achieved by substituting $\gamma = f$ which results in

$$\overline{ET_i} = (-f^2 + f + 1)c \quad and \quad \overline{PT_i} = (-f^2 + 2f + 1)c \tag{6}$$

Using the assumption in Section 4 the processor utilization and the value function for this approach are respectively given by

$$UTIL_i = \frac{1}{-f^2 + 2f + 1} \quad and \quad VAL_i = \frac{1}{(-f^2 + f + 1)} \tag{7}$$

The SR_i index for this approach (Figure 2a) is then $SR_i = \frac{1}{(-f^2+2f+1)(-f^2+f+1)}$

5.2   Primary-Backup OVERlap SWITCH (PB-OVER-SWITCH)

In PB-OVER-SWITCH, the scheduler switches from PB-EXCL to PB-CONCUR depending on the value of f. If f is less than a threshold f_o, then the PB-OVER-SWITCH approach behaves like PB-EXCL, else it behaves like PB-CONCUR. The threshold f_o is the value of f at which the SR index of PB-EXCL is equal to that of PB-CONCUR. Thus, $\overline{ET_i}$ and $\overline{PT_i}$ of the task become

$$\overline{ET_i} = \begin{cases} c \times (1 + f) & \text{for } 0 \leq f \leq f_o \\ c & \text{for } f_o < f \leq 1 \end{cases} \quad and \quad \overline{PT_i} = \begin{cases} c \times (1 + f) & \text{for } 0 \leq f \leq f_o \\ 2c & \text{for } f_o < f \leq 1 \end{cases} \tag{8}$$

For the given assumption the value of f_o is the value of f that satisfies $\frac{1}{(1+f)^2} = 0.5$, which is $f = \sqrt{2} - 1$. The SR index for this approach (Figure 2a) is given by

$$SR_i = \begin{cases} \frac{1}{(1+f)^2} & \text{for } 0 \leq f \leq \sqrt{2} - 1 \\ 0.5 & \text{for } \sqrt{2} - 1 \leq f \leq 1 \end{cases} \tag{9}$$

5.3   Effect of Task's Soft Laxity on Performance

In this section, we first study the effect of task's soft laxity on the performance of the PB-based fault-tolerant approaches. Next, we propose a mechanism that allow the fault probability threshold (f_o) to be adapted with task's soft laxity.

In Sections 4, 5.1, and 5.2 (Figure 2a), we assumed that the task's soft deadline (d^s) is equal to c. This assumption allows the PB-EXCL approach to schedule only one version of a task (Pr) within its soft deadline and the other version (Bk) must be scheduled after the soft deadline. This degrades the performance of the PB-EXCL approach as f increases because the correct output is always produced by the backup which has less value. When tasks have a large soft deadline, then both the primary and the backup versions of the task can be scheduled in an exclusive manner within their soft deadline. Therefore, all PB-based fault-tolerant approaches offer the same output value (VAL) for the finished tasks which is equal to one and does not depend on f. Hence, the SR index will be 0.5, $\frac{1}{1+f}$, and $\frac{1}{-f^2+2f+1}$, respectively for the PB-CONCUR, PB-EXCL, and the PB-OVER-CONT approaches (Figure 2b).

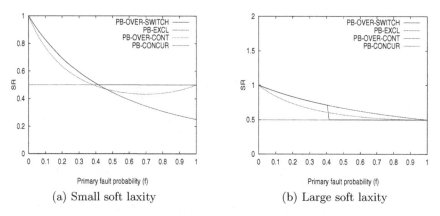

(a) Small soft laxity   (b) Large soft laxity

Fig. 2. SR index for the PB approaches

From Figure 2b, we notice that the PB-EXCL offers the best SR index for all values of f. The PB-SWITCH offers the best SR index only when $f < \sqrt{2} - 1$ and it offers the lowest SR index when $f > \sqrt{2} - 1$. This is because the threshold value (f_o) that is used by the PB-SWITCH approach to switch between the PB-EXCL and PB-CONCUR approaches is constant $(f_0 = \sqrt{2} - 1)$ and does not change with changing task's soft deadline. In Section 5.2, the threshold f_0 is defined as the value of f at which the SR index of PB-EXCL is equal to that of PB-CONCUR. Therefore, for the given task's deadline the value of f_0 is the value of f that satisfies $\frac{1}{(1+f)} = 0.5$, which is $f = 1$. By using this new threshold value $(f_0 = 1)$, the PB-OVER-SWITCH always behaves like PB-EXCL approach which offers the best performance in this case.

It is evident from the above discussion that the threshold value (f_0) have to be adapted with the task's soft laxity to be able to determine the correct overlap interval that maximizes the performance of the system. To do so, the scheduler behaves as follows for each task (T_i) that arrives in the system:

1. Schedules the primary (Pr_i) version of the task.
2. If the primary was scheduled within the soft deadline, then
   a) Determine the overlap interval ($\sigma_i c_i$) between the primary and the backup versions of the task so that both versions will be scheduled within the soft deadline.
   $$\sigma_i = \frac{c_i - (d_i^s - ft(Pr_i))}{c_i} \tag{10}$$
   where $ft(Pr_i)$ is the relative finish time of the primary version.
   b) If $\sigma_i < 0$ then, $\sigma_i = 0$.
3. Else if the primary was scheduled to finish after the soft deadline, then $\sigma_i = 1$.

Using this value (σ_i) the scheduler calculates the threshold value (f_0^i) that is used by the PB-OVER-SWITCH approach to select the correct overlap interval (i.e., if ($f < f_0^i$) : $\gamma = 0$? $\gamma = 1$) between the primary and the backup versions for this task (T_i). To calculate f_0^i for a given task T_i the scheduler uses the following equation:

$$f_0^i = (\sqrt{2} - 2)\sigma_i + 1 \tag{11}$$

The above equation is derived from the fact that f_0 change from $\sqrt{2} - 1$ to 1, when the task's soft laxity changes from the case where the scheduler is only able to schedule one version of the task within its soft deadline to the case where it is able to schedule both versions in an exclusive manner within the soft deadline.

5.4   Performance Studies

In this section, we compare the performance of the proposed adaptive fault-tolerant approaches with that of the non-adaptive PB-based fault-tolerant approaches, through simulation studies. The SR index has been used as the performance metric. For each point in the performance plots the system was simulated with 20,000 tasks. This number of tasks has been chosen to have a 99% confidence interval within ± 0.0035 around each value of SR.

Figures 3(a) and 3(b) show the effect of primary fault probability on the SR index for all PB approaches. Figure 3(a) shows the behavior of the PB approaches when f varies for tasks that have relative soft deadline equals to $2c_i$ and relative firm deadline equals to $5c_i$. Figure 3(b) shows the case for tasks that have relative soft deadline equals to $4c_i$ and relative firm deadline equals to $5c_i$.

From the figures, it can be seen that when the relative soft deadline is large (Figure 3(b)) the PB-OVER-SWITCH and the PB-EXCL approaches behave similar and offer the best SR index. This is because when d^s is large, the tasks

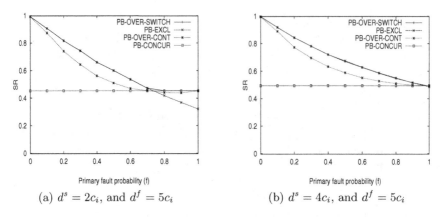

(a) $d^s = 2c_i$, and $d^f = 5c_i$ (b) $d^s = 4c_i$, and $d^f = 5c_i$

Fig. 3. Effect of primary fault probability (f) on SR index

have enough soft laxity to be scheduled in an exclusive manner within the soft deadline. The threshold value f_0 used by the PB-OVER-SWITCH approach is approximately equal to 1. From Figure 3(a), we can notice that the PB-EXCL approach has the highest SR index when $f < 0.75$, and the PB-CONCUR approach has the highest SR index when $f > 0.75$. The SR index offered by the PB-OVER-CONT lies between the PB-CONCUR and PB-EXCL for all values of primary fault probability. Since the PB-OVER-SWITCH behaves like PB-EXCL when $f \leq 0.75$ and like PB-CONCUR when $f > 0.75$, it offers the best SR index for all values of f.

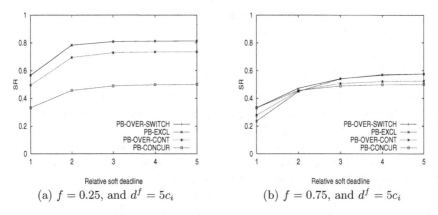

(a) $f = 0.25$, and $d^f = 5c_i$ (b) $f = 0.75$, and $d^f = 5c_i$

Fig. 4. Effect of task's soft laxity on SR index

Figures 4(a) and 4(b) show the effect of task's soft laxity on the SR index for all PB approaches. Figure 4(a) shows the behavior of the PB approaches when d^s varies for the system in which the primary fault probability is 0.25 and tasks' relative firm deadline is $5c_i$. Figure 4(b) shows the case for system in which the primary fault probability is 0.75 and tasks' relative firm deadline is $5c_i$.

From the figures, it can be seen that when the primary fault probability (f) is low (Figure 4(a)) the PB-OVER-SWITCH and the PB-EXCL approaches behave similar and offer the best SR index for all values of d^s. This is because when f is small, the PB-OVER-SWITCH switch to an overlap interval equal to zero. The SR index offered by the PB-OVER-CONT lies between the PB-CONCUR and PB-EXCL for all values of d^s. From Figure 4(b), we can notice that the PB-CONCUR and PB-OVER-SWITCH approaches have the highest SR index when $d^s \leq 2$. This is because, when f is high and tasks have small soft laxity, the threshold value that is used by the PB-OVER-SWITCH for each task is smaller than the fault probability in the system $(f = 0.75)$. Thus, PB-OVER-SWITCH behaves like PB-CONCUR for this region. It can be noticed that when $2 \leq d^s \leq 3$, the PB-OVER-SWITCH approach has the highest SR index. This is because, in this interval, the tasks have medium values of soft deadline. Thus, PB-OVER-SWITCH maximizes the performance by adapting the task's fault threshold (f_o^i) based on soft deadline. Finally, in Figure 4(b), it can be noticed that the PB-EXCL and PB-OVER-SWITCH approaches have the highest SR index when $d^s > 3$.

6   Conclusions

In this paper, we have proposed an adaptive scheme for the PB-based fault-tolerant scheduling of tasks on multiprocessor real-time systems. The key idea used is to control the overlap interval between the primary and backup versions of a task based on an estimated value of primary fault probability and task's soft laxity. Two variants (PB-OVER-CONT and PB-OVER-SWITCH) of the adaptive scheme have been proposed and studied. We have also proposed a new metric, called schedulability-reliability (SR) index. Our studies show that the proposed PB-OVER-SWITCH adaptive scheme always performs better than the other PB-based approaches for the SR index performance metric. The feedback based adaptation mechanism, such as the one developed in this paper, opens up many avenues for further research in value-based scheduling in real-time systems.

References

1. K.G. Shin and P. Ramanathan, "Real-time computing: A new discipline of computer science and engineering," *Proc. of IEEE,* vol. 82, no. 1, pp. 6-24, Jan. 1994.
2. L. Chen and A. Avizienis, "N-version programming: A fault tolerant approach to reliability of software operation", In Proc. *IEEE Fault-Tolerant Computing Symposium,* pp. 3-9., 1978.
3. B. Randell, "System structure for software fault-tolerance," *IEEE Transactions on Software Engineering,* vol. 1, no. 2, pp. 220-232, June 1975.
4. S. Tridandapani, A. K. Somani, and U. Reddy, "Low overhead multiprocessor allocation strategies exploiting system spare capacity for fault detection and location," *IEEE Transactions on Computers,* vol. 44, no. 7, pp. 865-877, July 1995.

5. S. Ghosh, R. Melhem, and D. Mosse, "Fault-tolerance through scheduling of aperiodic tasks in hard real-time multiprocessor systems," *IEEE Transactions on Parallel and Distributed Systems*, vol. 8, no. 3, pp. 272-284, Mar. 1997.
6. O. González, H. Shrikumar, J.A. Stankovic, and K. Ramamritham, "Adaptive fault-tolerance and graceful degradation under dynamic hard real-time scheduling," In Proc. *IEEE Real-Time Systems Symposium*, 1997.
7. G. Manimaran and C. Siva Ram Murthy, " A fault-tolerant dynamic scheduling algorithm for multiprocessor real-time systems and its analysis," *IEEE Transactions on Parallel and Distributed Systems*, vol. 9, no. 11, pp. 1137-1152, Nov. 1998.
8. I. Gupta, G. Manimaran, and C. Siva Ram Murthy, "Primary-Backup based fault-tolerant dynamic scheduling of object-based tasks in multiprocessor real-time systems," in *Dependable Network Computing*, D.R. Avresky (editor), Kluwer Academic Publishers, 1999.
9. F. Wang, K. Ramamritham and J. A. Stankovic, "Determining redundancy levels for fault tolerant real-time systems," *IEEE Transactions on Computers*, vol. 44, no. 2, pp. 292-301, Feb. 1995.

Learning from the Success of MPI

William D. Gropp

Mathematics and Computer Science Division, Argonne National Laboratory,
Argonne, Illinois 60439, gropp@mcs.anl.gov,
www.mcs.anl.gov/~gropp

Abstract. The Message Passing Interface (MPI) has been extremely successful as a portable way to program high-performance parallel computers. This success has occurred in spite of the view of many that message passing is difficult and that other approaches, including automatic parallelization and directive-based parallelism, are easier to use. This paper argues that MPI has succeeded because it addresses *all* of the important issues in providing a parallel programming model.

1 Introduction

The Message Passing Interface (MPI) is a very successful approach for writing parallel programs. Implementations of MPI exist for most parallel computers, and many applications are now using MPI as the way to express parallelism (see [1] for a list of papers describing applications that use MPI). The reasons for the success of MPI are not obvious. In fact, many users and researchers complain about the difficulty of using MPI. Commonly raised issues include the complexity of MPI (often as measured by the number of functions), performance issues (particularly the latency or cost of communicating short messages), and the lack of compile or runtime help (e.g., compiler transformations for performance; integration with the underlying language to simplify the handling of arrays, structures, and native datatypes; and debugging). More subtle issues, such as the complexity of nonblocking communication and the lack of elegance relative to a parallel programming language, are also raised [2]. With all of these criticisms, why has MPI enjoyed such success?

One might claim that MPI has succeeded simply because of its *portability*, that is, the ability to run an MPI program on most parallel platforms. But while portability was certainly a necessary condition, it was not sufficient. After all, there were other, equally portable programming models, including many message-passing and communication-based models. For example, the socket interface was (and remains) widely available and was used as an underlying communication layer by other parallel programming packages, such as PVM [3] and p4 [4]. An obvious second requirement is that of *performance*: the ability of the programming model to deliver the available performance of the underlying hardware. This clearly distinguishes MPI from interfaces such as sockets. However, even this is not enough. This paper argues that six requirements must *all* be satisfied for a parallel programming model to succeed, that is, to be widely

B. Monien, V.K. Prasanna, S. Vajapeyam (Eds.): HiPC 2001, LNCS 2228, pp. 81–92, 2001.

adopted. Programming models that address a subset of these issues can be successfully applied to a subset of applications, but such models will not reach a wide audience in high-performance computing.

2   Necessary Properties

The MPI programming model describes how separate *processes* communicate. In MPI-1 [5], communication occurs either through point-to-point (two-party) message passing or through collective (multiparty) communication. Each MPI process executes a program in an address space that is private to that process.

2.1   Portability

Portability is the most important property of a programming model for high-performance parallel computing. The high-performance computing community is too small to dictate solutions and, in particular, to significantly influence the direction of commodity computing. Further, the lifetime of an application (often ten to twenty years, rarely less than five years) greatly exceeds the lifetime of any particularly parallel hardware. Hence, any application must be prepared to run effectively on many generations of parallel computer, and that goal is most easily achieved by using a portable programming model.

Portability, however, does not require taking a "lowest common denominator" approach. A good design allows the use of performance-enhancing features without mandating them. For example, the message-passing semantics of MPI allows for the direct copy of data from the user's send buffer to the receive buffer without any other copies.[1] However, systems that can't provide this direct copy (because of hardware limitations or operating system restrictions) are permitted, under the MPI model, to make one or more copies. Thus MPI programs remain portable while exploiting hardware capabilities.

Unfortuately, portability does not imply portability with performance, often called *performance portability*. Providing a way to achieve performance while maintaining portability is the second requirement.

2.2   Performance

MPI enables performance of applications in two ways. For small numbers of processors, MPI provides an effective way to manage the use of memory. To understand this, consider a typical parallel computer as shown in Figure 1.

The memory near the CPU, whether it is a large cache (symmetric multiprocessor) or cache and memory (cluster or NUMA), may be accessed more rapidly than far-away memory. Even for shared-memory computers, the ratio of the number of cycles needed to access memory in L1 cache and main memory is roughly a hundred; for large, more loosely connected systems the ratio can exceed ten to one hundred thousand. This large ratio, even between the cache and

---

[1] This is sometimes called a zero-copy transfer.

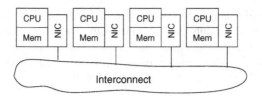

Fig. 1. A typical parallel computer

local memory, means that applications must carefully manage memory locality if they are to achieve high performance.

The separate processes of the MPI programming model provide a natural and effective match to this property of the hardware.

This is not a new approach. The C language provides `register`, originally intended to aid compilers in coping with a two-level memory hierarchy (registers and main memory). Some parallel languages, such as HPF [6], UPC [7], or CoArray Fortran [8], distiguish between local and shared data. Even programming models that do not recognize a distinction between local and remote memory, such as OpenMP, have implementations that often require techniques such as "first touch" to ensure that operations make effective use of cache. The MPI model, based on communicating processes, each with its own memory, is a good match to current hardware.

For large numbers of processors, MPI also provides effective means to develop scalable algorithms and programs. In particular, the collective communication and computing routines such as `MPI_Allreduce` provide a way to express scalable operations without exposing system-specific features to the programmer. Also important for supporting scalability is the ability to express the most powerful scalable algorithms; this is discussed in Section 2.4.

Another contribution to MPI's performance comes from its ability to work with the best compilers; this is discussed in Section 2.5. Also discussed there is how MPI addresses the performance-tradeoffs in using threads with MPI programs.

Unfortunately, while MPI achieves both portability and performance, it does not achieve perfect performance portability, defined as providing a single source that runs at (near) acheivable peak performance on all platforms. This lack is sometimes given as a criticism of MPI, but it is a criticism that most other programming models also share. For example, Dongarra et al [9] describe six different ways to implement matrix-matrix multiply in Fortran for a single processor; not only is no one of the six optimal for all platforms but *none* of the six are optimal on modern cache-based systems. Another example is the very existence of vendor-optimized implementations of the Basic Linear Algebra Subroutines (BLAS). These are functionally simple and have implementations in Fortran and C; if compilers (good as they are) were capable of producing optimal code for these relatively simple routines, the hand-tuned (or machined-tuned [10]) versions would not be necessary. Thus, while performance portability is a desirable goal, it is unreasonable to expect parallel programming models to provide it when uniprocessor models cannot. This difficulty also explains why relying on compiler-discovered parallelism has usually failed: the problem remains too difficult. Thus a successful programming model must allow the programmer to help.

2.3   Simplicity and Symmetry

The MPI model is often criticized as being large and complex, based on the number of routines (128 in MPI-1 with another 194 in MPI-2). The number of routines is not a relevant measure, however. Fortran, for example, has a large number of intrinsic functions; C and Java rely on a large suite of library routines to achieve external effects such as I/O and graphics; and common development frameworks have hundreds to thousands of methods.

A better measure of complexity is the number of concepts that the user must learn, along with the number of exceptions and special cases. Measured in these terms, MPI is actually very simple.

Using MPI requires learning only a few concepts. Many MPI programs can be written with only a few routines; several subsets of routines are commonly recommended, including ones with as few as six functions. Note the plural: for different purposes, different subsets of MPI are used. For example, some recommend using only collective communiation routines; others recommend only a few of the point-to-point routines. One key to the success of MPI is that these subsets can be used without learning the rest of MPI; in this sense, MPI is simple. Note that a smaller set of routines would *not* have provided this simplicity because, while some applications would find the routines that they needed, others would not.

Another sign of the effective design in MPI is the use of a single concept to solve multiple problems. This reduces both the number of items that a user must learn and the complexity of the implementation. For example, the MPI communicator both describes the group of communicating processes and provides a separate communication context that supports component-oriented software, described in more detail in Section 2.4. Another example is the MPI datatype; datatypes describe both the type (e.g., integer, real, or character) and layout (e.g., contiguous, strided, or indexed) of data. The MPI datatype solves the two problems of describing the types of data to allow for communication between systems with different data representations and of describing noncontiguous data layouts to allow an MPI implementation to implement zero-copy data transfers of noncontiguous data.

MPI also followed the principle of *symmetry*: wherever possible, routines were added to eliminate any exceptions. An example is the routine MPI_Issend. MPI provides a number of different send modes that correspond to different, well-established communication approaches. Three of these modes are the regular send (MPI_Send) and its nonblocking versions (MPI_Isend), and the synchronous send (MPI_Ssend). To maintain symmetry, MPI also provides the nonblocking synchronous send MPI_Issend. This send mode is meaningful (see [11, Section

7.6.1]) but is rarely used. Eliminating it would have removed a routine, slightly simplifying the MPI documentation and implementation. It would have created an exception, however. Instead of each MPI send mode having a nonblocking version, only some send modes would have nonblocking versions. Each such exception adds to the burden on the user and adds complexity: it is easy to forget about a routine that you never use; it is harder to remember arbitrary decisions on what is and is not available.

A place where MPI may have followed the principle of symmetry too far is in the large collection of routines for manipulating groups of processes. Particularly in MPI-1, the single routine MPI_Comm_split is all that is needed; few users need to manipulate groups at all. Once a routine working with MPI groups was introduced, however, symmetry required completing the set. Another place is in canceling of sends, where significant implementation complexity is required for an operation of dubious use.

Of course, more can be done to simplify the use of MPI. Some possible approaches are discussed in Section 3.1.

2.4   Modularity

Component-oriented software is becoming increasingly important. In commecial software, software components implementing a particular function are used to implement a clean, maintainable service. In high-performance computing, components are less common, with many applications being built as a monolithic code. However, as computational algorithms become more complex, the need to exploit software components embodying these algorithms increases.

For example, many modern numerical algorithms for the solution of partial differential equations are hierarchical, exploiting the structure of the underlying solution to provide a superior and scalable solution algorithm. Each level in that hierarchy may require a different solution algorithm; it is not unusual to have each level require a different decomposition of processes. Other examples are intelligent design automation programs that run application components such as fluid solvers and structural analysis codes under the control of a optimization algorithm.

MPI supports component-oriented software. Both describe the subset of processes participating in a component and to ensure that all MPI communication is kept within the component, MPI introduced the *communicator*.[2] Without something like a communicator, it is possible for a message sent by one component and intended for that component to be received by another component or by user code. MPI made reliable libraries possible.

Supporting modularity also means that certain powerful variable layout tricks (such as assuming that the variable a in an SPMD program is at the same address on all processors) must be modified to handle the case where each process may have a different stack-use history and variables may be dynamically allocated with different base addresses. Some programming models have assumed that all

---

[2] The context part of the communicator was inspired by Zipcode [12].

processes have the same layout of local variables, making it difficult or impossible to use those programming models with modern adaptive algorithms.

Modularity also deals with the complexity of MPI. Many tools have been built using MPI to provide the communication substrate; these tools and libraries provide the kind of easy-to-use interface for domain-specific applications that some developers feel are important; for example, some of these tools eliminate all evidence of MPI from the user program. MPI makes those tools possible. Note that the user base of these domain-specific codes may be too small to justify vendor-support of a parallel programming model.

2.5   Composability

One of the reasons for the continued success of Unix is the ease with which new solutions can be built by composing existing applications.

MPI was designed to work with other tools. This capability is vital, because the complexity of programs and hardware continues to increase. For example, the MPI specification was designed from the beginning to be thread-safe, since threaded parallelism was seen by the MPI Forum as a likely approach to systems built from a collection of SMP nodes. MPI-2 took this feature even further, acknowledging that there are performance tradeoffs in different degrees of threadedness and providing a mechanism for the user to request a particular level of thread support from the MPI library. Specifically, MPI defines several degrees of thread support. The first, called MPI_THREAD_SINGLE, specifies that there is a single thread of execution. This allows an MPI implementation to avoid the use of thread-locks or other techniques necessary to ensure correct behavior with multithreaded codes. Another level of thread support, MPI_THREAD_FUNNELLED, specifies that the process may have multiple threads but all MPI calls are made by one thread. This matches the common use of threads for loop parallelism, such as the most common uses of OpenMP. A third level, MPI_THREAD_MULTIPLE, allows multiple threads to make MPI calls. While these levels of thread support do introduce a small degree of complexity, they reflect MPI's pragmatic approach to providing a workable tool for high-performance computing.

The design of MPI as a library means that MPI operations cannot be optimized by a compiler. However, it also means that any MPI library can exploit the newest and best compilers and that the compiler can be developed without worrying about the impact of MPI on the generated code—from the compiler's point of view, MPI calls are simply generic function calls.[3] The ability of MPI to exploit improvements in other tools is called *composability*. Another example is in debuggers; because MPI is simply a library, any debugger can be used with MPI programs. Debuggers that are capable of handling multiple processes, such as TotalView [14], can immediately be used to debug MPI programs. Additional refinements, such as an interface to an abstraction of message passing that is

---

[3] There are some conflicts between the MPI model and the Fortran language; these are discussed in [13, Section 10.2.2]. The issues are also not unique to MPI; for example, any asynchronous I/O library faces the same issues with Fortran.

described in [15], allows users to use the debugger to discover information about pending communication and unreceived messages.

More integrated approaches, such as language extensions, have the obvious benefits, but they also have significant costs. A major cost is the difficulty of exploiting advances in other tools and of developing and maintaining a large, integrated system.

OpenMP is an example of a programming model that achieves the effect of composability with the compilers because OpenMP requires essentially orthogonal changes to the compiler; that is, most of the compiler development can ignore the addition of OpenMP in a way that more integrated languages cannot.

2.6   Completeness

MPI provides a complete programming model. Any parallel algorithm can be implemented with MPI. Some parallel programming models have sacrificed completeness for simplicity. For example, a number of programming models have required that synchronization happens only collectively for all processes or tasks. This requirement significantly simplifies the programming model and allows the use of special hardware affecting all processes. Many existing programs also fit into this model; data-parallel programs are natural candidates for this model. But as discussed in Section 2.4, many programs are becoming more complex and are exploiting software components. Such applications are difficult, if not impossible, to build using restrictive programming models.

Another way to look at this is that while many programs may not be easy under MPI, no program is impossible. MPI is sometimes called the "assembly language" of parallel programming. Those making this statement forget that C and Fortran have also been described as portable assembly languages. The generality of the approach should not be mistaken for an unnecessary complexity.

2.7   Summary

Six different requirements have been discussed, along with how MPI addresses each. Each of these is *necessary* in a general-purpose parallel programming system.

Portability and performance are clearly required. Simplicity and symmetry cater to the *user* and make it easy to learn and use safely. Composibility is required to prevent the approach from being left behind by the advance of other tools such as compilers and debuggers.

Modularity, like completeness, is required to ensure that tools can be built on top of the programming model. Without modularity, a programming model is suitable only for turnkey applications. While those may be important and easy to identify as customers, they represent the past rather than the future.

Completeness, like modularity, is required to ensure that the model supports a large enough community. While this does not mean that everyone uses every function, it means that the functionality that a user may need is likely to be present. An early poll of MPI users [16] in fact found that while no one was using all of the MPI-1 routines, essentially all MPI-1 routines were in use by someone.

The open standards process (see [17] for a description of the process used to develop MPI) was an important component in its success. Similar processes are being adopted by others; see [18] for a description of the principles and advantages of an open standards process.

3   Where Next?

MPI is not perfect. But any replacement will need to improve on all that MPI offers, particularly with respect to performance and modularity, without sacrificing the ability to express any parallel program. Three directions are open to investigation: improvements in the MPI programming model, better MPI implementations, and fundamentally new approaches to parallel computing.

3.1   Improving MPI

Where can MPI be improved? A number of evolutionary enhancements are possible, many of which can be made by creating tools that make it easier to build and maintain MPI programs.

1. Simpler interfaces. A compiler (or a preprocessor) could provide a simpler, integrated syntax. For example, Fortran 90 array syntax could be supported without requiring the user to create special MPI datatypes. Similarly, the MPI datatype for a C structure could be created automatically. Some tools for the latter already exist. Note that support for array syntax is an example of support for a subset of the MPI community, many of whom use data structures that do not map easily onto Fortran 90 arrays. A precompiler approach would maintain the composability of the tools, particularly if debuggers understood preprocessed code.
2. Elimination of function calls. There is no reason why a sophisticated system cannot remove the MPI routine calls and replace them with inline operations, including handling message matching. Such optimizations have been performed for Linda programs [19] and for MPI subsets [20]. Many compilers already perform similar operations for simple numerical functions like abs and sin. This enhancement can be achieved by using preprocessors or precompilers and thus can maintain the composability of MPI with the best compilers.
3. Additional tools and support for correctness and performance debugging. Such tools include editors that can connect send and receive operations so that both ends of the operation are presented to the programmer, or performance tools for massively parallel programs. (Tools such as Vampir and Jumpshot [21] are a good start, but much more can be done to integrate the performance tool with source-code editors and performance predictors.)

4. Changes to MPI itself, such as read-modify-write additions to the remote memory access operations in MPI-2. It turns out to be surprisingly difficult to implement an atomic fetch-and-increment operation [22, Section 6.5.4] in MPI-2 using remote memory operations (it is quite easy using threads, but that usually entails a performance penalty).

3.2   Improving MPI Implementations

Having an implementation of MPI is just the beginning. Just as the first compilers stimulated work in creating better compilers by finding better ways to produce quality code, MPI implementations are stimulating work on better approaches for implementing the features of MPI. Early work along this line looked at better ways to implement the MPI datatypes [23,24]. Other interesting work includes the use of threads to provide a lightweight MPI implementation [25,26]. This work is particularly interesting because it involves code transformations to ensure that the MPI process model is preserved within a single, multithreaded Unix process.

In fact, several implementations of MPI fail to achieve the available asymptotic bandwidth or latency. For example, at least two implementations from different vendors perform unnecessary copies (in one case because of layering MPI over a lower-level software that does not match MPI's message-passing semantics). These implementations can be significantly improved. They also underscore the risk in evaluating the design of a programming model based on a particular implementation.

1. Improvement of the implementation of collective routines for most platforms. One reason, ironically, is that the MPI point-to-point communication routines on which most MPI implementations build their collective routines are too *high* level. An alternative approach is to build the collective routines on top of stream-oriented methods that understand MPI datatypes.
2. Optimization for new hardware, such as implementations of VIA or Infiniband. Work in this direction is already taking place, but more can be done, particularly for collective (as opposed to point-to-point) communication.
3. Wide area networks (1000 km and more). In this situation, the steps used to send a message can be tuned to this high-latency situation. In particular, approaches that implement speculative receives [27], strategies that make use of quality of service [28], or alternatives to IP/TCP may be able to achieve better performance.
4. Scaling to more than 10,000 processes. Among other things, this requires better handling of internal buffers; also, some of the routines for managing process mappings (e.g., MPI_Graph_create) do not have scalable definitions.
5. Parallel I/O, particularly for clusters. While parallel file systems such as PVFS [29] provide support for I/O on clusters, much more needs to be done, particularly in the areas of communication aggregation and in reliability in the presence of faults.

6. Fault tolerance. The MPI intercommunicator (providing for communication between two groups of processes) provides an elegant mechanism for generalizing the usual "two party" approach to fault tolerance. Few MPI implementations support fault tolerance in this situation, and little has been done to develop intercommunicator collective routines that provide a well-specified behavior in the presence of faults.
7. Thread-safe and efficient implementations for the support of "mixed model" (message-passing plus threads) programming. The need to ensure thread-safety of an MPI implementation used with threads can significantly increase latency. Architecting an MPI implementation to avoid or reduce these penalties remains a challenge.

3.3   New Directions

In addition to improving MPI and enhancing MPI implementations, more revolutionary efforts should be explored.

One major need is for a better match of programming models to the multilevel memory hierarchies that the speed of light imposes, without adding unmanageable complexity. Instead of denying the importance of hierarchical memory, we need a memory centric view of computing.

MPI's performance comes partly by accident; the two-level memory model is better than a one-level memory model at allowing the programmer to work with the system to achieve performance. But a better approach needs to be found.

Two branches seem promising. One is to develop programming models targeted at hardware similar in organization to what we have today (see Figure 1). The other is to codevelop both new hardware and new programming models. For example, hardware built from processor-in-memory, together with hardware support for rapid communication of functions might be combined with a programming model that assumed distributed control. The Tera MTA architecture may be a step in such a direction, by providing extensive hardware support for latency hiding by extensive use of hardware threads. In either case, better techniques must be provided for both data transfer and data synchronization.

Another major need is to make it harder to write incorrect programs. A strength of MPI is that incorrect programs are usually deterministic, simplifying the debugging process compared to the race conditions that plague shared-memory programming. The synchronous send modes (e.g., MPI_Ssend) may also be used to ensure that a program has no dependence on message buffering.

4   Conclusion

The lessons from MPI can be summed up as follows: It is more important to make the hard things possible than it is to make the easy things easy. Future programming models must concentrate on helping programmers with what is hard, including the realities of memory hierarchies and the difficulties in reasoning about concurrent threads of control.

Acknowledgment. This work was supported by the Mathematical, Information, and Computational Sciences Division subprogram of the Office of Advanced Scientific Computing Research, U.S. Department of Energy, under Contract W-31-109-Eng-38.

References

1. Papers about MPI (2001) www.mcs.anl.gov/mpi/papers.
2. Hansen, P.B.: An evaluation of the Message-Passing Interface. ACM SIGPLAN Notices **33** (1998) 65–72
3. Geist, A., Beguelin, A., Dongarra, J., Jiang, W., Manchek, B., Sunderam, V.: PVM: Parallel Virtual Machine—A User's Guide and Tutorial for Network Parallel Computing. MIT Press, Cambridge, MA. (1994)
4. Boyle, J., Butler, R., Disz, T., Glickfeld, B., Lusk, E., Overbeek, R., Patterson, J., Stevens, R.: Portable Programs for Parallel Processors. Holt, Rinehart, and Winston, New York (1987)
5. Message Passing Interface Forum: MPI: A Message-Passing Interface standard. International Journal of Supercomputer Applications **8** (1994) 165–414
6. Koelbel, C.H., Loveman, D.B., Schreiber, R.S., Jr., G.L.S., Zosel, M.E.: The High Performance Fortran Handbook. MIT Press, Cambridge, MA (1993)
7. Carlson, W.W., Draper, J.M., Culler, D., Yelick, K., Brooks, E., Warren, K.: Introduction to UPC and language specification. Technical Report CCS-TR-99-157, Center for Computing Sciences, IDA, Bowie, MD (1999)
8. Numrich, R.W., Reid, J.: Co-Array Fortran for parallel programming. ACM SIGPLAN FORTRAN Forum **17** (1998) 1–31
9. Dongarra, J., Gustavson, F., Karp, A.: Implementing linear algebra algorithms for dense matrices on a vector pipeline machine. SIAM Review **26** (1984) 91–112
10. Whaley, R.C., Petitet, A., Dongarra, J.J.: Automated empirical optimizations of software and the ATLAS project. Parallel Computing **27** (2001) 3–35
11. Gropp, W., Lusk, E., Skjellum, A.: Using MPI: Portable Parallel Programming with the Message Passing Interface, 2nd edition. MIT Press, Cambridge, MA (1999)
12. Skjellum, A., Smith, S.G., Doss, N.E., Leung, A.P., Morari, M.: The design and evolution of Zipcode. Parallel Computing **20** (1994) 565–596
13. Message Passing Interface Forum: MPI2: A message passing interface standard. International Journal of High Performance Computing Applications **12** (1998) 1–299
14. TotalView Multiprocess Debugger/Analyzer (2000) http://www.etnus.com/Products/TotalView.
15. Cownie, J., Gropp, W.: A standard interface for debugger access to message queue information in MPI. In Dongarra, J., Luque, E., Margalef, T., eds.: Recent Advances in Parallel Virtual Machine and Message Passing Interface. Volume 1697 of Lecture Notes in Computer Science., Berlin, Springer (1999) 51–58
16. MPI poll '95 (1995) http://www.dcs.ed.ac.uk/home/trollius/www.osc.edu/Lam/mpi/mpi_poll.html.
17. Hempel, R., Walker, D.W.: The emergence of the MPI message passing standard for parallel computing. Computer Standards and Interfaces **21** (1999) 51–62
18. Krechmer, K.: The need for openness in standards. IEEE Computer **34** (2001) 100–101

19. Carriero, N., Gelernter, D.: A foundation for advanced compile–time analysis of linda programs. In Banerjee, U., Gelernter, D., Nicolau, A., Padua, D., eds.: Proceedings of Languages and Compilers for Parallel Computing. Volume 589 of Lecture Notes in Computer Science., Berlin, Springer (1992) 389–404

20. Ogawa, H., Matsuoka, S.: OMPI: Optimizing MPI programs using partial evaluation. In: Supercomputing'96. (1996)
    http://www.bib.informatik.th-darmstadt.de/sc96/OGAWA.

21. Zaki, O., Lusk, E., Gropp, W., Swider, D.: Toward scalable performance visualization with Jumpshot. High Performance Computing Applications **13** (1999) 277–288

22. Gropp, W., Lusk, E., Thakur, R.: Using MPI-2: Advanced Features of the Message-Passing Interface. MIT Press, Cambridge, MA (1999)

23. Gropp, W., Lusk, E., Swider, D.: Improving the performance of MPI derived datatypes. In Skjellum, A., Bangalore, P.V., Dandass, Y.S., eds.: Proceedings of the Third MPI Developer's and User's Conference, MPI Software Technology Press (1999) 25–30

24. Traeff, J.L., Hempel, R., Ritzdoff, H., Zimmermann, F.: Flattening on the fly: Efficient handling of MPI derived datatypes. Volume 1697 of Lecture Notes in Computer Science., Berlin, Springer (1999) 109–116

25. Demaine, E.D.: A threads-only MPI implementation for the development of parallel programs. In: Proceedings of the 11th International Symposium on High Performance Computing Systems. (1997) 153–163

26. Tang, H., Shen, K., Yang, T.: Compile/run-time support for threaded MPI execution on multiprogrammed shared memory machines. In Chien, A.A., Snir, M., eds.: Proceedings of the 1999 ACM Sigplan Symposium on Principles and Practice of Parallel Programming (PPoPP'99). Volume 34.8 of ACM Sigplan Notices., New York, ACM Press (1999) 107–118

27. Tatebe, O., Kodama, Y., Sekiguchi, S., Yamaguchi, Y.: Highly efficient implementation of MPI point-to-point communication using remote memory operations. In: Proceedings of the International Conference on Supercomputing (ICS-98), New York, ACM press (1998) 267–273

28. Roy, A., Foster, I., Gropp, W., Karonis, N., Sander, V., Toonen, B.: MPICH-GQ: Quality of service for message passing programs. Technical Report ANL/MCS-P838-0700, Mathematics and Computer Science Division, Argonne National Laboratory (2000)

29. Carns, P.H., Ligon III, W.B., Ross, R.B., Thakur, R.: PVFS: A parallel file system for Linux clusters. In: Proceedings of the 4th Annual Linux Showcase and Conference, Atlanta, GA, USENIX Association (2000) 317–327

Gyrokinetic Simulations of Plasma Turbulence on Massively Parallel Computers

J. L. V. Lewandowski, W.W. Lee and Z. Lin

Princeton University
Plasma Physics Laboratory, Princeton, NJ 08543, USA

Abstract. With the rapid development of massively parallel computers, the particle-in-cell (PIC) approach to plasma microturbulence in
toroidal geometry has become an increasingly important tool. Global,
self-consistent simulations with up to 100 millions particles yield valuable insights in the dynamics of the turbulence. The inclusion of the
fast-moving electrons in the dynamics represents a major challenge.

1 Introduction

It is now generally accepted in the fusion community that even if fast, large-scale
magnetohydrodynamic (MHD) instabilities can be suppressed, magnetically-
confined plasmas always contain sufficient free energy to drive slow, short-scale
instabilities. These slow, short-scale instabilities, often called microinstabilities [1],
are a major concern as confinement is concerned. The cross-field (perpendicular)
transport associated with microinstabilities is often one to two orders magnitude
larger than the neoclassical transport, and it is called 'anomalous' for this reason [2]. The source of free energy for driving these instabilities lies either in the
non-Maxwellian nature of the distribution functions in velocity space or in the
spatial gradients of the density or temperature of locally Maxwellian distributions. These instabilities have a characteristic wavelength of the order of Lar-
mor radii of the charges species and a typical frequency much smaller than the
the gyro-frequency associated with the species. There are two almost universal
features of the microinstabilities; first, because of the widely different masses of
electrons and ions $\left(m_i/m_e \sim 10^3 \right)$, and therefore widely different thermal speeds
and collision frequencies, the dynamical responses of ions and electrons during
the linear grwoth of the modes are quite different; second, the instability mechanism involves either collisional dissipation or a Landau resonance phenomenon.
Geometric effects complicate the mechanisms which drive the various microinstabilities. Shear of the magnetic field and modulation of the magnetic field strength
causing particle trapping are the most important of these. In summary, transport in high-temperature, magnetized plasmas is dominated by the long-range
collective electric field part of the Coulomb interactions between the charged
particles. In such plasmas the low-frequency modes $[\omega \ll \omega_{ci} \equiv eB/\left(m_i c \right)]$ dominate the cross-field transport [3]. There are various models that can be used
to study the transport generated by micro-turbulence in toroidal fusion plasmas. In this paper we describe the general aspects of the particle gyro-kinetic

B. Monien, V.K. Prasanna, S. Vajapeyam (Eds.): HiPC 2001, LNCS 2228, pp. 95–103, 2001.
Springer-Verlag Berlin Heidelberg 2001

algorithm. This algorithm, based on the Monte Carlo sampling of phase space, is particularly useful to include kinetic effects (missed or treated improperly in fluid approaches) and to treat a global real-space domain (rather than a subset of it). This paper is organized as follows; in section II, we describe briefly the basics of particle gyro-kinetic simulations and the model equations. The simulation domain and parameters are discussed in section III. Recent developments in gyro-kinetic turbulence and future directions will be discussed in section IV.

2   Gyrokinetic Algorithm & Model Equations

The basic aim for particle simulation is to solve the *self-consistent* Vlasov-Maxwell equations that describe the kinetic effects, including nonlinear wave-particle interactions, in six-dimensional phase space. To solve the Vlasov equation (for species j)

$$\frac{dF_j}{dt} \equiv \left[\frac{\partial}{\partial t} + \mathbf{V} \cdot \boldsymbol{\nabla} + \frac{q_j}{m_j} \left(\mathbf{E} + \frac{\mathbf{V} \times \mathbf{B}}{c} \right) \cdot \frac{\partial}{\partial \mathbf{V}} \right] F_j = 0 \qquad (1)$$

in Lagrangian coordinates, a sampling technique is used, whereas the self-consistent fields are obtained from Maxwell equations in the configuration space (Eulerian coordinates). The particle distribution function can be viewed as a collection of finite-size 'quasi-particles', $F_j (\mathbf{r}, \mathbf{v}, t) = \sum_{j=1}^{N} \delta \left[\mathbf{r} - \mathbf{r}_j(t) \right] \left[\mathbf{v} - \mathbf{v}_j(t) \right]$ (Klimontovich-Dupree representation). Each quasiparticle, as represented as a δ function in phase space, actually carries a finite area (which can contort and stretch) of the phase space. As it turns out, a direct simulation using the Klimontovich-Dupree F_j function requires extremely small time step and grid size given by ω_{pe}^{-1} (where ω_{pe} is the electron plasma frequency) and λ_D (the Debye shieding length), respectively. In order to bypass these stringent conditions on the time step and grid size, it is convenient to apply a coordinate transformation of the form [4]

$$\mathbf{r} = \mathbf{R} + \frac{\hat{\mathbf{b}} \times \mathbf{V}_\perp}{\Omega} \qquad (2)$$

where $\mathbf{r}$ is the (particle) position vector, $\mathbf{R}$ is the position vector of the *guiding center* of the particle, $\mathbf{V}_\perp$ is the component of the particle velocity in the plane perpendicular to $\hat{\mathbf{b}} \equiv \mathbf{B}/B$; and $\Omega \equiv qB/(mc)$ is the gyro-frequency. The coordinate transformation (2) is clearly adequate for the short-scale ($\lambda_\perp \sim \rho \sim V_\perp/\Omega$), low-frequency ($\omega/\Omega \ll 1$) modes of interest. As we have just mentioned, a set of quasiparticles is followed in phase space. In order to efficiently update ('push') the particle position in phase space, it is convenient to optimize the representation of the physical domain and of the confining magnetic field. In a toroidal domain, this can be done by introducing angle-like variables, θ and ζ, and a 'radial' coordinate ψ such that $\mathbf{B} \cdot \boldsymbol{\nabla} \psi \equiv 0$. Here θ is the poloidal angle (which labels the position the short way around the torus), ζ is the toroidal angle (which labels the position the long way around the torus), and ψ is the toroidal magnetic flux

(Fig. 1). In these coordinates, a magnetic surface is defined by $\psi = $ const or a square of side 2π is the $\theta - \zeta$ plane (Fig. 2). In the so-called magnetic coordinates $\{\psi, \theta, \zeta\}$ the confining magnetic field can be written in contravariant form

$$\mathbf{B} = \boldsymbol{\nabla}\psi \times \boldsymbol{\nabla}\left(\theta - \zeta/q\right) \tag{3}$$

or in covariant form as

$$\mathbf{B} = g\boldsymbol{\nabla}\zeta + I\boldsymbol{\nabla}\theta + F\boldsymbol{\nabla}\psi \tag{4}$$

where g and I are proportional to the currents flowing in the poloidal and toroidal directions, respectively; $F = F\left(\psi, \theta, \zeta\right)$ is related to the non-orthogonality of the coordinate system. It is clear from representation (3) that $\boldsymbol{\nabla}\cdot\mathbf{B} \equiv 0$, as it should be. In these coordinates, the product of the Jacobian of the transformation, $\mathcal{J} \equiv [\boldsymbol{\nabla}\psi\cdot(\boldsymbol{\nabla}\theta\times\boldsymbol{\nabla}\zeta)]^{-1}$, with the magnetic field strength squared depends on the radial label ψ only, that is $\mathcal{J}B^2 = g(\psi)q(\psi) + I(\psi)$; this is an important feature that improve the efficiency of the particle pushing algorithm. Setting the particle mass and charge to unity, the Hamiltonian guiding center equations of motions can be written as [5]

$$
\begin{aligned}
\frac{d\psi_p}{dt} &= I\mathcal{D}^{-1}\left[\frac{\partial\Phi}{\partial\zeta} - \rho_\|B^2\frac{\partial f}{\partial\zeta} + \left(\rho_\|^2 B + \mu\right)\frac{\partial B}{\partial\zeta}\right] \\
&\quad - g\mathcal{D}^{-1}\left[\frac{\partial\Phi}{\partial\theta} - \rho_\|B^2\frac{\partial f}{\partial\theta} + \left(\rho_\|^2 B + \mu\right)\frac{\partial B}{\partial\theta}\right] \\
\frac{d\theta}{dt} &= \mathcal{D}^{-1}\rho_\|B^2\left[1 - \frac{dg}{d\psi_p}\left(\rho_\| + f\right) - g\frac{\partial f}{\partial\psi_p}\right] \\
&\quad + g\mathcal{D}^{-1}\left[\frac{\partial\Phi}{\partial\psi_p} + \left(\rho_\|^2 B + \mu\right)\frac{\partial B}{\partial\psi_p}\right] \\
\frac{d\zeta}{dt} &= \mathcal{D}^{-1}\rho_\|B^2\left[q + \frac{dI}{d\psi_p}\left(\rho_\| + f\right) + I\frac{\partial f}{\partial\psi_p}\right] \\
&\quad - I\mathcal{D}^{-1}\left[\frac{\partial\Phi}{\partial\psi_p} + \left(\rho_\|^2 B + \mu\right)\frac{\partial B}{\partial\psi_p}\right] \\
\frac{d\rho_\|}{dt} &= \mathcal{D}^{-1}\left[\left(\rho_\|^2 B + \mu\right)\frac{\partial B}{\partial\theta} + \frac{\partial\Phi}{\partial\theta}\right]\left[g\frac{\partial f}{\partial\psi_p} + \frac{dg}{d\psi_p}\left(\rho_\| + f\right) - 1\right] \\
&\quad - \mathcal{D}^{-1}\left[\left(\rho_\|^2 B + \mu\right)\frac{\partial B}{\partial\zeta} + \frac{\partial\Phi}{\partial\zeta}\right]\left[q + \frac{dI}{d\psi_p}\left(\rho_\| + f\right) + I\frac{\partial f}{\partial\psi_p}\right] \\
&\quad - \mathcal{D}^{-1}\left[\left(\rho_\|^2 B + \mu\right)\frac{\partial B}{\partial\psi_p} + \frac{\partial\Phi}{\partial\psi_p}\right]\left(g\frac{\partial f}{\partial\theta} - I\frac{\partial f}{\partial\zeta}\right) - \frac{\partial f}{\partial t}
\end{aligned}
\tag{5}
$$

where $\mathcal{D} \equiv qg + I + \left(\rho_\| + f\right)\left[g\left(dI/d\psi_p\right) - I\left(dg/d\psi_p\right)\right]$. Since for the modes of interest the equilibrium distribution is known, it is advantageous to use a perturbative approach. The probability distribution function for the guiding centers is written as $F = F_0(\text{equilibrium}) + \delta f$ (perturbation) and

$$\mathcal{L}\left(F\right) = 0 \tag{6}$$

where we have defined the operator [6]

$$
\mathcal{L} \equiv \frac{\partial}{\partial t} + \left\{ V_{\|} \widehat{\mathbf{b}} + \frac{c}{B} \langle \mathbf{E} \rangle \times \widehat{\mathbf{b}} + \widehat{\mathbf{b}} \times \left[\frac{V_{\|}^2}{\Omega} \left(\widehat{\mathbf{b}} \cdot \nabla \right) \widehat{\mathbf{b}} + \frac{V_{\perp}^2}{2\Omega} \frac{1}{B} \nabla B \right] \right\} \cdot \nabla
$$
$$
+ \left\{ \left(\frac{q}{m} \langle \mathbf{E} \rangle - \frac{V_{\perp}^2}{2} \frac{1}{B} \nabla B \right) \cdot \left(\widehat{\mathbf{b}} + \frac{V_{\|}}{\Omega} \widehat{\mathbf{b}} \times \widehat{\mathbf{b}} \cdot \nabla \widehat{\mathbf{b}} \right) \right\} \frac{\partial}{\partial V_{\|}} \qquad (7)
$$

Apart from the geometrical effects arising due to the magnetic field inhomo-geneity (the terms in ∇B), the richness of the gyro-kinetic plasma is due to the turbulent electric field (averaged over the fast gyro-motion)

$$
\langle \mathbf{E} \rangle (\mathbf{R}) \equiv \frac{1}{2\pi} \int_0^{2\pi} \mathbf{E} \left[\mathbf{R} + \Omega^{-1} \widehat{\mathbf{b}} \times \mathbf{V}_{\perp} (\varphi) \right] d\varphi \qquad (8)
$$

Since $\langle \mathbf{E} \rangle = -\nabla \langle \Phi \rangle$, the problem is reduced to calculate the gyro-averaged elec-trostatic $\langle \Phi \rangle$. This is accomplished by solving the so-called gyrokinetic Poisson equation [4]

$$
\frac{\tau}{\lambda_D^2} \left(\Phi - \widetilde{\Phi} \right) = 4\pi e \left(\overline{n_i} - \overline{n_e} \right) \qquad (9)
$$

$$
\overline{\Phi} (\mathbf{R}) \equiv \frac{1}{2\pi} \int \Phi (\mathbf{r}) \, \delta \left(\mathbf{r} - \mathbf{R} - \boldsymbol{\rho} \right) d^3 \mathbf{r} d\varphi \qquad (10)
$$

and

$$
\widetilde{\Phi} (\mathbf{r}) \equiv \frac{1}{2\pi} \int \overline{\Phi} (\mathbf{R}) \, F_0 \left(\mathbf{R}, \mu, V_{\|} \right) \delta \left(\mathbf{R} - \mathbf{r} + \boldsymbol{\rho} \right) d^3 \mathbf{R} d\mu dv_{\|} d\varphi \qquad (11)
$$

where $\delta (\mathbf{x})$ is the generalized Kronecker delta function. The term $\Phi - \widetilde{\Phi}$ in the gyrokinetic Poisson equation (9) represents the ion polarization density. From the expressions for $\overline{\Phi}$ [Eq.(10)] and for $\widetilde{\Phi}$ [Eq.(11)] it is clear that a transformation between guiding center space ($\mathbf{R}$) and configuration space ($\mathbf{r}$) must be carried out in order to solve the gyrokinetic Poisson equation. This is done by performing the first-order and second-order gyro-phase averages in configuration space, as illustrated in Fig. 3. This method uses *local* operations in configuration space and are amenable to massively parallel algorithms [7].

3   Simulations

The model equations for micro-turbulence in magnetized plasmas are Eqs.(5) for the trajectories (quasi) particles, Eq.(6) for the sampling of phase-space and Eq.(9) for the self-consistent electrostatic potential. The simulation domain is illustrated in Fig. 4 for a Tokamak configuration with circular, concentric mag-netic surfaces; an example of a more general configuration is the stellarator illustrated in Fig. 5 [8]. The toroidal volume is divided in N segments. A one-dimensional decomposition [using the Message Passing Interface (MPI)] is used for which each processor is allocated a specific segment of the torus (Fig. 4). A second-order Runge-Kutta method is used to advance the particles in phase space. After each push, the contribution of each particle to scalar fields (e.g. density) is deposited onto the grid (gather) and the self-consistent, turbulent electrostatic potential is computed using the parallelized Poisson solver. The electric field can then be calculated and interpolated at the particle position (scatter). The particles are redistributed across the various processors and some diagnostics can be carried out. This is followed by a particle push, and the cycle is repeated. A typical turbulence run is carried out with up to 100 millions particles on a (ψ, θ, ζ) grid with 20 millions points. The specific MPI implementation used in the gyro-kinetic code is based on the underlying physics. A direct (and crude) implementation would divide the particles across the processors, so that each processor handles a fixed number of particles. However, because each particle samples a substantial fraction of the physical domain, a direct approach would require non-local communications. It is therefore preferable to allow the number of particles to 'fluctuate': this ensures a favorable load-balancing. The specific MPI implementation is as follows

- **Step 0**: Allocate memory for 1D domain decomposition for each PE; each PE handles $\sim N_p/N_{PE}$ particles, where N_{PE} is the total number of processors and N_p is the total number of (marker) particles.
- **Step 1**: The particle density on each PE are computed, and the self-consistent, turbulent electrostatic field is determined (this is not a communication intensive step).
- **Step 2**: the electric field, used to follow the particle trajectories, is determined on each PE.
- **Step 3**: The particles positions and velocities are updated on each PE.
- **Step 4:** The particles are redistributed across the various PEs (communication intensive step)
- **Step 5:** Collisional effects are included perturbatively (some MPI collective operations are required here).
- **Step 6:** Various diagnostics are carried out (global operations are needed here, to obtain radial profiles, for examples)
- **Step 7:** Go to Step 1 (the sequence 1-6 is repeated for N time steps).

In this talk, we will present recent computational studies of plasma microturbulence in magnetized toroidal plasmas carried out on the T3E at NERSC (Fig. 6). It is will be shown that although the turbulent activity occurs at very small scales (of the order of the radius of gyration of the particles), there is a spontaneous formulation of a large-scale global electric field which, in turn, regulates the turbulence. In addition, the collisional processes (which act as a sink of energy) strongly modify the characteristics of the turbulence, and can yield to a bursty behavior.

4   Future Directions

Most of simulations reported so far have been carried out in a toroidal plasma with circular surfaces (Fig. 4); equilibrium quantities (such the magnetic field strength) are then invariant under a rotation around the axis of symmetry. The stellarator (Fig. 5) configuration is *not* axi-symmetric and modification of the turbulence can be expected [9]. Almost all computational studies involved a single species (i.e. ions). The turbulence can be modified by the inclusion of the fast-moving electrons. However, because of the mass disparity between ions and electrons (see Introduction), the proper treatment of electron dynamics in a global gyro-kinetic code is a challenging problem. There has been substantial recent progress in addressing the electron dynamics for low-dimensionality problems, for both the electrostatic regime [10] and the electromagnetic regime [11]. The most promising avenue is the so-called slipt-weight scheme [10]; the key idea is to write the electron distribution function as (in the electrostatic case)

$$F = F_0 + \left[\exp\left(\frac{e\Phi}{T_e}\right) - 1\right] F_0 + h \tag{12}$$

where F_0 is the equilibrium distribution function (usually a Maxwellian distribution function), Φ is the turbulent electrostatic potential, T_e is the electron temperature and h is the so-called nonadiabatic electron response. If the electrons had zero mass, they would respond adiabatically to the turbulent electrostatic potential in which case $h \approx 0$. However the mass ratio m_e/m_i is small but finite; knowing that the bulk of the electrons do respond adiabatically, it is sufficient (and numerically convenient) to follow the non-adiabatic response, h, of the remaining electrons. As it turns out, the electromagnetic generalization of the split-weight scheme (12) displays good *thermodynamic* properties [11].

References

1. P. C. Liewer, Nucl. Fusion **25**, 543 (1985).
2. W. Horton, Review of Modern Physics **71**, 735 (1999).
3. B. B. Kadomtsev, *Plasma Turbulence* (Academic Press, New York, 1965).
4. W.W. Lee, Physics of Fluids **26**, 556 (1983).
5. R.B. White, Physics of Fluids **B2**, 845 (1990).
6. T.S. Hahm, Physics of Fluids **31**, 2670 (1988).
7. Z. Lin and W.W. Lee, Physical Review E **52**, 5646 (1995).
8. J.L.V. Lewandowski, Computing and Visualization in Science (Springer-Verlag), **1**(4), p. 183 (1999).
9. J.L.V. Lewandowski, Physics of Plasmas **7**, 3360 (2000).
10. I. Manuilskiy and W.W. Lee, Physics of Plasmas **7**, 1381 (2000).
11. W.W. Lee, J.L.V. Lewandowski and Z. Lin [accepted for publication, to appear in Physics of Plasmas (2001)].

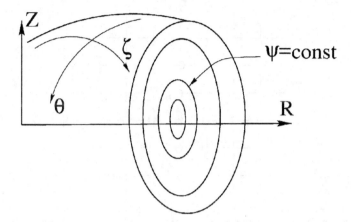

Fig. 1. Illustration of magnetic coordinates in a toroidal plasma; a magnetic surface is defined as $\psi = $ const, where the radial coordinate ψ satisfies $\mathbf{B} \cdot \nabla \psi \equiv 0$.

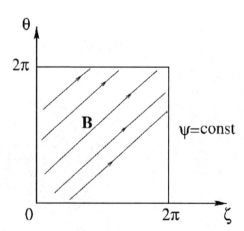

Fig. 2. Local magnetic field lines $\mathbf{B}$ on a given magnetic $\psi = $ const surface. The equation for a magnetic field line is $\alpha - \zeta - q(\psi)\theta = $ const, where $q(\psi)$ is the so-called safety factor. Note that any physical quantity F must be satisfy the periodicity conditions $F(\psi, \theta, \zeta) = F(\psi, \theta + 2\pi M, \zeta + 2\pi N)$, for any integers M and N.

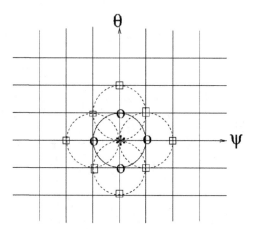

Fig. 3. Illustration of the first order and second order gyro-phase procedure used to calculate the self-consistent turbulent potential on a plane ϕ = const. The star * symbol denotes the guiding center position; the gyro-averaged potential is obtained by a weighted, linear combination of the 4 grid points located at 'O'. In turn, the contribution at point 'O' can be calculated from another average procedure involving the points '□'.

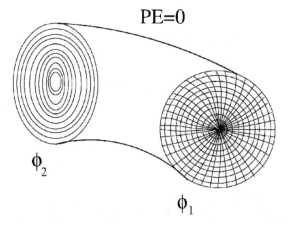

Fig. 4. Illustration of the one-dimensional domain decomposition in toroidal geometry. Each processor is allocated a section of the torus in the interval $[\phi, \phi + \Delta\phi)$, where $\Delta\phi = 2\pi/N$ and N is the total number of processors.

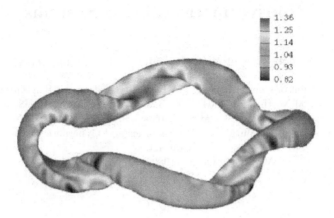

Fig. 5. The use of magnetic coordinates greatly simplify the treatment of complex magnetic geometries such as the stellarator configuration; here the magnetic field strength on a magnetic surface of the stellarator TJ-2 (in operation in Madrid, Spain) is shown [J.L.V. Lewandowski, Computing & Visualization in Science (Springer-Verlag), 1(4), p. 183 (1999).]

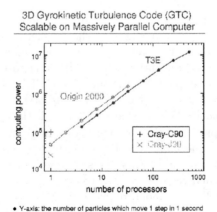

Fig. 6. Scalability of the 3-dimensional gyrokinetic code on the Origin 2000 and on the T3E.

A Parallel Krylov-Type Method for Nonsymmetric Linear Systems

Anthony T. Chronopoulos[1] and Andrey B. Kucherov[2]

[1] Division of Computer Science, University of Texas at San Antonio,
6900 North Loop 1604 West, San Antonio, TX 78249, USA
atc@cs.utsa.edu
[2] Department of Computational Mathematics and Cybernetics, Moscow State
University,
Vorobjevy Goru, 119899, Moscow, Russia

Abstract. Parallel Krylov (**S-step** and **block**) iterative methods for
linear systems have been studied and implemented in the past. In this article we present a parallel Krylov method based on **block s-step** method
for nonsymmetric linear systems. We derive two new averaging algorithm
to combine several approximations to the solution of a single linear system using the block method with multiple initial guesses. We implement
the new methods with ILU preconditioners on a parallel computer. We
test the accuracy and present performance results.

1 Introduction

The s-Step Orthomin forms, at each iteration, s independent direction vectors
using repeated matrix-vector products of the coefficient matrix with a single
residual vector ([4]). Then the solution is advanced simultaneously using the s
direction vectors. The **orthogonal s-Step GCR/Orthomin** was introduced in
[4]. In the Orthogonal s-Step GCR/Orthomin (OSGCR(s)/OSOmin(k,s)) ([5]), a
Modified Gram-Schmidt method (**MGS**) is used to orthonormalize the direction
vectors within each block. We note that OSGCR(s) is the same as OSOmin(0,s).

An alternative approach to the s-step methods, in terms of parallel properties,
is offered by the block methods. The block methods use a number of **linearly
independent** *initial residual* vectors. This number is called the **blocksize** of
the method. The residual vectors are used to compute simultaneously a block of
direction vectors which are orthonormalized via MGS (much like in OSOmin).
These direction vectors are then used to advance the solution and compute the
new residual vector. Several authors have studied the block methods see for
example [2] , [3] , [8] , [9] , [10] , [12] , [14] and the references therein.

A block OSGCR/OSOmin for the solution of nonsymmetric linear systems
if obtained by turning the OSGCR/OSOmin into a block algorithm as in the
references above. We want to use this method to solve a linear system with a
single right handside on a parallel computer. One can use many initial guesses

B. Monien, V.K. Prasanna, S. Vajapeyam (Eds.): HiPC 2001, LNCS 2228, pp. 104–114, 2001.

and then combine the final solution approximations into one by an averaging scheme. Such a scheme was mentioned in [6], for the Conjugate Gradient method.

The convergence study of the s-step methods can be found in [4], [5]. Here we simply derived block algorithms in order to increase the parallelism of the methods. The block algorithms are expected to exhibit similar convergence properties in reference to (the class of) the matrix of coefficients as in the s-step methods, assuming that the initial residual vectors are linearly independent. By varying the parameters k, s in the OSGCR(s)/OSOMin(k,s) method we obtain methods mathematically equivalent to other widely used Krylov methods (e.g. OSOmin(k,1) is equivalent to Omin(k), OSGCR(s) is equivalent to Odir(s) and GMRES(s)) (see [4]). Thus, we only test the block OSOmin method for various k, s and we expect these comparisons to hold for other widely used methods.

The article follows the following structure. In section 2, the block s-step method is presented. In section 3, new solution averaging block methods for solving a linear system with a single right hand side are proposed. In section 4, a partial differential equation problem discretization which generates a large sparse matrix and a parallel preconditioner are described. In section 5, implementation and test results are presented. In section 6, we draw conclusions.

2   Block OSGCR/OSOmin

For integers i, k, s and $1 \leq i, k$, let $j_i = 1$ for OSGCR(s) $j_i = \max(1, i-k+1)$ for OSOMin(k,s). In OSGCR and OSOmin, each iteration generates a *block* of s direction vectors, which are denoted by the *matrix* $P_i = [p_i^1, \ldots, p_i^s]$. Firstly, AP_i is obtained from the column vectors $[Ar_i, A^2 r_i, \ldots, A^s r_i]$, by simultaneously $A^T A$ -orthogonalizing them against the preceding blocks of direction vectors and then orthogonalizing them amongst themselves. Then the direction vectors P_i are formed using the same linear combinations (as in AP_i) starting with the vectors $[r_i, Ar_i, \ldots, A^{s-1} r_i]$. The norm of the residual $\|r_{i+1}\|_2$ is minimized simultaneously in all s new direction vectors in order to obtain x_{i+1}. All orthogonalizations apply the Modified Gramm Schmidt algorithm (MGS) [5].

Let us assume that we are to solve a single linear system $Ax = f$ of dimension n with b righthand sides. We next present the block OSGCR(s)/OSOmin(k,s) (BOSGCR(b,s)/BOSOmin(b,k,s)). We note that BOSGCR(b,s) is the same as BOSOMIN(b,0,s). The following notation facilitates the description of the algorithm. We omit here the iteration subscripts which appear in the algorithm. Let b denote the block size, for example the number of initial solutions. The matrices F (right hand sides), X (solution vectors), R (residual vectors) are of dimensions $n \times b$. The matrices P (direction vectors), AP (matrix times direction vectors), Q, S (used in the orthonormalization of AP and similar transformations on P, respectively) are of dimension $n \times bs$. And U is upper triangular of dimension $bs \times bs$. The (parameter) matrix of steplengths α is of dimension $bs \times b$. Note that if (the b righthand sides are distinct) then $f_i \neq f_j$, for $i \neq j$; else (for a single righthand side $f_i = f$, for $i = 1, \ldots, b$.

Algorithm 1 BOSGCR(b,s)/BOSOmin(b,k,s)
Initialization
Set $F = [f_1, \ldots, f_b]$, Compute the initial residuals $R_1 = F - AX_1$ and set
$\quad Q_0 = S_0 = 0$.
Iterations
For $i = 1, 2, \ldots$ **until** convergence **do**
$\quad$ 1. Compute $P_i = [R_i, AR_i, \ldots, A^{s-1}R_i]$
$\qquad\qquad$ and (set) $AP_i = [AR_i, A^2R_i, \ldots, A^sR_i]$
$\quad$ 2. Orthogonalize AP_i against the matrices
$\qquad Q_{i-1}, \ldots, Q_{j_i}$ and update P_i
$\qquad$ **If** $(i > 1)$ **then**
$\qquad AP_i := AP_i - \sum_{j=j_i}^{i-1} Q_j(Q_j^T AP_i)$
$\qquad P_i := P_i - \sum_{j=j_i}^{i-1} S_j(Q_j^T AP_i)$
$\qquad$ **EndIf**
$\quad$ 3. Compute QR-decomposition (via MGS) of AP_i and obtain S_i
$\qquad\qquad AP_i = Q_iU_i$
$\qquad\qquad P_i = S_iU_i$
$\quad$ 4. Compute the steplengths/Residuals/Solutions
$\qquad\qquad \alpha_i = Q_i^T R_i$
$\qquad\qquad R_{i+1} = R_i - Q_i\alpha_i$
$\qquad\qquad X_{i+1} = X_i + S_i\alpha_i$
EndFor
$\quad$ 5. If(b distinct righthand sides) **then** exit **Else** Combine the b solutions X_{i+1}
$\quad$ to obtain a single solutions $\bar{x}$

3 Solution Averaging Algorithms

In this section we discuss how a single system could be solved using Algorithm1.
In order to apply some block method, in which b vectors are iterated, to solving
of a single right hand side system

$$Ax = f \tag{1}$$

one must obtain b linearly independent initial guesses. Output of block iterations
gives us b approximate solutions. It is reasonable that on input we have only one
initial guess $\hat{x}_0$, and want to get only one solution (approximation) at the end.
If we use a preconditioned method, in which a preconditioning matrix depends
upon some parameter and easy to construct, it would be natural to obtain b
initial guesses, choosing b different parameter values. For an application of this
approach see the end of the next section.
$\quad$ We now derive a new Solution Averaging algorithm which extracts a single
approximate solution with a optimal from b final iterates. Let

$$X_i = [x_i^{(1)}, x_i^{(2)}, \ldots, x_i^{(b)}] \tag{2}$$

$$R_i = [r_i^{(1)}, r_i^{(2)}, \ldots, r_i^{(b)}] \tag{3}$$

be final block solution and block residual vector respectively. Since the right hand side is the same we can write

$$R_i = f e^T - A X_i, \tag{4}$$

where $e^T = [1, 1, \ldots, 1]$ is a unit vector of size b. For brevity we shall omit the index i. Multiplying the last equation by some b-vector one can obtain

$$\frac{1}{e^T \xi} R \xi = f - A(\frac{1}{e^T \xi} X \xi) \tag{5}$$

So, a residual $\bar{r} = f - A \bar{x}$ of a vector

$$\bar{x} = \frac{1}{e^T \xi} X \xi \tag{6}$$

equals to

$$\bar{r} = \frac{1}{e^T \xi} R \xi. \tag{7}$$

Now we are ready to formulate an optimization problem for an averaging vector ξ.

$$\text{Find} \quad \xi = argmin_\xi ||\bar{r}||. \tag{8}$$

By direct computations we get

$$||\frac{1}{e^T \xi} R \xi||^2 = \frac{1}{(e^T \xi)^2} \xi^T R^T R \xi. \tag{9}$$

Then using the Cauchy-Schwarz inequality

$$(e^T \xi)^2 \le (\xi^T (R^T R) \xi)(e^T (R^T R)^{-1} e) \tag{10}$$

we obtain

$$||\frac{1}{e^T \xi} R \xi||^2 \ge \frac{\xi^T R^T R \xi}{(\xi^T R^T R \xi)(e^T (R^T R)^{-1} e)} \tag{11}$$

It is seen now that a minimal value of $||\bar{r}||$ equals

$$min||\bar{r}|| = \frac{1}{\sqrt{e^T (R^T R)^{-1} e}} \tag{12}$$

which is attainable if and only if

$$\xi = c(R^T R)^{-1} e, \quad c \ne 0 \tag{13}$$

for any nonzero scalar c. We choose $c = 1$ for simplicity. An equation $R^T R \xi = e$ should be solved with a special care, since usually columns of matrix R are very close to linearly dependent vectors and $R^T R$ is badly conditioned. QR-decomposition with column pivoting is an appropriate tool for factorizing the matrix $R^T R$. We apply MGS with column pivoting to compute $R = Q_R W$, where $Q_R^T Q_R = I_{b \times b}$, W is an upper triangular $b \times b$-matrix, and then find ξ from the equation $W^T W \xi = e$.

Therefore we obtain a Solution Averaging algorithm as follows

Algorithm 2 Solution Averaging I

 For b solution guesses $X = [x^{(1)}, \ldots, x^{(b)}]^T$ compute
 residuals $R = [r^{(1)}, \ldots, r^{(b)}]^T = fe^T - AX$
 where: $e = [1, \ldots, 1]^T$
 Iterate with Algorithm 1 and compute R and X
 Compute the QR-decomposition $R = Q_R W$
 (by applying MGS with column pivoting)
 Back-Solve two $b \times b$ linear systems $W^T W \xi = e$,
 Compute:

  an average solution: $\bar{x} = \frac{1}{e^t \xi} X \xi$

  an average residual: $\bar{r} = \frac{1}{e^t \xi} R \xi$

  an estimate for the average residual:

  $||\bar{r}|| = \frac{1}{\sqrt{(e^T \xi)}}$

Another algorithm for combining the residuals can be derived from an averaging scheme proposed for the CG method with multiple initial guesses [6]. There, it was mentioned (without providing an algorithm) that the average residual can be obtained by optimally chosing coefficients $\bar{\xi}$ of the expansion

$$\bar{r} = r^{(b)} + \sum_{m=1}^{b-1} (r^{(m)} - r^{(b)}) \bar{\xi}[m] \tag{14}$$

This condition implies that the 2-norm of the average solution $\bar{r}_i$, must be minimized. We use QR Decomposition with column pivoting to solve this problem and we obtain and implement the following residual averaging algorithm. We omit the iteration subscripts for simplicity.

Algorithm 3 Solution Averaging II

 For b solution guesses $X = [x^{(1)}, \ldots, x^{(b)}]^T$ compute
 residuals $R = [r^{(1)}, \ldots, r^{(b)}]^T = fe^T - AX$
 where: $e = [1, \ldots, 1]^T$
 Iterate with Algorithm 1 and compute R and X
 Compute QR-decomposition $\bar{R} = Q_{\bar{R}} \bar{W}$
 where: $\bar{R} = [r^{(1)} - r^{(b)}, \ldots, r^{(b-1)} - r^{(b)}]^T$
 (by applying MGS with column pivoting)
 Back-Solve a $(b-1) \times (b-1)$ system
 $\bar{W} \bar{\xi} = -Q_{\bar{R}}^T r^{(b)}$
 Compute: an average residual/solution:
 $\bar{r} = r^{(b)} + \sum_{m=1}^{b-1} (r^{(m)} - r^{(b)}) \bar{\xi}[m]$
 $\bar{x} = x^{(b)} + \sum_{m=1}^{b-1} (x^{(m)} - x^{(b)}) \bar{\xi}[m]$

4   Test Problem and Preconditioning

We consider right preconditioning because it minimizes the norm of the residual error. Left preconditioning is similar (see [7]), but it minimizes the norm of the preconditioned residuals. Let B^{-1} be the preconditioning matrix. The transformed system is

$$(AB^{-1})(Bx) = f, \tag{15}$$

which is then solved by the iterative process. Either B is a close approximation to the A, i.e.

$$AB^{-1} \approx I \tag{16}$$

or AB^{-1} has clustered eigenvalues. The preconditioner B must be easily invertible, so that the system $By = \phi$ is easy to solve. Following [5], in combining right preconditioning with the block OSGCR/OSOmin, we only need to modify *Step 1*, as follows

1. Compute $P_i = [B^{-1}R_i, B^{-1}(AB^{-1})R_i, \ldots, B^{-1}(AB^{-1})^{s-1}R_i]$,
   $AP_i = [AB^{-1}R_i, (AB^{-1})^2 R_i, \ldots, (AB^{-1})^s R_i]$,

For a nonsymmetric linear system we consider the linear system arising from the discretization of the three-dimensional elliptic equation

$$\Delta u + \gamma(xu_x + yu_y + zu_z) = f(x, y, z) \tag{17}$$

on the unit cube with homogeneous Dirichlet boundary conditions. The degree of nonsymmetry of the problem is controlled by γ. We chose $\gamma = 50$. A discretization by centered finite differences on a uniform $n_h \times n_h \times n_h$ grid with mesh size $1/(n_h + 1)$ yields a linear system

$$Au_h = f_h^{(m)} \tag{18}$$

of the size $n = n_h^3$. As an example one could consider solving $b = 4$ linear systems. For $m = 1, 2, \ldots, b$, we could choose the following exact solutions

$$u_h^{[2m-1]} = w(x, y, z) \sin(\pi m xyz), \tag{19}$$

$$u_h^{[2m]} = w(x, y, z) \cos(\pi m xyz), \quad m = 1, 2 \tag{20}$$

where

$$w(x, y, z) = x(1 - x)y(1 - y)z(1 - z) \exp(xyz), \tag{21}$$

and x, y, z are taken at grid points. Righthand sides $f_h^{(m)}$ could be obtained by matrix-vector product $Au_h^{(m)}$, $m = 1, 2, 3, 4$. In the tests (in the next section) we only consider the first linear system with $b = 4$ initial guesses.

We use a parallel incomplete decomposition preconditioner based on a domain overlapping. Let p be the number of processors. We split the domain into p overlapping subdomains. Firstly, we split a computational domain along the z-axis into p subdomains, each of them contains $(n_h/p + 2)$ xy-planes. So, each two neighbour subdomains share 2 overlapped xy-planes. Secondly, we consider a restriction of the original operator on each subdomain. The loss of connection between subdomains is partially compensated for by introducing overlapping planes. Thus we obtain the following restricted matrix on each subdomain Ω

$$A_\Omega y(j) = a_s(j)y(j-n_h) + a_w(j)y(j-1) + a_0(j)y(j)+$$
$$+a_e(j)y(j+1) + a_n(j)y(j+n_h) + a_b(j)y(j-n_h^2) + a_t y(j+n_h^2), \qquad (22)$$
$$-n_h^2 + 1 \le j \le n_h^3/p + n_h^2,$$

$$a_b(j) = 0, \quad -n_h^2 + 1 \le j \le 0; \qquad (23)$$

$$a_t(j) = 0, \quad n_h^3/p + 1 \le j \le n_h^3/p + n_h^2. \qquad (24)$$

Thirdly, on each subdomain we construct an Incomplete LU decomposition (ILU) preconditioner B as follows $B = B_\Omega(\theta) = LDU$ where L/U - lower/upper triangular and D - diagonal matrices respectively, and $0 \le \theta \le 1$. In our case we chose L and U with the same sparsity pattern as A. In this case, one can easily see that only the diagonal of the matrix is modified by the elimination. We show the sparsity patterns of these matrices by their action on a vector v in the matrix times vector product.

$$Lv(j) = \frac{1}{d(j)}v(j) + a_w(j)v(j-1) + a_s(j)v(j-n_h) + a_b(j)v(j-n_h^2), \qquad (25)$$

$$Dv(j) = d(j)v(j), \qquad (26)$$

$$Uv(j) = \frac{1}{d(j)}v(j) + a_e(j)v(j+1) + a_n(j)v(j+n_h) + a_b(j)v(j+n_h^2), \qquad (27)$$

$$-n_h^2 + 1 \le j \le n_h^3/p + n_h^2, \qquad (28)$$

where

$$1/d(j) = a_0(j) - a_w(j)d(j-1)(a_e(j-1) + \theta a_n(j-1) + \theta a_t(j-1)) -$$
$$-a_s(j)d(j-n_h)(\theta a_e(j-n_h) + a_n(j-n_h) + \theta a_t(j-n_h)) - \qquad (29)$$
$$-a_b(j)d(j-n_h^2)(\theta a_e(j-n_h^2) + \theta a_n(j-n_h^2) + a_t(j-n_h^2))$$

with $j = -n_h^2 + 1, \ldots, n_h^3/p + n_h^2$.

We use the parallel $ILU(\theta)$ preconditioner described above. We take $\theta = 1$ for the iteration process (which provides the fastest convergence), and use other values $0 \le \theta < 1$ to generate initial multiple guesses. We compute b initial guesses as follows

$$x_0^{(m)} = B^{-1}(\theta_m)f, \quad \theta_m = \frac{b-m}{b}, m = 1, 2, \ldots, b \qquad (30)$$

5   Implementation and Test Results

We ran our tests on the CRAY T3E at the University of Texas Austin. We used MPI and Fortran 90 and BLAS 1-3 for our implementation.

Table 1. BOSGCR/BOSOMin , $b = 1$, $b = 4$.

Block Method	(k, s)	$Iter's$	$MError1$	$MRRES1$	$MError4$	$MRRES4$
Omin	$(4, 1)$	61	$1.5E - 12$	$6.55E - 11$	$3.5E - 10$	$5.12E - 8$
	$(8, 1)$	55	$1.9E - 12$	$9.76E - 11$	$8.1E - 10$	$1.27E - 7$
	$(16, 1)$	55	$1.0E - 12$	$5.99E - 11$	$4.5E - 10$	$7.72E - 8$
OSGCR	$(0, 4)$	15	$6.9E - 13$	$4.88E - 11$	$3.3E - 12$	$5.84E - 10$
	$(0, 8)$	8	$3.0E - 13$	$1.63E - 11$	$4.3E - 11$	$1.10E - 8$
OSOmin	$(3, 2)$	30	$1.2E - 12$	$6.08E - 11$	$9.8E - 12$	$2.13E - 9$
	$(1, 4)$	15	$1.0E - 12$	$4.60E - 11$	$1.2E - 10$	$1.98E - 8$
	$(1, 8)$	7	$8.7E - 13$	$6.49E - 11$	$7.9E - 10$	$1.77E - 7$

We see that the algorithms consist of the following operations: (i) Sparse Matrix x vector; (ii) Inner Products; (iii) Linear Combinations; (iv) MGS. Our implementation is outlined as follows: (1) Map A and vectors to a logical linear array of PEs (processors); (2) on each PE use BLAS-2 (e.g. SDOT, SGER for (ii)-(iv)); (3) use ALLREDUCE for global communication (e.g. for (ii)); (4) use local PE communication (e.g. for (i)). For our experiments $n_h = 64$ so the size of a matrix is $n = n_h^3 = 262,144$. The smallest number of PEs that have enough memory to run the problem is $p = 4$.

Convergence tests: We ran Algorithm 1 for $b = 1$ and $b = 4$ with the Solution Averaging methods I - II, on $p = 4$ PEs, for convergence. We report the number of iterations (Iter's). The Maximum Relative Residual Error in 2-norm (MRRES1 and MRRES4 for $b = 1$ and $b = 4$ respectively). The Maximum True Error in ∞-norm (MError1 and MError4 for $b = 1$ and $b = 4$ respectively). We use (MRRES =) $Max_{m=1}^{b}(\|r_i^m\|_2/\|r_0^m\|_2) < \epsilon$, as stopping criterion. The results are in Table 1. The methods are: the standard Omin(k) or Orthomin(k) for $s = 1$; the OSGCR for $k = 0$; the OSOMIN(k,s) for $k = 1$, $s > 0$. *Iter* is the total number of iterations. Algorithms 2 and 3 gave the same results for this test problem. Our aim in this test was to obtain Maximum-True-Error of order 10^{-10}. We used $\epsilon = 10^{-10}$ at first in all the runs. Then we observed that we can use $\epsilon = 10^{-6}$ in the Algorithms 2, 3 with $b = 4$ (thus with fewer iterations) and obtain Maximum-True-Error less than 10^{-10}.

Table 2. BOSGCR/BOSOMin, $b = 4$, $(k, s) = (16, 1)$; Execution times (secs)

p	Av	Prv	DotPr	LinComb	LocCom	GlobCom	T_p/T_{com}
64	$4.91E - 2$	$4.77E - 2$	$2.1E - 3$	$5.2E - 3$	$1.46E - 2$	$6.7E - 3$	$.1661/2.13E - 2$
32	$9.85E - 2$	$6.22E - 2$	$3.5E - 3$	$1.00E - 2$	$1.46E - 2$	$7.0E - 3$	$.2862/2.16E - 2$
16	$.1934$	$9.23E - 2$	$5.9E - 3$	$1.93E - 2$	$1.47E - 2$	$7.1E - 3$	$.5218/2.18E - 2$
8	$.3815$	$.1522$	$1.07E - 2$	$3.81E - 2$	$1.71E - 2$	$5.9E - 3$	$.9932/2.27E - 2$
4	$.7572$	$.2667$	$2.05E - 2$	$7.55E - 2$	$1.53E - 2$	$4.9E - 3$	$1.9280/2.02E - 2$

Table 3. BOSGCR/BOSOMin, $b = 4$, $(k, s) = (0, 8)$; Execution times (secs)

p	Av	Prv	DotPr	LinComb	LocCom	GlobCom	T_p/T_{com}
64	.3581	.3330	$1.22E - 2$	$3.62E - 2$	$9.79E - 2$	$3.54E - 2$	$1.107/.1334$
32	.7144	.4300	$2.11E - 2$	$7.07E - 2$	$9.68E - 2$	$2.60E - 2$	$1.950/.1228$
16	1.400	.6282	$3.55E - 2$	.1365	.1005	$2.34E - 2$	$3.576/.1239$
8	2.718	1.002	$8.04E - 2$	.3216	.1041	$2.22E - 2$	$6.715/.1263$
4	5.211	1.707	.1934	.7870	$9.66E - 2$	$1.80E - 2$	$12.66/.1146$

Performance tests: We ran the algorithms on $p = 4$, 8, 16, 32, 64 processors and measured the different computation/communication parts for each iteration. We report in Table 2-4 the times for: Matrix times vector product (Av), Preconditioning step (Prv), Dotproducts (DotPr), Linear Combinations (LinComb), Local Communication (LocCom), Global Communication (GlobCom), Parallel Execution Time (T_p), Total Communication Time (T_{com}). We plot the speedup in Figure 1, which shows that the algorithms are scalable.

Table 4. BOSGCR/BOSOMin , $b = 4$, $(k, s) = (1, 8)$; Execution times (secs)

p	Av	Prv	DotPr	LinComb	LocCom	GlobCom	T_p/T_{com}
64	.3593	.3306	$2.40E - 2$	$8.73E - 2$	$9.56E - 2$	$3.71E - 2$	$1.683/.1328$
32	.7193	.4295	$4.75E - 2$	.1759	$9.66E - 2$	$3.97E - 2$	$3.142/.1363$
16	1.365	.6039	.1101	.4392	$9.72E - 2$	$3.32E - 2$	$5.616/.1305$
8	2.589	.9397	.2601	1.055	$9.91E - 2$	$3.15E - 2$	$10.16/.1306$
4	5.277	1.713	.4605	1.912	$9.92E - 2$	$2.84E - 2$	$20.91/.1276$

6   Conclusions

We have derived two new parallel algorithms for solving linear systems with a single right hand side. We have implemented the algorithms with ILU preconditioning to approximate the solution of a large sparse system. We have studied the convergence and the parallel performance of the algorithms. Our results show that that methods are convergent and they are scalable.

Acknowledgement. Some reviewers' comments helped enhance the quality of presentation. This research was supported, in part, by research grants from (1) NASA NAG 2-1383 (1999-2000), (2) State of Texas Higher Education Coordinating Board through the Texas Advanced Research/Advanced Technology Program ATP 003658-0442-1999 (3) This research was also supported in part by NSF cooperative agreement ACI-9619020 through computing resources provided by the National Partnership for Advanced Computational Infrastructure at the University of California San Diego.

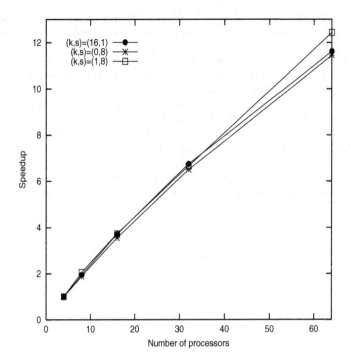

Fig. 1. BOSGCR/BOSOMin, $b = 4$; Speedup

References

[1] Axelsson, O.: Iterative Solution Methods. Cambridge University Press, (1996)

[2] Broyden, C.G.: Block Conjugate Gradient Methods. Optimization methods and Software, **2**(1993) 1-17

[3] Calvetti, D., Reichel L.: Application of a block modified Chebyshev algorithm to the iterative solution of symmetric linear systems with multiple right-hand side vectors. Numer. Math. **68**(1994), 3-16

[4] Chronopoulos, A.T.: S-step Iterative Methods for (Non)symmetric (In)definite Linear Systems. SIAM J. on Num. Analysis, No. 6, **28**(1991) 1776-1789.

[5] Chronopoulos, A.T., Swanson, C.D.: Parallel Iterative S-step Methods for Unsymmetric Linear Systems. Parallel Computing. Volume 22/5, (1996) 623-641

[6] Hackbush, W.: A parallel variant of the conjugate gradient method. Applied Mathematical Sciences, 95. Springer-Verlag, New-York, (1994) 111-119

[7] Meurant, G.: Computer Solution of Large Linear Systems. Elsevier (1999)

[8] Nikishin, A.A., Yeremin, A.Y.: Variable block CG algorithms for solving large sparse symmetric positive definite linear systems on parallel computers, I: General iterative scheme. SIAM J. Matrix Anal. Appl. **16**(1995) 1135-1153

[9] Papadrakakis M., Smerou, S.: A new implementation of the Lanczos method in linear problems. Internat. J. Numer. Methods Engrg., **29**(1990) 141-159

[10] Parlett, B.N.: A new look at the Lanczos and the block-Lanczos methods. SIAM J. Numer. Anal., **17**(1980) 687-706

[11] Radicati di Brozolo, G., Robert, Y.: Parallel conjugate gradient-like algorithms for sparse nonsymmetric systems on a vector multiprocessor. Parallel Computing, **11**(1989) 223-239

[12] Sadkane, M., Vital, B.: Davidson's Method for Linear Systems of Equations. Implementation of a Block Algorithm on a Multi-processor. Tech. Rept. TR/PA/91/60, CERFACS, Toulouse, (1991)

[13] Simon, H., Yeremin, A.: New Approach to Construction of Efficient Iterative Schemes for Massively Parallel Applications: Variable Block CG and BiCG Methods and Variable Block Arnoldi Procedure. Proc. of the 6th SIAM Conference on Parallel Proc. for Scientific Computing, ed. by R. Sincovec et al., SIAM, Philadelphia, (1993) 57-60

[14] Simoncini, V., Gallopoulos, E.: An iterative method for nonsymmetric systems with multiple right hand sides. SIAM J. Sci. Stat. Comput., **16**(1995) 917-933

Evolving Cellular Automata Based Associative Memory for Pattern Recognition

Niloy Ganguly[1], Arijit Das[2], Pradipta Maji[2], Biplab K. Sikdar[2], and
P. Pal Chaudhuri[1]

[1] Computer centre, IISWBM, Calcutta, West Bengal, India 700073,
n_ganguly@hotmail.com
[2] Department of Computer Science & Technology, Bengal Engineering College (D.U),
Botanic Garden, Howrah, West Bengal, India 711103,
{arij@,pradipta@,biplab@,ppc@ppc.}becs.ac.in

Abstract. This paper reports a *Cellular Automata* (*CA*) model for pattern recognition. The special class of *CA*, referred to as *GMACA* (*Generalized Multiple Attractor Cellular Automata*), is employed to design the *CA* based associative memory for pattern recognition. The desired *GMACA* are evolved through the implementation of genetic algorithm (*GA*). An efficient scheme to ensure fast convergence of *GA* is also reported. Experimental results confirm the fact that the *GMACA* based pattern recognizer is more powerful than the Hopfield network for memorizing arbitrary patterns.

Keywords: Cellular Automata, Genetic Algorithm, Pattern Recognition.

1   Introduction

Pattern recognition is the study as to how the machines can learn to distinguish patterns of interest from their background. The training enables a machine to learn/store the patterns and recognize them with reasonable certainty even from a noisy environment. The performance of pattern recognition depends on the depth of training. In conventional approach of recognition, the machine compares the given input pattern with each of the stored patterns and identifies the closest match. The search requires the time proportionate to the number of patterns learnt/stored. Obviously, with the growth of the number of patterns stored, the recognition process becomes too slow.

The *Associative Memory* provides the solution to the problem where the time to recognize a pattern is independent of the number of patterns stored. In such a design, the entire state space gets divided into some pivotal points (a,b,c of *Fig.1*). The states close to a pivotal point get associated with that specific point. Identification of a noisy pattern (original pattern distorted due to noise) amounts to traversing the transient path (*Fig.1*) from the given noisy pattern to the closest pivotal point learnt. As a result, the process of recognition becomes independent of the number of patterns learnt/stored.

B. Monien, V.K. Prasanna, S. Vajapeyam (Eds.): HiPC 2001, LNCS 2228, pp. 115–124, 2001.
© Springer-Verlag Berlin Heidelberg 2001

In the above background, since early 80's the model of associative memory has attracted considerable interests among the researchers [1]. Both sparsely connected machine (Cellular Automata) and densely connected network (Neural Net) [1,2] have been explored to solve this problem. The seminal work of Hopfield [3] made a breakthrough by modeling a *recurrent, asynchronous, neural net* as an *associative memory* system. However, the complex structure of neural net model has partially restricted its application. Search for alternative model around the simple structure of Cellular Automata (*CA*) continued [4].

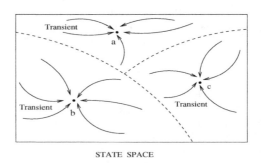

STATE SPACE

Fig. 1. Model of associative memory with 3 pivotal points

The design of *CA* based associative memory has so far mostly concentrated around uniform *CA* [2,4] with same rule applied to each of the *CA* cells. Some works on non-uniform *CA* has also been reported in [5]. But none reported the evolution of *CA* to a general purpose pattern recognizer although it has been able to recognize some patterns [2,6]. In this paper, we propose the evolution of associative memory for pattern recognition with more general class of *CA* referred to as *GMACA*. This specific class of *CA* displays very encouraging result in pattern recognition. The memorizing capacity of such a class of *CA*, as established in this paper, can be found to be better than that of Hopfield Network.

The genetic algorithm (*GA*) has been employed to arrive at the desired *GMACA* configurations. An innovative scheme to identify the initial population, which significantly improves the *GA* convergence rate, has been also reported.

In order to present the underlying principle of *CA* based model for pattern recognition, in *Section II* we introduce an overview on *Cellular Automata*. *GMACA* based associative memory and its application for pattern recognition is next outlined in *Section III*. Application of *genetic algorithm* to evolve *CA* based associative memory is subsequently presented in *Section IV*. In *Section V*, we report the experimental observations that establish the *CA* as an efficient pattern recognizer. A comparative study with Neural Net model has been also reported in this section.

2   Cellular Automata Preliminaries

A *Cellular Automaton* (*CA*) is a discrete system which evolves in discrete space and time. It consists of a large number of cells, organized in the form of a lattice (*Fig.2*). Each cell acts as a single processor and communicates only with the neighboring cells. The update of state (next state) for a cell depends on its own state and the states of its neighboring cells. In this paper we will discuss only 3-neighborhood (left neighbor, self and right neighbor) one dimensional *CA* and each *CA* cell having only two states - 0 or 1.

In a two state 3-neighborhood *CA*, there can be a total of 2^{2^3} i.e, 256 distinct next state functions of a cell referred to as the *rule* of *CA* cell [7]. If the next state function of a cell is expressed in the form of a truth table, then the decimal equivalent of the output column is conventionally called the *rule number* for the cell. The following is an illustration for two such rules, 90 and 150:

Neighborhood : 111 110 101 100 011 010 001 000 *RuleNo*
(i) NextState :   0   1   0   1   1   0   1   0   90
(ii) NextState : 1   0   0   1   0   1   1   0   150

The first row lists the possible combinations of present states (left, self and right) for a 3-neighborhood *CA* cell at time t. The next two rows list the next states for the ith cell at time instant $(t + 1)$ - the decimal equivalent of the 8 bit binary number is referred to as Rule Number. The number of 1's in a rule is quantified by the Langton's parameter λ [8]. For example, λ value for both the rules 90 and 150 is 0·5 (λ=number of 1's in the rule number/8). If the same rule is applied to all the cells, then the *CA* is referred to as *uniform CA*, else it is a *hybrid CA*. The *CA* rule vector shown in *Fig.2(a)* <90,90,90,90,90> is of a uniform *CA*, where rule 90 is applied to each of the cells. By contrast, the *Fig.2(b)* represents hybrid *CA* rule with rule vector <100,130,150,170,180>. The average value of λ for all the cells in a hybrid *CA* is referred to as λ_{av}. In *Fig.2(b)*, the value of λ_{av} for the hybrid *CA* is 0·425.

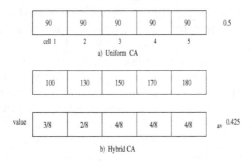

Fig. 2. Rules and Langton's parameter values of uniform and hybrid *CA*

The attractive feature of the *CA* is its global complex behavior [7]. Moreover, the locality of inter-connection among the *CA* cells and its synchronous update scheme makes it ideally suitable for $VLSI$ implementation. As a result, studies of cellular automata space [2,4,6] has generated considerable interests. Exhaustive study of its state transition behavior is also reported in [9].

A $GMACA$ (*Generalized Multiple Attractor Cellular Automata*) is a hybrid CA [9] that employs non-linear rules with AND/OR logic with attractor cycle length greater than 1 and can perform pattern recognition task. *Fig.3* illustrates a 5-cell hybrid $GMACA$ with rule vector $< 89, 39, 87, 145, 91 >$. Its state space is divided into two attractor basins, attractor-1 and attractor-2. The states not covered by the attractor cycles can be viewed as *Transient States* [10] in the sense that the $GMACA$ finally settles down in one of its attractors after passing through such transient states. The state transition behavior of this class of CA has been partially characterized in [9].

Characterization of the uniform CA state transition, based on the value of λ is reported in [8]. As the λ value for a rule R is shifted from 0 to 0·5 the average behavior of the uniform CA having rule R passes through the transitions : homogeneous $\rightarrow$ periodic $\rightarrow$ complex $\rightarrow$ chaotic. In [8] Langton claimed that each regime can be associated with a range of λ value. The range of λ value for which the rule exhibit complex function is referred to as λ_c.

In the present work, we have concentrated on the $GMACA$ with hybrid rule vector and its attractive feature to memorize the large number of patterns for a specific range of λ_{av}. The $GMACA$ is modeled to design the CA based associative memory which in turn has been effectively utilized as the pattern recognizer.

3 $GMACA$ Based Associative Memory for Pattern Recognition

The pattern recognizer is trained to get familiarized with some specific pattern set. It can recognize an incoming pattern even if the pattern is corrupted with limited noise. If the entire pattern is viewed as a binary string, then the hamming distance between the incoming pattern $\acute{P}_i$ and the original one P_i measures the amount of noise. A pattern recognizing system typically memorizes k patterns $\{P_1, \cdots, P_i, \cdots P_k\}$. When a new pattern $\acute{P}_i$ to be recognized is input to the system, the patern recognizer identifies it as P_i, where $\acute{P}_i$ is the closest to P_i -that is, the hamming distance between $\acute{P}_i$ and P_i is the least among all P's.

A $GMACA$ having its state space distributed into disjoint basins (*Fig.3*) with transient and attractor states models an associative memory and is effectively employed for pattern recognition. The entire state space around some pivotal points (*Fig.1*) can be viewed as the state space generated by a $GMACA$ with its attractors and attractor basins containing the transient states. Memorizing the set of pivotal points $\{P_1, \cdots, P_i, \cdots P_k\}$ is equivalent to design of CA with the pivotal points as attractor states. Any other transient point $\acute{P}_i$ in close vicinity of a pivotal point P_i can be considered as a transient state of the $GMACA$ in the attractor basin of P_i. Thus $\acute{P}_i$ can be viewed as a pattern P_i with some noise.

Therefore, the state space of the $GMACA$ employed as pattern recognizer of a particular input set gets divided in such a way that each attractor cycle contains one of the patterns ($\mathcal{P}_i$) to be learnt. Moreover, the hamming distance between any transient state $\acute{\mathcal{P}}_i$ in the attractor basin of $\mathcal{P}_i$ and the attractor $\mathcal{P}_i$ is the least compared to the hamming distances of $\acute{\mathcal{P}}_i$ with other $\mathcal{P}$s . This $GMACA$ thus produces the correct output in time proportionate to the traversal of the input pattern to an attractor, and thus independent of the number of patterns (k) learnt/stored. The essential design goals for the $GMACA$ based associative memory model are next summarized.

Design Goals : The $GMACA$ designed for a particular pattern set maintains following two relations :

R1: Each attractor basin of the CA should contain one pattern ($\mathcal{P}_i$) to be learnt in its attractor cycle; and

R2: Hamming distance of each state $\mathcal{S}_i$ ($\in \mathcal{P}_i$-basin) and $\mathcal{P}_i$ (a state in the attractor cycle) is less than the hamming distance of $\mathcal{S}_i$ with any other $\mathcal{P}$s. The $GMACA$ of *Fig.3* maintains both R1 and R2. It learns two patterns, $\mathcal{P}_1 = 10000$ and $\mathcal{P}_2 = 00111$. The state $\mathcal{P} = 11010$ has the hamming distances 2 and 3 with $\mathcal{P}_1$ & $\mathcal{P}_2$ respectively. If $\mathcal{P}$ is given as the input to be recognized, then the recognizer must return $\mathcal{P}_1$. The $GMACA$ of *Fig.3* loaded with the pattern $\mathcal{P}=$ 11010 returns the desired pattern $\mathcal{P}_1 = 10000$ after two time steps.

The next section describes the GA based searching scheme to find the desired $GMACA$, tuned for a set of patterns to be recognized.

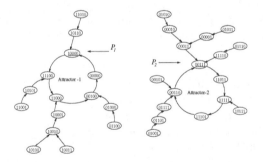

Fig. 3. State space of a GMACA based 5-bit pattern recognizer with two attractor basins (Transient States : All states other than the ones lying on attractor cycles)

4   Evolving $GMACA$ Based Associative Memory

The aim of this evolution scheme is to arrive at the desired $GMACA$ (rule vector) that can perform pattern recognition task. This subsection describes the GA based solution to evolve the $GMACA$ rules with the desired functionality noted in the previous subsection. The fitness function and the method of selecting population for the next generation of GA are noted below.

Fitness function: The fitness $\mathcal{F}(C_r)$ of a particular chromosome C_r (that is, CA rule) in a population is determined by the hamming distance between the attractor $\acute{\mathcal{P}}_i$, evolved for a state from the run (generation), and the desired attractor $\mathcal{P}_i$. The chromosome (CA) is run with 300 randomly chosen initial configurations (ICs) and the fitness of the CA is determined by averaging the fitness for each individual IC.

Let us assume that, a chromosome C_r is run for the maximum iterations ($\mathcal{L}_{max}$) for an IC and it reaches to a state $\acute{\mathcal{P}}_i$. If $\acute{\mathcal{P}}_i \notin$ any attractor cycle, then the fitness value of the chromosome C_r is considered as zero. On the other hand, if $\acute{\mathcal{P}}_i \in$ an attractor cycle containing $\mathcal{P}_i$, then the fitness of C_r is $\frac{n-|\mathcal{P}_i-\acute{\mathcal{P}}_i|}{n}$, where n is the length of the CA rule (chromosome). Therefore, the fitness function can be written as

$$\mathcal{F}(C_r) = \frac{1}{k} \sum_{i=1}^{k} \frac{n-\mid \mathcal{P}_i - \acute{\mathcal{P}}_i \mid}{n} \tag{1}$$

where k is the number of random ICs.

Selection, Crossover and Mutation: From the exhaustive experimentation, we have set the ratio of selection, mutation and crossover to be performed on the present population (PP) to form the population (NP) for next generation. To construct NP the top 10% CA rules (elite rules) [11] of PP are selected without any modifications. The 80% chromosomes of NP are constructed by taking crossover operations among the chromosomes of PP. The rest 10 % population of NP are generated from the mutations of the elite rules. In the formation of NP the single point crossover between randomly chosen pairs of rules and single point mutations of elite rules are employed. The algorithm for evolving $GMACA$ next follows.

Algorithm 1 Evolve-$GMACA$

  Input : $\mathcal{N}$: *Number of CA cell,*
  $\{P\}$: *Patterns $\{P_1, P_2, \cdots, P_\mathcal{M}\}$ to be trained,*
  $\mathcal{G}_{max}$: *Maximum number of generations permitted*
  Output : *Rule table of different $GMACA$*
  begin
   Step 1. Generate 50 new rules (i.e, chromosomes) for initial population (IP)
   Step 2. Initialize generation counter GC=zero; $PP \leftarrow IP$
   Step 3. Generate a new set of 300 IC
   Step 4. Compute fitness values $\mathcal{F}(C_r)$ for each chromosome of the present population (PP)
   Step 5. Store the rule and the corresponding generation number for which the fitness value $\mathcal{F}(C_r) = 1$
    If $\mathcal{F}(C_r) = 1$ for at least one chromosome then go to stop
   Step 6. Increment generation counter (GC)
   Step 7. If $GC > \mathcal{G}_{max}$ then goto Step 1
    else rank the chromosomes in order of fitness
   Step 8. Form next population (NP) by selection, crossover and mutation

Step 9. $PP \leftarrow NP$; *Go to* Step 3
Step 10. *Stop*

We next outline a scheme for quick convergence to the the desired solution.
Selection of Initial Population (IP)**:** The number of generations needed to evolve an n-cell $GMACA$, for associative memory model, increases with the value of n. The quick convergence of *Algorithm 1* demands the intelligent selection of the IP. The selection schemes aim to form the IP by maintaining the diversity in initial population and simultaneously enhancing the mean fitness.

Scheme 1: Initial population from λ_{pr} region
From exhaustive experimentation, it is found that the n-cell $GMACA$ rules, acting as an efficient pattern recognizer lie within a specific range of λ_{av} value and referred to as λ_{pr} region. This fact basically extends the Langton's observation, applicable to the uniform CA to the domain of hybrid CA.

To find λ_{pr} region, the experiment has been conducted for n=10 to 30. For each n, 15 different sets S_i (i=1,2,..., 15) of patterns, to be learnt, are selected randomly. The number of patterns in a set S_i is taken as $\lceil .15n \rceil$ - that is, the maximum number of n-bit patterns that a Hopfield network can memorize [3]. *Algorithm 1* starts with the initial population (IP) of 50 chromosomes chosen at random from an unbiased distribution and then undergoes evolution till the 100% fit rules are obtained. The λ_{av} values for each of the fit rules are computed for an n. *Table 1* depicts the mean and standard deviation of λ_{av} to arrive at the desired $GMACA$ rules. *Column III* indicates the mean value below 0·5 while *column V* indicates the mean above 0·5 for different CA size. The result reports that the λ_{av} values of the desired $GMACA$ rules are clustered around in the areas that are roughly equidistant from 0.5 and, therefore, λ_{pr} can be chosen as 0.45± or 0.55±. This observation leads to form the initial population (IP) with λ_{av} value close to 0·45 or 0·55. The average convergence rate of the population is much higher than the pure random populations.

Table 1. Mean and standard deviations of λ_{av} value of fit GMACA rules

No of	No of gene^n	λ_{av} value less than 0.5		λ_{av} value greater than 0.5	
CA cell	to converge	Mean	Standard Deviation	Mean	Standard Deviation
10	140	0.464423	0.025872	0.530833	0.015723
15	138	0.457576	0.025365	0.524074	0.020580
20	486	0.363194	0.006875	0.560937	0.013532
25	612	0.486250	0.005995	0.517500	0.007500
30	703	0.465000	0.024840	0.509375	0.004541

Scheme 2: Mismatch pair Algorithm
In selecting the initial population, the mismatch pair algorithm presented by Myer [12] is effectively used in the current design. Mismatch pair algorithm constructs a rule table from the state transition diagram of a CA. The state transition diagram (G) is empirically constructed (*Fig.4*) for the patterns

$P_1, P_2, \cdots, P_{\mathcal{M}}$ which are to be learnt, considering them as the attractors of a $GMACA$. We apply a modified version of the algorithm to construct a CA rule table in 3-neighborhood. The derived CA approximates the state transition diagram G constructed from the attractors $P_1, P_2, \cdots, P_{\mathcal{M}}$. These CA rules are considered as the member of the initial population (IP). The average fitness of the rules produced from this process (*Table 3, column VII*) are approximately 15% better than the random population and convergence rate of *Algorithm 1* is also increased.

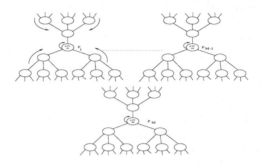

Fig. 4. Empirical basins created for Mismatch pair algorithm

Scheme 3: Mixed Rules
The initial population (IP) in this scheme comprises of the (i) chromosomes from λ_{pr} region, (ii) chromosomes produced through mismatch pair algorithm, (iii) randomly generated chromosomes and (iv) chromosomes that are produced from the concatenation of fit CA of smaller sizes - that is, if a design demands the evolution of a 25-cell CA capable of recognizing 5 given patterns, then the chromosome is formed by concatenating a 10-cell and a 15-cell CA rule vectors, where the corresponding 10-cell and 15-cell CA are capable of recognizing two and three patterns respectively.

5   Experimental Observations

The experiments to evolve pattern recognizable n-cell $GMACA$ for different values of n are carried out. For each n, 15 different sets of patterns to be trained are selected randomly. The number of patterns to be learnt by the CA is progressively increased. In case of mixed rules, out of the 50 chromosomes of IP the 35 are selected from mismatch pair algorithm and the rest 15 are generated from the other three processes.

Table 2 demonstrates the potential of $GMACA$ based pattern recognition scheme. *Column II* of *Table 2* depicts the maximum number of patterns that an n-cell $GMACA$ can memorize. The average number of generations required to converge *Algorithm 1* are noted in *Column III*. The results of Hopfield Net

Table 2. Performance of the $GMACA$ based pattern recognizer

Pattern size (n)	CA based Patt. Recog		No of patterns memorized in Hopfield net
	No of patterns memorized	No of $gene^n$ required	
10	4	128	2
15	4	66	2
20	5	172	3
25	6	210	4
30	7	280	5
35	8	340	5
40	9	410	6
45	10	485	7

on the same data set are provided in *Column IV* for the sake of comparison. The experimental result clearly indicates that the $GMACA$ have much higher capacity to store or learn patterns in comparison to Hopfield net.

The performance evaluation of the four *IP* selection schemes (random, λ_{pr} region, mismatch pair and mixed rules) is provided in *Table 3*. The graph showing the number of generations required to converge *Algorithm 1* for different number of CA cells with random and preselected initial populations are presented in *Fig.5*. It is observed that the convergence rate is much better for the initial population formed from the mixing of chromosomes. The entries '$*$' in *column II* of *Table 3* signify that *Algorithm 1* with random *IP* does not converge within 1000 generations for $n \geq 35$.

Table 3. Mean fitness and number of generations for random and preselected rules

No of CA cell	Random		λ_{pr} region		Mismatch pair		Mixed rule	
	No of $gene^n$	Mean Fitness	No of $gene^n$	Mean Fitness	No of $gene^n$	Mean Fitness	No of $gene^n$	Mean Fitness
10	166	65.35	141	67.15	118	72.86	128	73.12
15	156	62.28	253	65.05	148	72.67	66	74.81
20	512	61.16	413	61.78	364	70.66	172	72.84
25	721	60.44	690	61.15	376	73.46	210	74.38
30	852	59.33	718	62.10	468	77.03	280	76.88
35	$*$	59.75	785	60.62	525	67.96	340	70.44
40	$*$	59.71	826	59.81	575	67.21	410	71.28
45	$*$	58.48	935	58.50	612	66.78	485	71.17

6   Conclusion

This paper has established the Cellular Automata as a powerful machine in designing the pattern recognition tool. The pattern recognizer is designed with

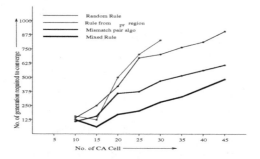

Fig. 5. Graph showing the number of generations required for random and preselected rules

the hybrid CA based associative memory. It is proved experimentally that the storage capacity of CA based associative memory is far better than that of Hopfield Net.

References

1. J. Hertz, A. Krogh and R. G. Palmer, *"Introduction to the theory of Neural computation"*, Santa Fe institute studies in the sciences of complexity, Addison Wesley, 1991.
2. M. Chady and R. Poli, *"Evoluation of Cellular-automaton-based Associative Memories"*, Technical Report no. CSRP-97-15, May 1997.
3. J. J. Hopfield, *"Pattern Recognition computation using action potential timings for stimulus representations"*, Nature, 376: 33-36; 1995.
4. E. Jen, *"Invariant strings and Pattern Recognizing properties of 1D CA"*, Journal of statistical physics, 43, 1986.
5. M. Sipper, *"Co-evolving Non-Uniform Cellular Automata to Perform Computations"*, Physica D, 92: 193-208, 1996
6. M. Mitchell, P. T. Hraber and J. P. Crutchfield, *"Revisiting the Edge of Chaos: Evolving Cellular Automata to Perform Computations"*, Complex Systems, 7:89-130, 1993.
7. S. Wolfram, *"Theory and application of Cellular Automata"*, World Scientific, 1986.
8. W. Li, N. H. Packard and C. G. Langton, *"Transition Phenomena in Cellular Automata Rule Space"*, PhysicaD, 45; 1990.
9. P. Pal Chaudhuri, D Roy Chowdhury , S. Nandi and S. Chatterjee, *"Additive Cellular Automata, Theory and Applications, VOL. 1"*, IEEE Computer Society Press, Los Alamitos, California.
10. Wuensche,A., and M.J.Lesser, *"The Global Dynamics of Cellular Automata"*, Santa Fe Institute Studies in the Sciences of Complexity, Addison-Wesley, 1992.
11. J. H. Holland, *"Adaptation in natural and artificial systems"*, The University of Michigan Press, Ann Arbor, MI, 1975.
12. J. E. Myers, *"Random boolean Networks - Three recent results"*, Private communication; 1997.

Efficient Parallel Algorithms and Software for Compressed Octrees with Applications to Hierarchical Methods*

Bhanu Hariharan and Srinivas Aluru

Dept. of Electrical and Computer Engineering
Iowa State University, Ames, IA, USA
{bhanu,aluru}@iastate.edu

Abstract. We describe the design and implementation of efficient parallel algorithms, and a software library for the parallel implementation of compressed octree data structures. Octrees are widely used in supporting hierarchical methods for scientific applications such as the N-body problem, molecular dynamics and smoothed particle hydrodynamics. The primary goal of our work is to identify and abstract the commonalities present in various hierarchical methods using octrees, design efficient parallel algorithms for them, and encapsulate them in a software library. We designed provably efficient parallel algorithms and implementation strategies that perform well irrespective of data distribution. The library will enable rapid development of applications, allowing application developers to use efficient parallel algorithms developed for this purpose, without the necessity of having detailed knowledge of the algorithms or of implementing them. The software is developed in C using the Message Passing Interface (MPI). We report experimental results on an IBM SP and a Pentium cluster.

1   Introduction

Octrees [6] are hierarchical tree data structures used in a number of scientific applications that require accessing and manipulating information on points and clusters of points in space. A significant portion of the effort involved in realizing such applications on parallel computers is directed towards the design of parallel algorithms for constructing and querying octrees and their implementation. There is a large body of literature on sequential and parallel algorithms for the construction of octrees, and on applications using octrees. But often, this work is carried out in the context of developing the best possible algorithms for a particular problem or application, such as, parallel octree algorithms for solving the N-body problem [3].

The primary motivation behind this work is the large number of scientific applications that use octrees. There has been recent interest in studying parallel algorithms for octrees in the context of supporting hierarchical applications.

---

* Research supported by NSF Career under CCR-0096288 and NSF CCR-9988347.

B. Monien, V.K. Prasanna, S. Vajapeyam (Eds.): HiPC 2001, LNCS 2228, pp. 125–136, 2001.

Shan *et al.* study the construction of octrees on shared address multiprocessors [8]. Liu *et al.* discuss parallel algorithms for constructing octrees and for incrementally updating octrees [3]. The main distinguishing feature of our work is the design of a parallel software library using radically different algorithms and implementation strategies that ensure theoretically provable and practically efficient performance. In order to allow the design of provably efficient algorithms, we use compressed octrees instead of regular octrees. It is easy to migrate octree based applications to use compressed octrees, as will become evident later. Also the software can be easily modified to construct octrees as well, and going through compressed octrees allows for the design of more efficient algorithms. Our contributions in this paper are the following: Provably efficient algorithms for compressed octree construction and queries using compressed octrees, implementation strategies to minimize the constant overheads hidden in complexity analysis, and development of a software library suitable for use in applications. The library is developed in the C programming language using MPI.

The rest of the paper is organized as follows: Section 2 contains a description of octrees and compressed octrees. In Section 3, we outline some parallel communication primitives used in describing our algorithms. Parallel algorithms and implementation strategies for tree construction and queries are discussed in Sections 4 through 6. Each section also describes experimental results on both uniform and non-uniform distributions. Section 7 consists of future work and conclusions.

2   Octrees and Compressed Octrees

Octrees and compressed octrees are hierarchical tree data structures used to represent point data. The root node of the tree represents all points and a cubic region containing them. The children of a node in the tree represent subregions of the region represented by the node and the leaf nodes in the tree represent individual points. We use the term *cell* to denote a cubic region, the term *subcell* to denote a cell that is contained in another and the term *immediate subcell* to denote a cell obtained as a result of immediate subdivision of another cell into 8 cells having half the side length of the original cell. The *length of a cell* is the span of the cell along any dimension. Let n denote the number of points. Consider a cell large enough to contain all points. Recursive subdivision of this cell into subcells is naturally represented by a tree, called *octree*. The subdivision process is stopped on cells having exactly one point or when the smallest allowable size for a cell (fixed based on machine precision or other considerations) is reached. A cell that is not further subdivided is a *leaf cell*.

In an octree, when the points that lie within a cell also lie within one of its immediate subcells, *chains* are formed in the tree. Each node on a chain represents the same set of points and hence contains the same information duplicated, perhaps in a different form. A chain can be arbitrarily large irrespective of the total number of points, making the size of the octree dependent upon the distribution of the points. In a compressed octree, each such chain is replaced by a single node

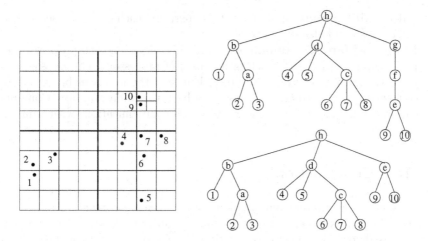

Fig. 1. A two dimensional domain and its corresponding quadtree and compressed quadtree. The term *quadtree* is used in two-dimensions because each node has at most 4 children.

(see Figure 1). This not only avoids information duplication but also makes the size distribution independent. The size of a compressed octree is $O(n)$. It is not a height-balanced tree and its height could be as large as $\Omega(n)$. For each node in a compressed octree, we define two cells: The *small cell* at a node v, denoted $S(v)$, is the smallest subcell that contains all the points in its subtree. The *large cell* of the node, denoted $L(v)$, is the largest cell the node is responsible for. It is obtained by taking the appropriate immediate subcell of the small cell at its parent node. A compressed octree for n points can be constructed sequentially in $O(n \log n)$ time, which is optimal [1]. It can be directly used by applications coded to use octrees. If necessary, octrees can be constructed from compressed octrees. To do this, a chain can be created in place of a node whose large cell and small cell are not identical, such that each node in the chain represents a cell half the length of its parent cell.

3   Preliminaries

In this section, we briefly describe some primitive parallel operations used in our algorithms. For a detailed description and run-time analysis, see [2]. In what follows, p denotes the number of processors.

- **Parallel Prefix:** Given n data items $x_1, x_2, \ldots, x_n$ and a binary associative operator $\otimes$ that operates on these data items, the prefix sums problem is to compute the partial sums, $s_1, s_2, \ldots, s_n$, where $s_i = x_1 \otimes x_2 \otimes x_3 \otimes \ldots \otimes x_i$. Parallel prefix is the problem of computing the prefix sums in parallel.
- **Allgather:** Given a data item on each processor, the Allgather operation collects all the data and stores a copy of it in each processor.

- **All-to-All Communication:** In this operation each processor sends a distinct message of size m to every processor.
- **Many-to-Many Communication:** In this operation each processor has a personalized message for each of a subset of processors. The message sizes can have varying lengths. If the total length of the messages being sent out or received at any processor is bounded by t, the Many-to-Many Communication can be performed using two all-to-all communications with a uniform message size of $\frac{t}{p}$ [5].

4   Tree Construction

In this section, we present our parallel algorithm for constructing compressed octrees. The goal is to generate the tree in postorder traversal order, which will be later used to design efficient algorithms for queries. The algorithm takes as input the number of allowable levels (l) in the tree to determine the size of the smallest cell. In practice, l is quite small. For instance, 20 levels allows us to have a tree with potentially $8^{20} = 2^{60}$ leaf nodes, enough to capture a wide variety of distributions for even the largest simulations.

Definition 1. *Let D be the length of the domain and l be the length of a cell. The level of the cell is defined to be $\log_2 \frac{D}{l}$.*

Definition 2. *The level of a node in the compressed octree is defined to be the level of its small cell.*

Following [9], we represent cells using integer keys. Each cell can be represented by an integer triple (x, y, z), denoted by the x, y and z coordinates of the cell in the cell space. The key is formed by interleaving the bits of x, y and z in that order, and prepending it with a 1 bit. This unique and unambiguous representation of all the cells in the tree corresponds to ordering them according to Morton ordering, also known as Z-space filling curve ordering [4]. For any pair of cells, either the two cells are disjoint or one of them is completely contained in the other cell. We consider two cells to be disjoint if they merely touch at the boundaries. Define an ordering of a pair of cells as follows: If a cell is completely contained in another, then the cell with smaller size appears first in the ordering. If they are disjoint, find the smallest cell that contains both these cells. Each of the two cells will be contained in a distinct immediate subcell of this smallest cell. Order the two cells according to the value of the keys of the corresponding immediate subcells. These operations can be done using bit arithmetic, and are taken as $O(1)$ time operations. We order the nodes in a compressed octree according to their small cells. In drawing a compressed octree, we make use of this ordering. Ordering of all the nodes thus, corresponds to the postorder traversal of the nodes in the tree. Furthermore, given two nodes v_1 and v_2 in the compressed octree, the smallest cell containing $S(v_1)$ and $S(v_2)$ is the same as $S(u)$, where u is the lowest common ancestor of v_1 and v_2.

The compressed octree is constructed as follows: Each processor is initially given $\frac{n}{p}$ points. For each point, the corresponding leaf cell is generated. These are sorted in parallel, eliminating duplicates. Each processor borrows the first leaf cell from the next processor and runs a sequential algorithm to construct the compressed octree for its leaf cells together with the borrowed cell in $O\left(\frac{n}{p} \log \frac{n}{p}\right)$ run-time [1]. This local tree is stored in postorder traversal order in an array. Each node stores the indices of its parent and children in the array.

Lemma 1. *Consider the compressed octree for the n points, which we term the global compressed octree. Each node in this tree can be found in the local compressed octree of at least one processor.*

Proof. Every internal node is the lowest common ancestor of at least one consecutive pair of leaves. More specifically, let u be a node in the global compressed octree and let $v_1, v_2, \ldots, v_k$ be its children $(2 \le k \le 8)$. Consider any two consecutive children v_i and v_{i+1} $(1 \le i \le k - 1)$. Then, u is the lowest common ancestor of the rightmost leaf in v_i's subtree and the leftmost leaf in v_{i+1}'s subtree. If we generate the lowest common ancestors of every consecutive pair of leaf nodes, we are guaranteed to generate every internal node in the compressed octree. However, there may be duplicates — each internal node is generated at least once but at most seven times $(2^d - 1$ times in d dimensions$)$.

Each node in a local compressed octree is also a node in the global compressed octree. In order to generate the postorder traversal order of the global compressed octree, nodes which should actually appear later in the postorder traversal of the global tree, should be sent to the appropriate processors. Such nodes, termed *out of order nodes*, appear consecutively after the borrowed leaf in the postorder traversal of the local tree and hence can be easily identified. Also, the borrowed leaf makes sure that the lowest common ancestor of every consecutive pair of leaves is generated; it is not considered a part of the local tree. An Allgather operation is used to collect the first leaf cell in each processor into an array of size p. Using a binary search in this array, the destination processor for each out of order node can be found. Nodes that should be routed to the same processor are collected together and sent in a single message. This uses Many-to-Many communication.

Lemma 2. *The number of out of order nodes in a processor is at most l.*

Proof. We show that there can be at most one node per level in the local tree that is out of order. Suppose this is not true. Consider any two out of order nodes in a level, say v_1 and v_2. Suppose v_2 occurs to the right of v_1 (v_2 comes after v_1 in cell order). Since v_1 and v_2 are generated, some leaves in their subtrees belong to the processor. Because the leaves allocated to a processor are consecutive, the rightmost leaf in the subtree of v_1 belongs to the processor. This means v_1 is not an out of order node, a contradiction.

Lemma 3. *The total number of nodes received by a processor is at most 7l.*

Proof. We first show that a processor cannot receive more than one distinct node per level. Suppose this is not true. Let v_1 and v_2 be two distinct nodes received at the same level. Without loss of generality, let v_2 be to the right of v_1. A node is received by a processor only if it contains the rightmost leaf in the subtree of the node. Therefore, all the leaf nodes between the rightmost leaf in the subtree of v_1 and the rightmost leaf in the subtree of v_2 must be contained in the same processor. In that case, the entire subtree under v_2 is generated on the processor itself and the processor could not have received v_2 from another processor, a contradiction. Hence, a processor can receive at most l distinct nodes from other processors. As there can be at most 7 ($2^d - 1$ in d dimensions) duplicate copies of each node, a processor may receive no more than $7l$ (($2^d - 1)l$ in d dimensions) nodes.

The received nodes are merged with the local postorder traversal array and their positions are communicated back to the sending processors. The net result is the postorder traversal of the global octree distributed across processors. Each node contains the position of its parent and each of its children in this array. From here onwards, we use the term *local tree* or *local array* to refer to the local portion of the global compressed octree stored in postorder traversal order. The size of the local tree is $O\left(\frac{n}{p} + l\right)$.

4.1   Experimental Results

We experimentally tested our tree construction algorithm using an IBM SP with 4-way SMP nodes and a Pentium Pro cluster connected by Fast Ethernet. To verify the distribution-independent nature of the algorithm, we tested it with both uniformly and non-uniformly distributed points. We used multiplicative congruential random number generator for uniform distribution and generated points that are uniformly distributed in all three dimensions. For the non-uniformly distributed sample, we chose three normal distributions with different means (0.4, 0.5, 0.7 for x, y, and z respectively) and standard deviations (0.2, 0.2, 0.1 for x, y, and z respectively), to generate the x, y, and z coordinates of the points.

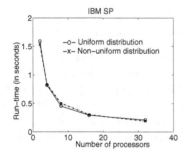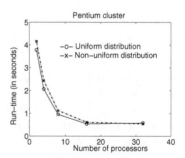

Fig. 2. Run-times of parallel tree construction as a function of the number of processors for 100,000 points on IBM SP and Pentium cluster.

Table 1. Run-times in seconds for various stages of tree construction for 100,000 points on IBM SP.

Number of procs	Uniform distribution					Non-uniform distribution				
	Parallel sort	Local tree	Comm	Merging	Total	Parallel sort	Local tree	Comm	Merging	Total
2	0.77	0.76	0.01	0.05	1.59	0.74	0.73	0.01	0.05	1.53
4	0.36	0.38	0.04	0.04	0.82	0.38	0.38	0.03	0.04	0.83
8	0.20	0.20	0.03	0.02	0.45	0.21	0.21	0.03	0.05	0.50
16	0.12	0.11	0.03	0.03	0.29	0.12	0.12	0.03	0.03	0.30
32	0.07	0.07	0.04	0.02	0.21	0.07	0.07	0.03	0.02	0.19

The run-times of parallel tree construction for $100,000$ points for both the distributions as a function of the number of processors are shown in Figure 2. From the graphs, it is easy to see that the algorithm works well even for non-uniform distributions, as predicted by the theoretical analysis. It also scales well with number of processors. The breakdown of the run-times on the SP into times spent in various stages in the algorithm is shown in Table 1. Parallel sorting and local tree construction dominate the run-time, and a negligible percentage of time is spent in communicating out of order nodes and incorporating them into the local tree. This is a direct consequence of the result that the number out of order nodes is a function of the height of the tree, which tends to be small. Hence, the algorithm has very little overhead due to parallelization.

5   Upward and Downward Accumulations

5.1   Upward Accumulation

In the upward accumulation query, typical of many hierarchical applications, the objective is to aggregate the information at all the descendants of each node in the tree. Typically, a function is supplied to aggregate the information at a node to its parent. To compute the upward accumulation at a node, this function is applied at each node in the tree for each of its children. Our algorithm for parallel upward accumulation is as follows: First, each processor scans its local array from left to right. During the scan, the information at each node is aggregated to its parent, provided the parent is local to the processor. As the tree is stored in postorder traversal order, if all the children of a node are present in the same processor, it is encountered only after all its children are. This ensures that such a node has received its upward accumulation when the scan reaches it. This computation takes $O\left(\frac{n}{p} + l\right)$ time.

During the scan, some nodes are labeled *residual nodes* based on the following rules: 1) If a node receives its upward accumulation but its parent lies in a different processor, it is labeled a *residual leaf node*. 2) If a node does not have its upward accumulation when it is visited, it is labeled a *residual internal node*. Each processor copies its residual nodes into an array. It is easy to see that the residual nodes form a tree (termed the *residual tree*) and the tree is present in its postorder traversal order, distributed across processors.

Lemma 4. *The number of residual internal nodes in a processor is at most $l-1$.*

Proof. We show that there can be at most one residual internal node per level in a processor. Suppose this is not true. Consider any two residual internal nodes in a level, say v_1 and v_2. Suppose v_2 occurs to the right of v_1 (v_2 comes after v_1 in cell order). Because of postorder property, the right most leaf in v_1's subtree and the rightmost leaf in v_2's subtree should belong to this processor. Because the leaves allocated to a processor are consecutive, all the leaves in the subtree of v_2 belong to the processor. This means v_2 is not a residual internal node, a contradiction. Therefore, there can be only one residual internal node at each level in a processor. Furthermore, no leaf can be a residual internal node. Hence, the claim.

Corollary 1. *The size of the residual tree is at most $8pl$.*

Proof. From the previous lemma, the total number of residual internal nodes is at most $p(l-1)$. It follows that the maximum number of residual leaf nodes can at most be $7p(l-1)+1$ (or $p(2^d-1)(l-1)+1$ in d dimensions), for a total tree size of at most $8p(l-1)+1 < 8pl$ (or $2^d p(l-1)+1 < 2^d pl$ in d dimensions).

The residual tree is accumulated using a parallel upward tree accumulation algorithm [7]. The number of iterations of the algorithm is the logarithm of the size of the tree. During each iteration, the algorithm uses a parallel prefix and a Many-to-Many communication. The algorithm could have been directly applied to the global compressed octree but this would require $O(\log n)$ iterations. The residual tree can be accumulated in $O(\log p + \log l)$ iterations. Thus, we reduce the iterations from the logarithm of the size of the tree to the logarithm of the height of the tree, which is much smaller!

5.2   Downward Accumulation

In the downward accumulation query, the objective is to aggregate the information at all the ancestors of each node in the tree. The algorithm for this query is the reverse of the algorithm employed for upward accumulation. First, the downward accumulations for the residual tree are calculated. This requires $O(\log p + \log l)$ iterations. Then, the downward accumulations are computed for the local tree using a right-to-left scan of the postorder traversal of the local tree.

5.3   Experimental Results

The primary advantage of using the parallel accumulation algorithms on the residual tree instead of the global compressed octree is that the size of the

Table 2. Number of residual nodes for 100,000 (100K) and 1 million (1M) points. Also shown are the run-times in seconds for the upward accumulation query for 100K points. The accumulation function used is addition.

Table 2(a)

Number of procs	Number of residual nodes			
	Uniform		Non-uniform	
	100K	1M	100K	1M
2	13	44	27	26
4	71	65	97	83
8	160	193	200	211
16	281	365	329	364
32	502	693	547	732

Table 2(b)

Number of procs	Run-time for upward accumulation					
	Uniform			Non-uniform		
	Local	Parallel	Total	Local	Parallel	Total
2	0.03	0.01	0.04	0.03	0.01	0.04
4	0.02	0.01	0.03	0.02	0.02	0.04
8	0.01	0.02	0.03	0.01	0.05	0.06
16	0.01	0.04	0.05	0.01	0.04	0.05
32	0.01	0.04	0.05	0.01	0.04	0.05

residual tree is significantly smaller. The residual tree sizes as a function of the number of processors for $100,000$ and 1 million points and two different distributions are shown in Table 2 (a). The actual sizes are much less compared to the upper bound of $8pl$. This is because most of the nodes have their children in the same processor and only nodes at higher levels are more likely to be part of the residual tree. The results for the upward accumulation queries run using addition as the accumulation function are tabulated in Table 2 (b). The total run-times are significantly smaller when compared to tree construction run-times and do not show much improvement as the number of processors is increased. This is because: 1) the local run-times are small due to the simple accumulation function chosen, and 2) the parallel part, though has only few iterations, uses the time consuming Many-to-Many communication in each iteration. In any application, tree construction will be used prior to running any of these queries. Given that the query times for accumulations are smaller by an order of magnitude or more, the overall application run-time will still show near linear speed up.

Hierarchical applications typically use complex accumulation functions. For instance, in an application of the Fast Multipole Method to solve the 3D N-body problem, the upward accumulation function translates a truncated Taylor series expansion from a child to its parent in time proportional to the fourth power of the number of terms used. Studies show that run-time of the applications is dominated by this phase (for example, see [9]). To reflect real application behavior, we used an accumulation function consisting of a small delay of 0.25ms. This increased the overall query time but the time spent in communication remained almost the same. The run-times for upward and downward accumulation as a function of the number of processors are shown in Figure 3. The figure confirms that the algorithms scale well for complex accumulation functions.

6 Cell Queries

In a cell query, each processor generates an array of cells whose information should be retrieved. The query cells are directed to the processors owning these cells and the cell information is communicated back to the originating proces-

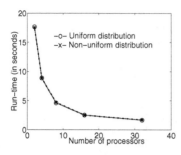 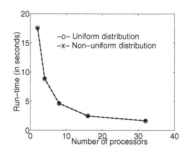

Fig. 3. Run-times of upward and downward accumulation queries for 100,000 points on IBM SP. The accumulation function uses a delay of a quarter of a millisecond.

sors. Consider two examples of such queries – In the Fast Multipole Method, for each cell, the information within cells that are in a "doughnut region" around the query cell is sought. In molecular dynamics, the information from points that are within a cut-off radius from each point is sought. If the smallest cell size is chosen such that its length is equal to the cut-off radius, this query translates to finding the neighboring cells of each cell. Specialized algorithms can be designed for each query type. However, they may each require a different tree representation and/or a different data distribution strategy. In order to preserve generality, we use a unified approach. The commonality between most such queries is that spatially local information is sought for each cell. The partitioning of the compressed octree using its postorder traversal order has the effect of partitioning the domain into regions, ordering the regions according to Morton ordering, and allocating them to processors. This ensures that most queries can be answered locally whereas communication is necessary only near the boundaries.

We classify queries into two categories – those that can be answered locally, and those that require remote access. Using an All-to-All communication, the non-local queries are sent to the appropriate processors. Another All-to-All communication is used to receive the answers. Here, communication for receiving remote queries can be overlapped with the computation used for answering local queries to virtually eliminate the parallel overhead.

6.1 Experimental Results

To experimentally evaluate the run-time performance of cell queries using our approach, we used the following query, typical of molecular dynamics applications: For each leaf cell in the compressed octree, all 26 of its neighboring leaf cells are sought. The times spent in answering queries locally and time spent in communication for 100,000 points on the IBM SP and on the Pentium cluster are shown in Table 3. The run-times scale well with the number of processors for both uniform and non-uniform distributions.

Table 3. Run-time in seconds for all leaf neighboring cell query for 100,000 points.

Number of procs	IBM SP				Pentium cluster			
	Uniform distribution		Non-uniform distribution		Uniform distribution		Non-uniform distribution	
	Comm	Total	Comm	Total	Comm	Total	Comm	Total
2	0.09	155.87	0.32	160.61	0.06	256.12	0.49	344.10
4	0.10	94.43	0.46	93.97	0.08	140.45	0.58	181.47
8	0.09	54.48	0.44	53.13	0.08	77.44	0.54	105.44
16	0.10	31.47	0.34	30.74	0.15	43.24	0.44	59.16
32	0.09	22.41	0.28	22.37	0.12	24.80	0.36	34.77

7   Conclusions and Future Work

We presented the design and development of a parallel tree library for compressed octrees. We designed efficient algorithms that perform well for non-uniform distributions and have little communication overhead. The library encapsulates most of the parallelization necessary in building applications using hierarchical methods. This makes it possible for application developers to develop parallel code with little effort. We are currently working on including more queries and developing sample applications to demonstrate the effectiveness of the library. The library can be obtained by contacting one of the authors.

Acknowledgements. We wish to acknowledge the Ira and Marylou Fulton Supercomputing Center at Brigham Young University for access to the IBM SP parallel computer.

References

1. S. Aluru and F. Sevilgen, Dynamic compressed hyperoctrees with applications to N-body problem, *Proc. Foundations of Software Technology and Theoretical Computer Science* (1999) 21-33.
2. V. Kumar, A. Grama, A. Gupta and G. Karypis, *Introduction to Parallel Computing*, The Benjamin/Cummings Publishing Co., 1994.
3. P. Liu and S. Bhatt, Experiences with parallel N-body simulation, *IEEE Transactions on Parallel and Distributed Systems, 11(12)* (2000) 1306-1323.
4. G.M. Morton, A computer oriented geodesic data base and a new technique in file sequencing, IBM, Ottawa, Canada, 1966.
5. S. Ranka, R.V. Shankar and K.A. Alsabti, Many-to-many communication with bounded traffic, *Prod. Frontiers of Massively Parallel Computation* (1995) 20-27.
6. H. Samet, Design and analysis of spatial data structures, Addison-Wesley Publishing Company, 1990.
7. F. Sevilgen, S. Aluru and N. Futamura, A provably optimal, distribution-independent, parallel fast multipole method, *Proc. International Parallel and Distributed Processing Symposium* (2000) 77-84.

8. H. Shan and J.P. Singh, Parallel tree building on a range of shared address space multiprocessors: algorithms and application performance, *Proc. International Parallel Processing Symposium and Symposium on Parallel and Distributed Processing* (1998) 475-484.

9. M.S. Warren and J.K. Salmon, A parallel hashed oct-tree N-body algorithm, *Proc. Supercomputing* (1993) 1-12.

A Case Study of Improving Memory Locality in Polygonal Model Simplification: Metrics and Performance

Victor Salamon[1], Paul Lu[1], Ben Watson[2], Dima Brodsky[3], and Dave Gomboc[1]

[1] Dept. of Computing Science, University of Alberta,
Edmonton, AB, Canada, T6G 2E8, {salamon,paullu,dave}@cs.ualberta.ca
[2] Dept. of Computer Science, Northwestern University,
Evanston, IL, USA, watsonb@cs.nwu.edu
[3] Dept. of Computer Science, University of British Columbia,
Vancouver, BC, Canada, dima@cs.ubc.ca

Abstract. Polygonal model simplification algorithms take a full-sized polygonal model as input and output a less-detailed version of the model with fewer polygons. When the internal data structures for the input model are larger than main memory, many simplification algorithms suffer from poor performance due to paging.

We present a case study of the recently-introduced *R-Simp* algorithm and how its data locality and performance can be substantially improved through an off-line spatial sort and an on-line reorganization of its internal data structures. When both techniques are used, *R-Simp*'s performance improves by up to 7-fold. We empirically characterize the data-access pattern of *R-Simp* and present an application-specific metric, called *cluster pagespan*, of *R-Simp*'s locality of memory reference.

1 Introduction

Trade-offs between quality and performance are important issues in real-time computer graphics and visualization. In general, the more polygons in a three-dimensional (3D) computer graphics model, the more detailed and realistic is the rendered image. However, the same level of detail and realism may not be required in all computer graphics applications. For real-time display, fewer polygons result in a faster rendering; rendering speed may be the most important criteria. For other applications, the speed of rendering may be traded off for improved image quality. The flexibility to select a version of the same model with a different level of detail (i.e., a different number of polygons) can be important when designing a graphics system.

Polygonal model simplification algorithms take a full-sized polygonal model as input and output a version of the model with fewer polygons. Although the simplified models are often of high quality, it is also clear that, upon close examination, some details have been sacrificed to reduce the size of the model (Figure 1). A number of model simplification algorithms have been proposed [3],

B. Monien, V.K. Prasanna, S. Vajapeyam (Eds.): HiPC 2001, LNCS 2228, pp. 137–148, 2001.
© Springer-Verlag Berlin Heidelberg 2001

| Original, 871,306 polygons | Simplified, R-$Simp$, 40,000 polygons |

Fig. 1. Original and Simplified Versions of Polygonal Models:*dragon*

each with their different strengths and weaknesses in terms of execution time and the resulting image quality. In general, the algorithms are used to pre-compute simplified versions of the models that can be selected for use at run-time.

However, models that have over a million polygons require tens of megabytes of disk storage, and require hundreds of megabytes of storage for their in-memory data structures (Table 1). For example, the *blade* model has almost 1.8 million polygons and requires 331.3 megabytes (MB) of virtual memory after being read in from an 80.1 MB disk file.[1]

Consequently, we examine the problem of data-access locality and its effect on performance for a model simplification algorithm. In particular, we make a case study of the recently-introduced *R-Simp* simplification algorithm. We focus on systems-oriented run-time performance and related metrics of *R-Simp*, as opposed to measures of model quality.

1.1   Motivation and Related Work: Large Models

As new model acquisition technologies have developed, the complexity and size of 3D polygonal models have increased. First, model producers have begun to use 3D scanners (for example, [5]). Second, as the speed and resolution of scientific instrumentation increases, so has the size of the data sets that must be visualized. Both of these technology changes have resulted in models with hundreds of millions to billions of polygons. Current hardware cannot come close to displaying these models in real time. Consequently, there is a large body of "model simplification" research addressing this problem [3].

Most of these model simplification algorithms (e.g., [9,4,2]) use a greedy search approach with a time complexity of $O(n \log n)$, where n is the number of polygons in the original model. Also, the greedy algorithms have poor locality of memory access, jumping around the surface of the input model, from puzzle piece to puzzle piece. One exception to this trend is the simplification algorithm

---

[1] Initial virtual memory size is read from the /proc file system's stat device under Linux 2.2.12 before any simplification computation is performed.

Table 1. Summary of Full-Sized Polygonal Models

Model	Faces (Polygons)	Vertices	Initial Virtual Memory Size (MB)	File Size for PLY Model (MB) [7]
hand	654,666	327,323	123.6	31.1
dragon	871,306	435,545	164.1	32.2
blade	1,765,388	882,954	331.3	80.1

described by Lindstrom [6], based on the algorithm by Rossignac and Borrel [8]. Lindstrom's algorithm is fast, but it produces simplified models of poor quality and, by its nature, it is difficult to control the exact number of polygons in the simplified model.

In contrast, the recently-introduced *R-Simp* algorithm [1] produces approximated models of substantially higher quality than those produced by Rossignac and Borrel's approach, and *R-Simp* allows for exact control of output size, all without a severe cost in execution speed. *R-Simp* is unique in that it iteratively refines an initial and poor approximation, rather than simplifying the full input model. Consequently, *R-Simp* has a time complexity of $O(n \log m)$, rather than $O(n \log n)$, where m is the number of polygons outputted in the simplified model. Since $m \ll n$ in practice, *R-Simp*'s advantage can be substantial.

2   The Problem: Locality of Memory Accesses

To understand the data-access patterns of *R-Simp*, we hand-instrumented a version of the code such that an on-line trace is produced of all the model-related memory accesses. Each memory access is timestamped using the value returned by the Unix system call gettimeofday(), where the first access is fixed at time 0. The ability to map and label a memory access to a specific *R-Simp* data structure is beyond the ability of most standard tools. The resulting trace file is processed off-line.

A scatter plot of the virtual address of each memory access versus the timestamp shows a "white noise" pattern (Figure 2), which indicates relatively poor locality of reference due to a large working set. The graph intuitively explains why paging is a performance problem for models that do *not* fit in main memory. We only show the scatter plot for *dragon* as the plots for the other two models are similar. The hardware platform is a 500 MHz Pentium III system, running Linux 2.2.12, with 128 MB of RAM, and a IDE swap disk. Although there are dual processors, *R-Simp* is a single-threaded application. All reported real times are the average of 5 runs.

The instrumentation and tracing adds significant overhead to *R-Simp*'s execution time. Consequently, we have scaled the X-axis. For example, the non-instrumented *R-Simp* requires 7,736 seconds to compute the simplified model for *dragon*. The instrumented *R-Simp* requires much more time to execute, but we have post-processed the trace information so that the X-axis appears to be 7,736

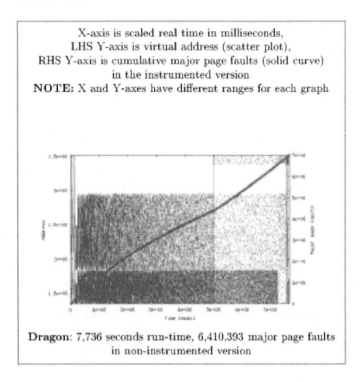

X-axis is scaled real time in milliseconds,
LHS Y-axis is virtual address (scatter plot),
RHS Y-axis is cumulative major page faults (solid curve)
in the instrumented version
NOTE: X and Y-axes have different ranges for each graph

Dragon: 7,736 seconds run-time, 6,410,393 major page faults
in non-instrumented version

Fig. 2. Memory-Access Pattern for Original *R-Simp* on *dragon*: Computing 40,000 polygon simplified model

seconds. Although the scaling may introduce some distortions to the figure, it still captures the basic nature of the memory-access patterns.

Figure 2 has some notable patterns. The darker horizontal band, between addresses 1.3×10^8 and 1.8×10^8 on the LHS Y-axis, is the region of virtual memory where the data structure for the vertex list is stored. The horizontal region above the vertices is where the face list is stored. The data structures and details of *R-Simp* are discussed in Section 3. There is a vertical line at time 5.0×10^6 milliseconds for *dragon*, which is when *R-Simp* finishes the simplification process and begins to create the new simplified model. The band of memory accesses at the top of the graph, and to the immediate right of the vertical line, is where the new model is stored.

Superimposed on Figure 2 is a solid curve representing the cumulative major page faults of *R-Simp* over time, as reported by the Unix function `getrusage()`. A major page fault may require a disk access to swap out the victim page and does require a disk access to swap in the needed page of data. Note that there is some discrepancy between the major page faults incurred by the instrumented *R-Simp* (Y-axis on the RHS) and the major page faults incurred by the non-instrumented *R-Simp* (noted in the captions of Figure 2 and Table 3). We have purposely not scaled the RHS Y-axis since that is more problematic than scaling real time. Again, the figures are only meant to give an intuitive picture of the data-access patterns inherent to *R-Simp*.

3   The *R-Simp* Algorithm

We briefly summarize the main data structures in *R-Simp* and the main phases of the computation [1].

3.1   Data Structures

R-Simp uses three main data structures to perform model simplification. The first two data structures are static and are used to store the original model. The last data structure, the *cluster*, is more dynamic.

First, the *vertex list* is a global array that stores the vertices from the input model. Each vertex contains (x, y, z) coordinates, an adjacency list of vertices, and an adjacency list of faces. Second, the *face list* is also a global array that stores the faces from the input model. Each face contains its normal, its area, and pointers to the vertices that make up the face. The global vertex and face lists are accessed throughout *R-Simp*'s execution. Third, the *cluster* structure represents a portion of the original model and is a node in an n-ary tree. A cluster contains a list of vertices from the original model and other supplementary data. Iteratively, a leaf cluster is selected from a priority queue and subdivided into 2, 4, or 8 sub-clusters. At the end of the simplification phase, each cluster represents one vertex in the simplified model.

3.2   Phases of *R-Simp*

The first phase is the input phase. The original model is read in from the file and the initial vertex list and face list are created. These data structures require a lot of virtual memory (Table 1). The blocks of memory that are allocated in this phase are used throughout the entire simplification process (Section 2, Figure 2).

In the second phase, the in-memory data structures are initialized. The initialization phase creates the vertex and face adjacency lists.

The third phase is the heart of *R-Simp*: the simplification phase. In this phase the original set of vertices is reduced to the desired size. The simplification phase starts with all the vertices in a single cluster. This cluster is then subdivided into a maximum of eight sub-clusters. These clusters are inserted into a priority queue and the main loop of *R-Simp* begins.

The priority queue holds references to the clusters and orders the clusters based on the surface variation (i.e., amount of curvature). The greater the surface variation, the higher the priority of the cluster. Therefore, the simplified model will have more vertices in regions of high surface variation. This process iterates until the priority queue contains the required number of clusters, which is the number of vertices in the simplified model. Each split causes more clusters to be created, thus the amount of memory used by this phase depends on the level

Table 2. Performance of Polygonal Model Simplification: Various Strategies, times in seconds

Model	Output (polygons)	Original *R-Simp*	w/Spatial Sort	w/Reorg.	w/Both
hand	10,000	1,401	854	305	270
hand	20,000	1,905	1,140	342	293
hand	40,000	2,597	1,582	410	353
dragon	10,000	4,375	1,524	1,409	892
dragon	20,000	5,829	2,058	1,778	1,183
dragon	40,000	7,736	2,781	2,484	1,563
blade	10,000	8,306	4,052	4,539	3,545
blade	20,000	10,877	5,289	5,700	4,792
blade	40,000	14,312	7,084	7,720	6,713

of detail required. In general, the coarser the level of detail required, the less memory is needed.

The fourth phase is the post-simplification phase: the final set of vertices for the simplified model are computed. Each cluster represents a single vertex, v_{so}, in the simplified model; for each cluster, *R-Simp* computes the optimal position of v_{so}. Finally, the algorithm changes all of the pointers from the original vertices in a cluster to v_{so}.

The fifth phase is the triangulation phase, where the algorithm creates the faces (i.e., polygons) of the new simplified model. The algorithm iterates through all the faces in the original model and examines where the vertices of these faces lie in the output cluster. If two or more vertices point to the same new vertex (i.e., two or more original vertices are within the same cluster) then the face has degenerated and only if all three original vertices point to different new vertices do we keep the face and add it to the new face list. The vertices of this new face point to the vertices in the new vertex list.

In the last phase, the output phase, the algorithm writes the new vertex list and the new face list to the output file.

4   Improving Memory Locality and Performance in *R-Simp*

In addition to experiments with an instrumented version of *R-Simp*, we have also benchmarked versions of *R-Simp* without the instrumentation and with compiler optimizations (-O) turned on. *R-Simp* is written in C++ and we use the egcs compiler, version 2.91.66, on our Linux platform. The hardware environment is the same as in Section 2.

When the data structures of a model fit within main memory, *R-Simp* is known to have low run times [1]. However, the *hand*, *dragon*, and *blade* models are large enough that they require more memory than is physically available

Table 3. Major Page Fault Count of Polygonal Model Simplification: Various Strategies

Model	Output (polygons)	Original *R-Simp*	w/Spatial Sort	w/Reorg.	w/Both
hand	10,000	1,042,444	701,789	265,328	249,849
hand	20,000	1,361,492	880,498	281,338	254,083
hand	40,000	1,812,649	1,167,593	297,028	268,952
dragon	10,000	3,670,516	1,409,489	1,070,466	702,050
dragon	20,000	4,892,932	1,808,060	1,329,220	875,622
dragon	40,000	6,410,393	2,386,238	1,767,490	1,084,346
blade	10,000	7,678,929	3,878,566	4,127,079	3,125,348
blade	20,000	9,988,417	5,002,567	5,195,597	4,149,098
blade	40,000	13,063,493	6,599,763	6,997,380	5,717,086

(Table 1).[2] Consequently, the baseline *R-Simp* (called "Original *R-Simp*") experiences long run times (Table 2) due to the high number of page faults that it incurs (Table 3). As expected, the larger the input model, the longer the run time and the higher the number of page faults. As the output model's size increases, the run time and page faults also increase.

As previously discussed (Section 3), the global vertex and face lists are large data structures that are accessed throughout the computation. Therefore, they are natural targets of our attempts to improve the locality of memory access. In particular, we have developed two different techniques that independently improve the memory locality of *R-Simp*; real-time performance improves by a factor of 2 to 6-fold, depending on the model and the size of the simplified model. When combined, the two techniques can improve performance by up to 7-fold.

4.1   Metrics: Cluster Pagespan and Resident Working Set

We introduce *cluster pagespan* as an application-specific metric of the expected locality of reference to a model's in-memory data structure. Cluster pagespan is defined as the number of *unique* virtual memory pages that have to be accessed in order to touch all of the vertices and faces, in the global vertex and face lists, in the cluster. For each iteration of the simplification phase, *R-Simp*'s computation is focussed on a single cluster from the front of the priority queue. Therefore, if the pagespan of the cluster is large, there is a greater chance that a page fault will be incurred. The smaller the cluster pagespan, the lower the chance that one of its pages has been paged out by the operating system.

Figure 3(a) shows the cluster pagespan of the cluster at the front of the priority queue during an execution of *R-Simp* with the *blade* model. Since the initial clusters are very large, the cluster pagespan is also large at the beginning

---

[2] The *hand* model by itself would initially fit within 128 MB, but the operating system also needs memory. Consequently, paging occurs.

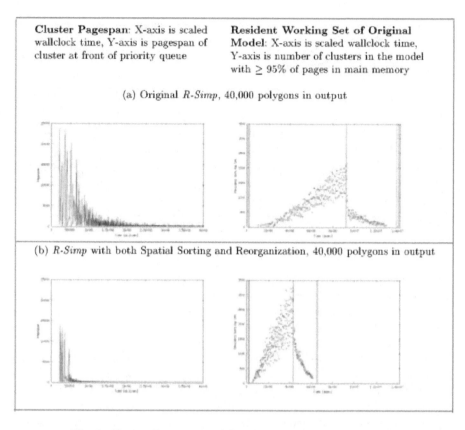

Cluster Pagespan: X-axis is scaled wallclock time, Y-axis is pagespan of cluster at front of priority queue

Resident Working Set of Original Model: X-axis is scaled wallclock time, Y-axis is number of clusters in the model with ≥ 95% of pages in main memory

(a) Original *R-Simp*, 40,000 polygons in output

(b) *R-Simp* with both Spatial Sorting and Reorganization, 40,000 polygons in output

Fig. 3. Cluster Pagespan and Resident Working Set for *blade*

of the execution. As clusters are split, the cluster pagespan decreases over time. The cluster pagespan data points are joined by lines in order to more clearly visualize the pattern. An instrumented version of *R-Simp* is used to gather the data. Furthermore, for each iteration of the simplification loop, *all* of the clusters in the priority queue are examined. If ≥ 95% of the pages containing the vertices and faces of a cluster are in physical memory (as opposed to being swapped out onto the swap disk), that cluster is considered to be resident in main memory. Therefore, Figure 3(a) also shows the *count* of how many of the clusters in the priority queue are considered to be in memory and part of the resident working set, over time.

As the clusters decrease in size, more clusters can simultaneously fit into main memory. Note that the vertices and faces are from the original (not simplified) model. The graph of the resident working set is not monotonically increasing because the operating system periodically (not continuously) reclaims pages. Also, the vertical lines represent important phase transitions. The resident working set count begins to decrease after the simplification phase because the post-simplification and triangulation phases dynamically create new vertex and face lists, which displace from memory the lists from the original model.

The cluster pagespan graph in Figure 3(a) indicates that there is poor memory locality in how the model is accessed for the initial portion of *R-Simp*'s execution. Consequently, the total number of clusters that reside in main memory remains under 2,000 for most of the simplification phase of the algorithm.

We now describe two techniques that measurably improve memory locality, according to cluster pagespan and the resident working set count, and also improve real time performance.

4.2   Offline Spatial Sorting

The models used in this case study are stored on disk in the PLY file format [7]. The file consists of a header, a list of the vertices, and a list of the faces. Each polygonal face is a triangle and is defined by a per-face list of three integers, which are index locations of vertices in the vertex list. There is no requirement that the order in which vertices appear in the list corresponds to their spatial locality. Two vertices that are spatial neighbours in the 3D geometry-space can be in contiguous indices in the vertex list, or they can be separated by an arbitrary number of other vertices. There is also no spatial locality constraint on the order of faces in the file. In *R-Simp*, the vertices and faces are stored in main memory in the same order in which they appear in the file, therefore the order of the vertices and faces in the file have a direct impact on the layout of the data structures in memory.

The large cluster pagespan values seen in the early portion of *R-Simp*'s execution (Figure 3(a)) suggests that perhaps the PLY models have not been optimized for spatial locality. Therefore, we decided to spatially sort the PLY file. The model itself is unchanged; it has the same number of vertices and faces at the same locations in geometry-space, but we change the order in which the vertices and faces appear in the file. Our spatial sort reads in the model from the file, sorts the vertices and faces, and then writes the same model back to disk in the PLY format. Therefore, the spatial sort is a preprocessing step that occurs before model simplification. The spatially-sorted version of the PLY file can then be re-used for different runs of the simplification program.

The spatial sort is a recursive Quicksort-like algorithm. After reading the model into memory, a 3D bounding box for the model is computed, as are three orthogonal dividing planes that partition the bounding box into eight isomorphic sub-boxes. Each sub-box is recursively partitioned; the stopping condition for the recursion is when a sub-box contains less than two vertices. As the sort recurses back up, the vertices in the vertex list are reordered so that vertices in the same sub-box at each level of recursion, which are spatial neighbours, have indices that are contiguous.

We spatially sort the faces according to their vertices to ensure that faces which are neighbours in geometry-space are also neighbours in the face list. Finally, the model is written back to disk in the PLY format with the vertex and face lists in the new, spatially-sorted order.

Our implementation of the spatial sort is written in C++. Spatially sorting the *hand*, *dragon*, and *blade* models require 28, 28, and 89 seconds, respectively,

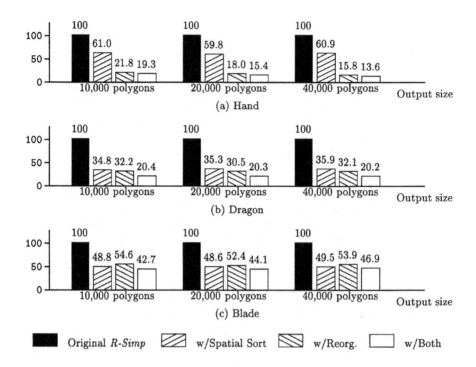

Fig. 4. Normalized Execution Times

on our 500 MHz Pentium III-based Linux platform. Again, the models only have to be sorted once since the new PLY files are retained on disk.

When *R-Simp* is given a spatially-sorted model for the input, cluster pagespan is reduced throughout the process's execution with a resulting improvement in the resident working set of the original model. For *blade*, there are significant improvements in both cluster pagespan and resident working set, which results in more than a 50% reduction in *R-Simp*'s execution time (Table 2). Spatial sorting also benefits the *hand* and *dragon* models.

Spatial sorting is a simple and fast procedure with substantial performance benefits for *R-Simp*. Although the spatial sort is currently an off-line preprocessing phase from *R-Simp*, we may integrate it into our implementation of *R-Simp* in the future. In the meantime, our experiments and measurements suggest that other researchers should consider spatially sorting the input models for their graphics systems.

4.3   Online Reorganization of Data Structures

The performance improvements due to a static spatial sort are substantial. However, as a model is iteratively simplified, there may be an opportunity to dynamically improve memory locality.

We have implemented a version of *R-Simp* that dynamically reorganizes its internal data structures in order to reduce cluster pagespan. Specifically, *before* a sub-cluster or cluster is inserted into the priority queue, the cluster may be selected for *cluster data structure reorganization* (or simply, *reorganization*). If selected for reorganization, the vertices and faces associated with the cluster are copied from the global vertex and face lists into new lists on new pages of virtual memory. The basic idea is similar to compacting memory to reduce fragmentation in memory management. Internal to the cluster data structure, the lists of vertices and faces now refer to the new vertex and face lists, thereby guaranteeing the minimal possible cluster pagespan.

Since there are copying and memory allocation overheads associated with reorganization, it is not done indiscriminately. Two criteria must be met before reorganization is performed:

1. The cluster is about to be inserted into the priority queue within the front 50% of clusters in the queue (i.e., the cluster is in the front half of the queue).
2. The cluster pagespan **after** reorganization must be less than n pages.

The first criteria tries to maximize the chances that a reorganized cluster will be accessed again (i.e., it will reach the front of the priority queue again and be re-used). Reducing the pagespan of a cluster that is not accessed again until the post-simplification phase produces fewer benefits. The value of "50%" was empirically determined.

The second criteria controls at what point reorganization is performed, in terms of cluster size. Reorganizing when clusters are large is expensive and inherently preserves large cluster pagespans. Reorganizing when clusters are small may delay reorganization until most of the simplification phase is completed, thus reducing the chances of benefitting from the reorganization. The specific value of n chosen as the threshold has, so far, been determined experimentally. The optimal value for n is found to be 1,024 pages (i.e., 4 MB given the 4 K pages on our platform) for *blade* (Figure 5). For the three models, the optimal value of n was empirically determined to be between 1,024 and 4,096 pages.

The benefits of reorganization are reflected in the reduced cluster pagespan and increased resident working set (for example, Figure 3(b)), reduced number of page faults (Table 3), and most importantly, in the lower run times (Table 2). When both spatial sorting and reorganization are applied to *R-Simp*, there is an additional benefit (Table 2 and Figure 4). Unfortunately, the two techniques have some overlap in how they improve data-access locality, so the benefits are not completely additive.

5   Concluding Remarks

In computer graphics and visualization, the complexity of the models and the size of the data sets have been increasing. Modern 3D scanners and rising standards for image quality have fueled a trend towards larger 3D polygonal models and also into model simplification algorithms. *R-Simp* is a new model simplification

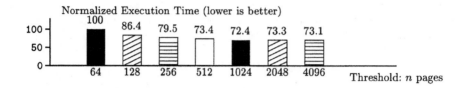

Fig. 5. Varying Reorganization Threshold, *blade*, Normalized Execution Time, 40,000 polygon output

algorithm with low run times, easy control of the output model size, and good model quality. However, no matter the amount of RAM that one can afford, there may be a model that is too large to fit in memory.

Therefore, we have developed spatial sorting and data structure reorganization techniques to improve the memory locality of *R-Simp* and experimentally shown it to improve performance by up to 7-fold. We have also introduced the cluster pagespan metric as one measure of memory locality in model simplification. For future work, we plan to study if spatial sorting and reorganization can also improve the performance of other simplification algorithms.

References

[1] D. Brodsky and B. Watson. Model simplification through refinement. In *Graphics Interface '00*, pages 221–228. Canadian Information Processing Society, Canadian Human-Computer Communications Society, May 2000.

[2] M. Garland and P.S. Heckbert. Surface simplification using quadric error metrics. In *Proc. ACM SIGGRAPH 1997*, pages 209–216, August 1997.

[3] P.S. Heckbert and M. Garland. Survey of polygonal surface simplification algorithms. Technical report, Carnegie Mellon University, 1997. Draft Version.

[4] H. Hoppe, T. DeRose, T. Duchamp, J. McDonald, and W. Stuetzle. Mesh optimization. In *Proc. ACM SIGGRAPH 1993*, volume 27, pages 19–26, August 1993.

[5] M. Levoy, K. Pulli, B. Curless, S. Rusinkiewicz, D. Koller, L. Pereira, M. Ginzton, S. Anderson, J. Davis, J. Ginsberg, J. Shade, and D. Fulk. The digital michaelangelo project: 3D scanning of large statues. In *Proc. ACM SIGGRAPH 2000*, pages 131–144, 2000.

[6] P. Lindstrom. Out-of-core simplification of large polygonal models. In *Proc. ACM SIGGRAPH 2000*, pages 259–262. ACM, 2000.

[7] PLY File Format. http://www.cc.gatech.edu/projects/large_models/ply.html.

[8] J. Rossignac and P. Borrel. Multi-resolution 3D approximations for rendering complex scenes. In *Modeling in Computer Graphics: Methods and Applications*, pages 455–465, Berlin, 1993. Springer-Verlag.

[9] W.J. Schroeder, J.A. Zarge, and W.E. Lorensen. Decimation of triangle meshes. *Computer Graphics*, 26(2):65–70, July 1992.

High-Performance Scalable Java Virtual Machines

Vivek Sarkar and Julian Dolby

IBM Thomas J. Watson Research Center
P. O. Box 704, Yorktown Heights, NY 10598, USA,
{vsarkar,dolby}@us.ibm.com

Abstract. In this paper, we discuss the requirements for building high-performance scalable Java virtual machines (JVMs) for server applications executing on symmetric multiprocessors (SMPs). We provide a survey of known performance issues, and outline future technology solutions. We also introduce a simple performance model for scalability of JVMs in server environments, and validate the model by presenting preliminary experimental results obtained by executing server applications on the Jalapeño virtual machine.

1 Introduction

Java^{TM} [4] has been rapidly gaining importance as a programming language initially for the client side of networked applications, and more recently, for the server side with the Java 2 Enterprise Edition (J2EE) programming model. Java programs are executed on a Java Virtual Machine (JVM) [21]. An important source of Java's appeal is its *portability i.e.,* the fact that it can be used as a common programming language across a heterogeneous mix of hardware and operating system (O/S) platforms. Recent evidence of Java's success on the server side includes the fact that two of the most popular application servers, IBM WebSphere and BEA WebLogic, are written in Java. Figure 1 illustrates the *four-tier* architecture that is emerging for current and future ebusiness software platforms. The *client tier* is used for direct interaction with the user, and can be lightweight enough to run on small devices with embedded processors. The *presentation/formatting tier* is used to format web pages for delivery to and presentation on the client tier. The presentation tier is usually located in the front-end of a web server, though it may also be located on *edge servers* [18]. The *application tier* provides session and transaction support to business applications; this tier is usually located on an *application server*. Finally, the *back-end services tier* provides access to data sources and transaction sources that are expected to be wrapped as *web services* in the future. These services are usually provided by enterprise-scale database and transaction servers.

The foundations for the four-tier architecture illustrated in Figure 1 lie in several portable and interoperable standards such as TCP/IP, HTML, XML, and Java. Even though there are many non-Java components in the ebusiness

B. Monien, V.K. Prasanna, S. Vajapeyam (Eds.): HiPC 2001, LNCS 2228, pp. 151–163, 2001.

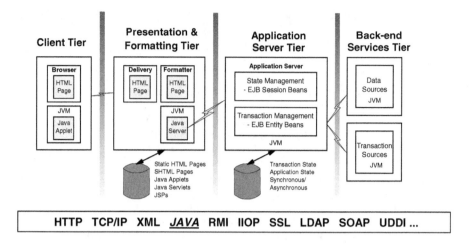

Fig. 1. Four-tier architecture for ebusiness software platforms

middleware outlined in Figure 1, we believe that the performance of the Java components have a significant impact on end-to-end performance and hence deserve special attention. In this paper, we focus on the JVM requirements for the server (non-client) tiers in Figure 1, provide a survey of known performance issues, and outline future technology solutions for building high-performance scalable JVMs for servers. Performance and scalability are important issues for server applications, because these applications need to be robust and effective in satisfying multiple requests from multiple clients. Further, server JVMs are usually executed on symmetric multiprocessor (SMP) systems, so that additional computing capacity can be obtained by using multiple processors in an SMP.

The rest of the paper is organized as follows. Section 2 identifies key JVM requirements for server machines. Section 3 summarizes how the Jalapeño research virtual machine addresses some of the key challenges in building high-performance scalable JVMs. Section 4 introduces a simple performance model for scalability, and presents experimental results to validate the model. Section 5 contains our conclusions.

2   JVM Requirements on Servers

Since Java is a portable language, it can be used across a range of hardware from embedded devices to large servers. However, the virtual machine (VM) and library requirements for this range of hardware can be quite varied. Specifically, the Java platform comes in three editions — Micro Edition, Standard Edition, and Enterprise Edition. There is already an established need for a special class of Micro Edition VMs that are tailored to obey the unique resource constraints of embedded devices. In this section, we identify key VM requirements for using the Standard and Enterprise editions of the Java platform on server hardware. These requirements are applicable to the use of Java in standard ebusiness scenarios, as well as scenarios that we anticipate will be important in future server configurations such as Java servers for grid computing [16]. In our opinion, the following critical requirements for servers need to be addressed more effectively by future VMs for Java:

1. *Multiprocessor scalability.*
   Multithreading in C and C++ is usually supported by a thread library in the operating system. In contrast, the Java language specification defines threads as part of the Java language. A Java thread is an object with special methods and operations for activation, termination and synchronization; specifically, a Java thread is an instance of a subtype of the class `Runnable`.
   As mentioned earlier, SMP configurations are very popular for server machines. However, many VMs for Java perform a direct 1:1 mapping of Java threads onto heavyweight operating system (O/S) threads for execution on SMPs. A 1:1 mapping can lead to poor scalability of multithreaded Java programs along two dimensions:
   a) *Capacity* — The large footprint size of O/S threads limits the number of Java threads that can be created on a single SMP. For typical server applications that create new threads for new requests, this can be a severe limitation on the throughput delivered by the application.
   b) *Speedup* — Further, the speedup obtained by using multiple processors in an SMP is degraded when the SMP is swamped with a large number of O/S threads.

2. *Efficient fine-grained synchronization.*
   Like other monitor-based languages, each dynamic object in a Java program can potentially be used as a synchronization lock via the MONITOR_ENTER and MONITOR_EXIT bytecode instructions. Server applications rely on the use of these locks to ensure correct multithreaded execution. It is important for JVMs to provide efficient support for fine-grained synchronizations, so that the monitor operations do not become a performance bottleneck. Further, it is important for different subsystems of the VM (*e.g.,* memory management, dynamic compilation, class loading, etc.) to execute as asynchronously as possible so as to avoid serialization bottlenecks during SMP execution.

3. *Efficient memory management.*
   To enhance application safety, the burden of memory management, *i.e.,* of freeing and recycling objects, is transferred from the programmer to the language runtime system in Java by including Garbage Collection (GC) [20] in the VM's memory management system. However, the presence of GC can cause performance problems for server applications due to the following reasons:
   a) A full GC requires *scanning* the entire heap. Therefore, GC overhead is larger for server SMP systems with larger memories (and larger heap sizes) than smaller uniprocessor systems.
   b) Many server applications have stringent *response-time* requirements *e.g.,* an application may have a requirement that 99% of requests must be served in < 1 second. GC can be a major obstacle in satisfying such response-time requirements because most current VMs perform non-incremental "stop-the-world" GC, usually leading to a response-time

failure each time a GC occurs (or, for generational GC systems, each time a major collection occurs).

4. *Exploitation of high-performance processors in servers.*

The latest features in high-performance processors are usually made available first on sever platforms, with accompanying extensions to C/C++ static compilers for exploiting the new hardware features *e.g.,* multiple functional units, branch prediction, prefetching, etc. Unfortunately, there still remains a gap between the effectiveness of hardware exploitation delivered by static compilers for Fortran and C compared to that delivered by Just-In-Time (JIT) compilers in current VMs for Java. There are two reasons for this performance gap. First, there are features in the Java language, such as threads, synchronization, precise exceptions, and dynamic class loading that make code optimization more challenging for Java than for C/C++. Second, since JIT compilation is performed dynamically, it is forced to restrict its optimization algorithms to those that can be performed efficiently so that compilation overhead does not become a major performance bottleneck at runtime.

5. *Efficient library and component usage.*

One of the major benefits offered by the Java platform is the availability of a large number of libraries and components that implement commonly-used functionalities. This is especially significant for the Enterprise Edition of the Java platform, which includes Enterprise Java Beans (EJBs) as a standard interface for accessing enterprise resources such as database and transaction systems. However, by necessity, all Java libraries and components are written to handle generic cases, and may therefore exhibit overheads that are unnecessary in a specific context *e.g.,* unnecessary synchronizations, unnecessary allocations, and unnecessary data format conversions. Further, interface calls are expensive in many current VMs, thus resulting in additional overheads for server applications which need to make frequent use of interface calls to access services provided by components and libraries.

6. *Continuously availability*

Server applications need to run for long durations with 7×24 availability. Two features of Java can help in this regard. First, compared to C/C++, the availability of garbage collection in Java greatly reduces the likelihood of storage leaks being present in the application. Second, the strong typing semantics of Java, combined with extra runtime checking, makes it impossible for a Java application to cause a hard failure (segmentation fault). Unfortunately, most current VMs are written in native code (C/C++) and do not enjoy the same benefits as the Java applications that they support. Hence, it is possible for bugs in current VMs to cause storage leaks and hard failures in the VM code, thus hindering the goal of continuous availability.

We now briefly summarize how these requirements for Java on servers are being addressed by current and future VM technologies being developed by the research community.

Some existing product JVMs address Requirement 1 on *uniprocessor* systems by providing the user the option of using a "green threads" package that supports an $m : 1$ mapping of multiple lightweight Java threads on a single O/S thread. Green threads deliver an improved *capacity scalability* compared to a $1 : 1$ mapping. Note that *speedup scalability* is not an issue for uniprocessors. In addition to being limited to uniprocessor use, green threads provide little support for pre-emption (other than priorities) and defer much of the scheduling burden to the programmer.

Further, all product JVMs that we are aware of revert to a $1 : 1$ mapping of Java threads on O/S threads when executing on SMPs, thus compromising both dimensions of scalability. There has been extensive past work on thread implementations in other languages, *e.g.,* Cilk [9], that demonstrates the scalability benefits that can be obtained by using lightweight threads instead of O/S threads for SMP execution. Section 3 includes a discussion of the $m : n$ mapping approach performed in the Jalapeño VM.

Requirement 2 has been partially addressed in recent JVMs by efficient runtime locking for the common case in which locks are uncontended [8,23,15]. *Escape analysis* has also been used to enable synchronization elimination as a compiler optimization [13]. However, when multiple application threads try to access JVM runtime services, the JVM often becomes a serialization bottleneck because of coarse-grained synchronization among subsystems of the JVM.

Requirement 3 has been partially addressed in recent JVMs by the use of *generational GC* algorithms [20], in which minor collections are performed with small pause times and major collections occur infrequently. A few VMs support *parallel GC* on a SMP, where multiple processors can be used to speed up the time taken by GC while the application threads are suspended [6]. Historically, there has also been extensive research on *concurrent GC* algorithms, which enable GC to be performed without requiring the application threads to be suspended. Practical experience with concurrent GC algorithms [7] suggests that the maximum pause times can indeed be reduced, but the reduction is usually accompanied by a 5%–10% increase in total execution time.

Significant progress has been made towards addressing Requirement 4 by increased levels of optimization performed by dynamic and just-in-time (JIT) compilers for Java [19,17]. Though optimized compilation of Java is challenging due to the presence of exceptions, threads, garbage collection, many features of Java (such as strong typing and restricted pointers) also make Java easier to optimize than native code. New technologies such as *adaptive optimization* [5] are further reducing the performance gap between Java and native code, and even offer the promise of turning the gap around so as to make Java programs run faster than native code.

A special-case solution to Requirement 5 would be to rewrite critical libraries so as to replace them by more efficient implementations that are well-suited to commonly-occurring idioms of library usage. The obvious drawbacks of the rewriting approach are that it is expensive, error-prone, and limited to known idioms of library usage. A more general approach would be to use *specialization* techniques to tailor the dynamically compiled code for library methods to the calling context of the application. Calling context information can be obtained by *context-sensitive profiling* (as in [3]), which goes beyond the context-insensitive profiling performed by current adaptive optimization systems.

Requirement 6 is a major challenge for current JVMs. One approach to avoiding storage leaks and hard failures in a JVM is to implement the JVM in Java. This approach has been undertaken in the Jalapeño VM, but not for any product JVM. However, the availability of GC for Java programs is not sufficient to guarantee the absence of storage leaks. Even though GC ensures that all space occupied by *unreachable* objects is reclaimed, it is possible to have a storage leak in a Java program (including VMs written in Java) that unwittingly causes unused storage to become *reachable*. Future technologies for automatic detection and correction of such storage leaks will be a big boon for long-running applications and VMs.

An additional consequence of the *continuous availability* requirement is that JVMs will need to be able to execute multiple Java applications without having to start up a new JVM for each application. This requirement will become increasingly important for future distributed computing environments such as *web services* and *grid computing* [16]. The Multiheap multiprocess JVM described in [14] provides a limited but effective solution to this requirement for transaction workloads. More general solutions will also need to include automatic support for class unloading.

3   Jalapeño Solution to Server Requirements

Jalapeño is a research virtual machine for Java built at the IBM T.J. Watson Research Center, and is currently being used by several research groups within IBM and outside IBM. A comprehensive description of Jalapeño can be found in [1]. In this section, we provide a brief summary of Jalapeño, and discuss how it addresses some of the key challenges in building high-performance scalable JVMs.

The Jalapeño VM has three key distinguishing features. First, the Jalapeño VM is itself implemented in Java. This design choice brings with it several advantages as well as technical challenges. The advantages include a uniform memory space for VM objects and application objects, and ease of portability. Second, the Jalapeño VM performs a lightweight m:n mapping of Java threads on OS-level threads, thus resulting in enhanced scalability on SMPs. Third, the Jalapeño VM takes a compile-only approach to program execution. Instead of providing both an interpreter and a JIT/dynamic compiler, it provides two dynamic compilers — a quick non-optimizing "baseline" compiler, and a slower but smarter optimizing compiler. An adaptive optimization system is used to dynamically select the subset of methods to be compiled by the optimizing compiler. The current Jalapeño system has been ported to SMPs running AIX/PowerPC, Linux/PowerPC, and Linux/IA-32.

As mentioned in the previous section, Jalapeño performs a $m : n$ mapping of Java threads onto virtual processors (O/S threads). Specifically, Jalapeño multiplexes all (m) Java threads created in a VM execution onto a smaller number (n) of O/S threads. The m Java threads include application threads, as well as VM threads created for garbage collection, compilation, adaptive optimization and other VM services. This $m : n$ mapping leads to improvements in both capacity and speedup dimensions of scalability (c.f. item 1 in Section 2).

Jalapeño threads are *quasi-preemptive*. This means that a thread can only be suspended (preempted) at predefined *yieldpoints* inserted in the code by Jalapeño's compilers. The code for a yieldpoint consists mainly of a test to check if a yield has been requested by some other thread in the VM. If so, the currently executing thread is suspended, and some other "ready" thread is selected for execution on the virtual processor. For this approach to work, it is necessary for the compilers to insert yieldpoints so as to guarantee that at most a bounded amount of time can elapse between successive dynamic instances of yieldpoints. The current approach in Jalapeño's compilers to deliver this guarantee is to insert a yieldpoint in the prologue of most methods, and in the header node of each loop. One of the major benefits of quasi-preemptive scheduling is that it enables a rapid transition between mutation and garbage collection [6], especially in the face of type-accurate GC. This capability helps reduce the response-time constraints for server JVMs discussed in item 3b in Section 2.

There are three kinds of locks available for use by threads in Jalapeño to address the requirements outlined in item 2 in Section 2. (Additional details can be found in [1].) *Processor locks* are low-level primitives used to support the other two locking mechanisms, and sometimes also used directly for efficient locking within the VM. Processor locks use a busy wait mechanism, and therefore should only be held for a short period of time. *Thin locks* [8,23] provide an efficient mechanism for locking in the absence of contention. *Thick locks* are used as a fallback to provide general locking when contention occurs. However, a challenge that still remains for Jalapeño is to further refine the granularity of locking among VM systems so as to avoid serialization bottlenecks when VM services such as dynamic class loading and compilation are performed concurrently with application threads.

As described in [6], the Jalapeño VM contains a modular GC framework which has been used to implement five different *parallel* garbage collectors: non-generational and generational versions of mark-and-sweep and semi-space copying collectors, as well as a hybrid version that uses mark-and-sweep for the mature space and copying for the nursery. These garbage collectors are type accurate both in the heap and thread stacks, stop-the-world (synchronous with respect to mutator execution), and support load-balancing among all available processors. Parallelization of GC has proved to be very effective for server applications with large heaps, thus addressing requirement 3a in Section 2.

When discussing requirement 4 in Section 2, we observed that the two challenges for effective exploitation of high-performance processors by Java programs were 1) Java language features that complicate code optimization, and 2) the difficulty of performing aggressive or time-consuming optimizations in a dynamic compilation framework. These two challenges are addressed by the Jalapeño optimizing compiler as follows. First, attention has been paid to relaxing dependence constraints as much as possible without compromising the language semantics [11,12]. Second, much emphasis has been placed on using *adaptive optimization* [5] to reduce the overhead of dynamic compilation, and on the use of dynamic optimistic analyses (*e.g.,* [24]) to obtain better code quality than might be possible with static compilation.

Requirement 5 in Section 2 highlighted the importance of reducing the overheads associated with using standard libraries. This area is yet to be explored for optimizing the use of J2EE libraries in Jalapeño. However, some attention has been paid to optimizing the implementation of the `String` class in Jalapeño. Specifically, a *semantic inlining* [27] approach is used along with flow-insensitive escape analysis [10] to remove unnecessary synchronizations in the `String` class. In addition, it has been demonstrated in the Jalapeño VM that the overhead of interface calls need not be large; instead, they and can be implemented as efficiently as virtual calls [2].

The last requirement discussed in Section 2 (item 6) has to do with continuous availability. The fact that the Jalapeño VM is itself written in Java is a big boon in this regard, since it eliminates the possibility of conventional memory leaks and hard failures. However, as described in [22], extra care is still necessary to guarantee safety for Jalapeño modules that use Jalapeño's special MAGIC class for low-level access to machine resources. Finally, an interesting extension to the standard VM model for Java would be to enable a single VM for multiple applications as in [14]; such a VM might be very useful for running applications in a *web services* or *grid computing* framework [16].

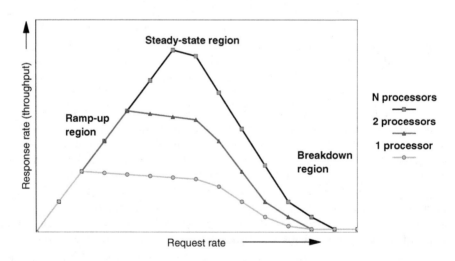

Fig. 2. Performance model for scalability of server applications

4   A Scalability Performance Model and Its Validation

In this section, we introduce a simple performance model for scalability of JVMs in server environments. At a coarse level, all server applications receive as input a number of requests at some variable *request rate*, and deliver as output a certain number of responses at a *response rate* (also referred to as the *throughput* of the server application).

Our performance model is outlined in Figure 2 by showing how we expect the response rate to vary as a function of the input request rate. The graph in Figure 2 is divided into three regions — *ramp-up*, *steady-state*, and *breakdown*. In the ramp-up region, the server is under-utilized, so the response rate increases monotonically with an increase in the request rate. In the steady-state region, the server is running at capacity and delivers a reasonably steady response rate with slight degradation as the request rate continues to increase. Finally, the request rate reaches a threshold that marks the start of the breakdown region. The throughput deteriorates significantly in the breakdown region, usually due to the working set of a key resource (*e.g.*, memory, threads, sockets) exceeding the available capacity on the underlying SMP platform.

The model in Figure 2 also indicates how we expect the response rate to vary with an increase in the number of processors used on an SMP. (Different values for the number of processors are represented by different curves.) The throughput obtained in the steady-state region should increase as the number of processors available increases, thus providing multiprocessor scalability. We expect the ramp-up regions in the different curves to coincide prior to reaching steady-state. Finally, since the amount of non-processor resources in an SMP (*e.g.*, memory) available is independent of the number of processors, we expect the breakdown region to start at approximately the same request rate for different numbers of processors.

To validate the model, we now present preliminary experimental results for two publicly available server benchmarks, Volano [26] and SPECjbb [25]. These experimental results were obtained using the Jalapeño virtual machine described in Section 3.

Figure 3 shows the throughput obtained by increasing the number of chat rooms in the Volano benchmark, while keeping the number of users per room fixed at 15. The number of rooms was increased from 1 to 50, as shown on the X-axis. Note that increasing the number of rooms leads to a proportional increase in the request rate for chat messages. The resulting throughput is shown on the Y-axis in units of messages/second. The performance measurements were obtained on an IBM RS/6000 Enterprise Server S80 running AIX v4.3, with 24 PowerPC processors and 4GB of main memory. The results in Figure 3 exhibit the ramp-up and steady-state phases outlined in our model, as well as scalable performance obtained by increasing the number of CPUs used from 1 to 4. Unfortunately, we were unable to run the benchmark for large numbers of rooms due to an inability to create sufficient socket connections. Therefore, no results were obtained to exhibit a breakdown region in Figure 3. (Alternatively, this

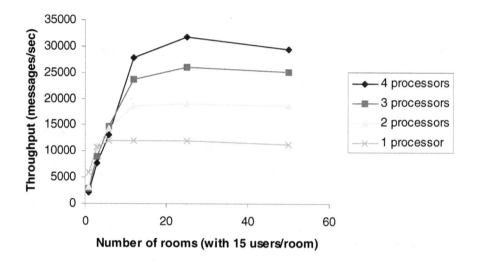

Fig. 3. Scalability results for Volano benchmark

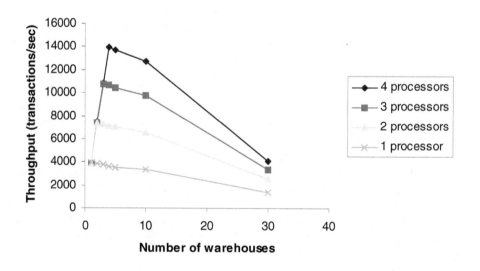

Fig. 4. Scalability results for SPECjbb benchmark

can be viewed as encountering a "breakdown" that prevented us from reporting results for the breakdown region!)

Figure 4 shows the throughput obtained by increasing the number of warehouses in the SPECjbb benchmark from 1 to 30. The Y-axis shows the throughput in units of transactions/second. It is important to note that SPECjbb is not a true server application, in that it does not have separate client threads and server threads as in Volano. The performance measurements were obtained on an IBM RS/6000 Enterprise Server F80 running AIX v4.3, with six 500MHz PowerPC processors and 4GB of main memory. These results clearly exhibit ramp-up and breakdown phases as predicted by the model, as well as scalable speedups obtained by increasing the number of CPUs used from 1 to 4. The ramp-up curves for the different number of CPUs coincide perfectly before reaching the steady-state and breakdown phases. However, the steady-state phases observed for SPECjbb are much shorter in duration than those observed for Volano. Further study is necessary to understand why these results differ from Jalapeño results obtained for an earlier version of the SPECjbb benchmark called pBOB [1].

5 Conclusions

In this paper, we discussed the requirements for building high-performance scalable Java virtual machines for server applications executing on symmetric multiprocessors. We provided a survey of known performance issues, and outlined future technology solutions. We also introduced a simple performance model for scalability of JVMs in server environments, and validated the model by presenting preliminary experimental results obtained by executing server applications on the Jalapeño virtual machine. Finally, we identified areas that need to be addressed in future work such as supporting continuous availability and improving the efficiency of library usage and component usage.

Acknowledgments. We would like to acknowledge the contributions of the entire Jalapeño team in two significant ways. First, the design discussions surrounding Jalapeño greatly influenced the ideas presented in this paper. Second, the Jalapeño VM infrastructure was extremely useful in obtaining the experimental results reported in this paper. We are also grateful to the OTI VAME team for providing an implementation of the Java class libraries for use in conjunction with Jalapeño. Finally, we would like to thank the following people at IBM for their input on performance issues and future trends for Java on servers: Rajiv Arora, Sandra Baylor, Peter Bunk, Curt Cotner, Robert Dimpsey, Stephen Fink, Larry Lowen, Bowie Snyder, and James West.

References

1. B. Alpern, C. R. Attanasio, J. J. Barton, M. G. Burke, P. Cheng, J.-D. Choi, A. Cocchi, S. J. Fink, D. Grove, M. Hind, S. F. Hummel, D. Lieber, V. Litvinov, M. F. Mergen, T. Ngo, J. R. Russell, V. Sarkar, M. J. Serrano, J. C. Shepherd, S. E. Smith, V. C. Sreedhar, H. Srinivasan, and J. Whaley. The Jalapeño virtual machine. *IBM Systems Journal*, 39(1), 2000.
2. B. Alpern, A. Cocchi, S. Fink, D. Grove, and D. Lieber. invokeinterface considered harmless. In *ACM Conference on Object-Oriented Programming Systems, Languages, and Applications*, Oct. 2001.

3. G. Ammons, T. Ball, and J. Larus. Exploiting hardware performance counters with flow and context sensitive profiling. In *SIGPLAN '97 Conf. on Programming Language Design and Implementation*, 1997.

4. K. Arnold and J. Gosling. *The Java Programming Language*. Addison-Wesley, 1996.

5. M. Arnold, S. Fink, D. Grove, M. Hind, and P. Sweeney. Adaptive optimization in the Jalapeño JVM. In *ACM Conference on Object-Oriented Programming Systems, Languages, and Applications*, Oct. 2000.

6. C. R. Attanasio, D. F. Bacon, A. Cocchi, and S. Smith. A comparative evaluation of parallel garbage collector implementations. In *14th International Workshop on Languages and Compilers for Parallel Computing*, Aug. 2001.

7. D. F. Bacon, C. R. Attanasio, H. B. Lee, V. T. Rajan, and S. Smith. Java without the coffee breaks: a nonintrusive multiprocessor garbage collector. In *SIGPLAN '01 Conf. on Programming Language Design and Implementation*, pages 92–103, June 2001.

8. D. F. Bacon, R. Konuru, C. Murthy, and M. Serrano. Thin locks: featherweight synchronization for Java. In *SIGPLAN '98 Conference on Programming Language Design and Implementation*, pages 258–268, June 1998.

9. R. Blumofe, C. Joerg, B. Kuszmaul, C. Leiserson, K. Randall, and Y. Zhou. Cilk: An Efficient Multithreaded Runtime System. In *Proceedings of the Fifth ACM SIGPLAN Symposium on Principles and Practice of Parallel Programming (PPOPP), Santa Barbara California, July 19-21*, pages 207–216, 1995.

10. M. Burke, J.-D. Choi, S. Fink, D. Grove, M. Hind, V. Sarkar, M. Serrano, V. Sreedhar, H. Srinivasan, and J. Whaley. The Jalapeño Dynamic Optimizing Compiler for Java. In *ACM Java Grande Conference*, June 1999.

11. C. Chambers, I. Pechtchanski, V. Sarkar, M. J. Serrano, and H. Srinivasan. Dependence analysis for Java. In *12th International Workshop on Languages and Compilers for Parallel Computing*, Aug. 1999.

12. J.-D. Choi, D. Grove, M. Hind, and V. Sarkar. Efficient and precise modeling of exceptions for the analysis of Java programs. In *ACM SIGPLAN-SIGSOFT Workshop on Program Analysis for Software Tools and Engineering*, pages 21–31, Sept. 1999.

13. J.-D. Choi, M. Gupta, M. Serrano, V. C. Sreedhar, and S. Midkiff. Escape analysis for Java. In *ACM Conference on Object-Oriented Programming Systems, Languages, and Applications*, pages 1–19, 1999.

14. D. Dillenberger, R. Bordawekar, C. Clark, D. Durand, D. Emmes, O. Gohda, S. Howard, M. Oliver, F. Samuel, and R. S. John. Building a Java Virtual Machine for Server Applications: The JVM on OS/390. *IBM Syst. J.*, 39(1), 2000.

15. R. Dimpsey, R. Arora, and K. Kuiper. Java Server Performance: A Case Study of Building Efficient, Scalable JVMs. *IBM Syst. J.*, 39(1), 2000.

16. I. Foster and C. Kesselman. *The Grid: Blueprint for a New Computing Infrastructure*. Morgan Kaufmann Publishers, Inc., San Francisco, California, July 1998.

17. The Java Hotspot Performance Engine Architecture. White paper available at http://java.sun.com/products/hotspot/whitepaper.html.

18. IBM. IBM WebSphere Edge Services Architecture. http://www-4.ibm.com/software/webservers/edgeserver/doc/esarchitecture.pdf.

19. K. Ishizaki, M. Kawahito, T. Yasue, M. Takeuchi, T. Ogasawara, T. Suganama, T. Onodera, H. Komatsu, and T. Nakatani. Design, implementation, and evaluation of optimizations in a just-in-time compiler. In *ACM Java Grande Conference*, San Fransisco, CA, June 1999.

20. R. Jones and R. Lins. *Garbage Collection: Algorithms for Automatic Dynamic Memory Management*. John Wiley and Sons, Chichester, England, 1996.
21. T. Lindholm and F. Yellin. *The Java Virtual Machine Specification*. The Java Series. Addison-Wesley, 1996.
22. J.-W. Maessen, V. Sarkar, and D. Grove. Program analysis for safety guarantees in a java virtual machine written in java. In *ACM SIGPLAN-SIGSOFT Workshop on Program Analysis for Software Tools and Engineering*, pages 21–31, June 2001.
23. T. Onodera and K. Kawachiya. A study of locking objects with bimodal fields. In *ACM Conference on Object-Oriented Programming Systems, Languages, and Applications*, Nov. 1999.
24. I. Pechtchanski and V. Sarkar. Dynamic optimistic whole program analysis: a framework and an application. In *ACM Conference on Object-Oriented Programming Systems, Languages, and Applications*, Oct. 2001.
25. The Standard Performance Evaluation Corporation. SPEC JBB 2000. http://www.spec.org/osg/jbb2000/, 2000.
26. VolanoMark 2.1. http://www.volano.com/benchmarks.html.
27. P. Wu, S. P. Midkiff, J. E. Moreira, and M. Gupta. Improving Java performance through semantic inlining. Technical Report 21313, IBM Research Division, 1998.

Shared Virtual Memory Clusters with Next-Generation Interconnection Networks and Wide Compute Nodes

Courtney R. Gibson and Angelos Bilas

Department of Electrical and Computer Engineering
University of Toronto
Toronto, Ontario M5S 3G4, Canada
{gibson,bilas}@eecg.toronto.edu

Abstract. Recently much effort has been spent on providing a shared address space abstraction on clusters of small–scale symmetric multi-processors. However, advances in technology will soon make it possible to construct these clusters with larger–scale cc-NUMA nodes, connected with non-coherent networks that offer latencies and bandwidth compa-rable to interconnection networks used in hardware cache–coherent sys-tems. The shared memory abstraction can be provided on these systems in software across nodes and in hardware within nodes.

In this work we investigate this approach to building future software shared memory clusters. We use an existing, large–scale hardware cache–coherent system with 64 processors to emulate a future cluster. We present results for both 32- and 64-processor system configurations. We quantify the effects of faster interconnects and wide, NUMA nodes on system design and identify the areas where more research is required for future SVM clusters. We find that current SVM protocols can only partially take advantage of faster interconnects and they need to be ad-justed to the new system features. In particular, unlike in today's clusters that employ SMP nodes, improving intra–node synchronization and data placement are key issues for future clusters. Data wait time and synchro-nization costs are not major issues, when not affected by the cost of page invalidations.

1   Introduction

Recently, there has been a lot of work on providing a shared address space ab-straction on clusters of commodity workstations interconnected with low-latency, high-bandwidth system area networks (SANs). The motivation for these efforts is two–fold. First, system area network (SAN) clusters have a number of ap-pealing features: They follow technology curves well since they are composed of commodity components. They exhibit shorter design cycles and lower costs than tightly–coupled multiprocessors. They can benefit from heterogeneity and there is potential for providing highly-available systems since component replication is not as costly as in other architectures. Second, the success of the shared address

B. Monien, V.K. Prasanna, S. Vajapeyam (Eds.): HiPC 2001, LNCS 2228, pp. 167–181, 2001.
© Springer-Verlag Berlin Heidelberg 2001

space abstraction: Previous work [13,6] has shown that a shared address space can be provided efficiently on tightly-coupled hardware DSM systems up to the 128 processor scale and most vendors are designing hardware cache-coherent machines, targeting both scientific as well as commercial applications.

Traditionally, SAN clusters have been built using small-scale symmetric multiprocessors (SMPs) that follow a uniform memory access (UMA) architecture. The interconnection networks used are faster than LANs and employ user-level communication that eliminates many of the overheads associated with the operating system and data copying. With current advances in technology it will be possible in the near future to construct commodity shared address space clusters out of larger–scale nodes connected with non-coherent networks that offer latencies and bandwidth comparable to interconnection networks used in hardware cache–coherent systems. With these changes in technology it is important to examine the impact that both the interconnection network performance and the node architecture have on the performance of shared memory clusters. The design space for these next generation SANs is not yet fixed; there is a need to determine the required levels of both functionality and performance. Similarly, it is not clear how using wider nodes will impact software shared memory protocol design and implementation as well as system performance.

In this paper we address the following questions: (i) To what degree will future shared memory clusters benefit from using interconnection networks that offer about one order of magnitude better latencies and bandwidth than todays SANs? Are protocol costs the dominating factor, or can faster interconnection networks improve system performance? If yes, what is the achievable improvement? (ii) How does the use of wider, potentially cc-NUMA, nodes affect system performance? Can existing protocols be used in these systems, or is it necessary to adjust them to new system characteristics? How can SVM protocols best take advantage of the new architecture? (iii) What are the remaining system bottlenecks and what are the areas where further research and improvements are necessary. To investigate these issues we build extensive emulation infrastructure. We use an existing, large–scale hardware cache–coherent system, an SGI Origin2000, to emulate such a cluster. We port an existing, low–level communication layer and a shared virtual memory (SVM) protocol on this system and study the behavior of a set of real applications.

Our high level conclusion is that existing SVM protocols need to be adjusted to the new system features. In particular, unlike in today's clusters that employ SMP nodes, improving intra–node synchronization and data placement are key issues for future clusters. Data wait time and synchronization costs are not major issues, when not affected by the cost of page invalidations. More specifically: (i) SVM protocols are able to take advantage of the faster interconnection network and reduce data wait time significantly. Thus, interconnection network overhead is still an important factor on today's clusters. Data wait time in many applications is reduced to less than 15% of the total execution time. With slight restructuring, the parallel speedup of FFT increases from 6 (on an existing cluster) to 23 (on the Origin2000) with 32 processors, and to 26 with 64 processors. (ii)

Using wide (8– and 16–processor) cc-NUMA nodes has an impact on intra-node synchronization and data placement. Most importantly, the NUMA features of the nodes result in imbalances, and the increased number of processes per node results in higher page invalidation (*mprotect*) costs. Overall, these effects combine to reduce system performance and need to be addressed in protocol design and implementation. (iii) The most important remaining protocol overheads are the cost of *mprotect*s and intra-node imbalance generated by data placement, which affect mainly synchronization overheads.

The rest of the paper is organized as follows: Section 2 presents our emulation infrastructure and our protocol extensions. Sections 3 and 4 present our results. Section 5 presents related work. Finally, Section 6 draws overall conclusions.

2   Emulation Infrastructure

To examine the behavior of next–generation clusters we use a state–of–the–art hardware cache–coherent DSM system, an SGI Origin 2000 to emulate future cluster architectures. Our approach has a number of advantages. It eliminates the need for simulation and allows direct execution of the same code used on the SAN cluster on top of the emulated communication layer. Using simulation, although advantageous in some cases, tends to overlook important system limitations that may shift bottlenecks to system components that are not modeled in detail. Using emulation on top of an existing, commercial multiprocessor results in a more precise model for future clusters. It uses commercial components that will most likely be used in future nodes and a commercial, off-the-shelf operating system. In fact, using a real, commercial OS reveals a number of issues usually hidden in simulation studies and/or custom built systems. Moreover, it allows us to use an SVM protocol and communication layer that have been designed and tuned for low-latency, high-bandwidth SANs and run on a Myrinet cluster.

To emulate a cluster on top of an SGI Origin2000 we follow a layered architecture. We implement a user-level communication layer that has been designed for low-latency, high-bandwidth cluster interconnection networks. We then port on this communication layer an existing SVM protocol that has been optimized for clusters with low-latency interconnection networks. At the highest level we use the SPLASH-2 applications. The versions we use include optimizations [10] for SVM systems (not necessarily clustered SVM systems). These optimizations are also useful for large scale hardware DSM systems. In the next few sections we describe our experimental platform in a bottom–up fashion.

2.1   Hardware Platform

The base system used in this study is an SGI Origin 2000 [12], containing sixty-four 300MHz R12000 processors and running Cellular IRIX 6.5.7f. The 64 processors are distributed in 32 nodes, each with 512 MBytes of main memory, for a total of 16 GBytes of system memory. The nodes are assembled in a full hypercube topology with fixed-path routing. Each processor has separate 32

KByte instruction and data caches and a 4 MByte unified 2-way set associative second-level cache. The main memory is organized into pages of 16 KBytes. The memory buses support a peak bandwidth of 780 MBytes/s for both local and remote memory accesses.

To place our results in context, we also present results from an actual cluster that has been built [8]. A direct comparison of the two platforms is not possible due to the large number of differences between the two systems. Our intention is to use the actual cluster statistics as a reference point for what today's systems can achieve.

2.2   Communication Layer

The communication layer we use is Virtual Memory Mapped Communication (VMMC) [3]. VMMC provides protected, reliable, low-latency, high-bandwidth user-level communication. The key feature of VMMC that we use is the remote deposit capability. The sender can directly deposit data into exported regions of the receiver's memory, without process (or processor) intervention on the receiving side. The communication layer has been extended to support a remote fetch operation and system-wide network locks in the network interface without involving remote processors, as described in [1]. It is important to note that these features are very general and can be used beyond SVM protocols.

We implement VMMC on the Origin2000 (VMMC-O2000) providing a message passing layer on top of the hardware cache-coherent interconnection network. VMMC-O2000 provides applications with the abstraction of variable-size, cache-coherent nodes connected with a non-coherent interconnect and the ability to place thread and memory resources within each cc-NUMA node. VMMC-O2000 differs from the the original, Myrinet implementation [3] in many ways:

The original implementation of VMMC makes use of the DMA features of the Myrinet NIs: data are moved transparently between the network and local memory without the need for intervention by the local processors. The Origin2000's *Block Transfer Engine* offers similar, DMA–like services but user-level applications do not have access to this functionality. VMMC-O2000 instead uses Unix shared–memory regions to accomplish the sharing of memory between the sender and receiver, and *bcopy(3C)* to transfer data between the emulated nodes.

Asynchronous send and receive operations are not implemented in VMMC-O2000. Asynchronous transmission on the cluster was managed by the dedicated processor on the Myrinet NI; the Origin offers no such dedicated unit. Although similar possibilities do exist—for instance using one of the two processors in each node as a communication processor—these are beyond the scope of this work and we do not explore them here. Asynchronous operations in the SVM protocol are replaced with synchronous ones.

VMMC-O2000 inter-node locks are implemented as ticket-based locks. Each node is the owner of a subset of the system–wide locks. At the implementation level, locks owned by one node are made available to other nodes by way of Unix shared–memory regions.

Since nodes within the Origin system are distributed cc-NUMA nodes, they do not exhibit the symmetry found in the SMP nodes that have been used so far in software shared memory systems. For this reason, we extend VMMC-O2000 to provide an interface for distributing threads and global memory within each cluster node. Compute threads are pinned to specific processors in each node. Also, global memory is placed across memory modules in the same node either in a round-robin fashion (default) or explicitly by the user.

Table 1 shows measurements for the basic operations of both VMMC and VMMC-O2000. We see that basic data movement operations are faster by about one order of magnitude in VMMC-O2000. Notification cost is about the same, but is not important in this context, as *GeNIMA* does not use interrupts. The cost for remote lock operations are significantly reduced under VMMC-O2000.

Table 1. Basic VMMC costs. All send and fetch operations are assumed to be synchronous, unless explicitly stated otherwise. These costs do not include contention in any part of the system.

VMMC Operation	SAN Cluster	Origin2000
1-word send (one-way lat)	$14\mu s$	$0.11\mu s$
1-word fetch (round-trip lat)	$31\mu s$	$0.61\mu s$
4 KByte send (one-way lat)	$46\mu s$	$7\mu s$
4 KByte fetch (round-trip lat)	$105\mu s$	$8\mu s$
Maximum ping-pong bandwidth	96 MBy/s	555 MBy/s
Maximum fetch bandwidth	95 MBy/s	578 MBy/s
Notification	$42\mu s$	$47\mu s$
Remote lock acquire	$53.8\mu s$	$8\mu s$
Local lock acquire	$12.7\mu s$	$7\mu s$
Remote lock release	$7.4\mu s$	$7\mu s$

Overall, VMMC-O2000 provides the illusion of a clustered system on top of the hardware cache-coherent Origin 2000. The system can be partitioned to any number of nodes, with each node having any number of processors. Communication within nodes is done using the hardware–coherent interconnect of the Origin without any VMMC-O2000 involvement. For instance, an 8-processor node consists of 4 Origin 2000 nodes, each with 2 processors that communicate using the hardware–coherent interconnect. Communication across nodes is performed by explicit VMMC-O2000 operations that access remote memory.

2.3   Protocol Layer

We use *GeNIMA* [1] as our base protocol. *GeNIMA* uses general-purpose network interface support to significantly improve protocol overheads and narrow the gap between SVM clusters and hardware DSM systems at the 16-processor scale. The version of the protocol we use here is the one presented in [8] with certain protocol-level optimizations to enhance performance and memory scalability. For the purpose of this work, we port *GeNIMA* on the Origin2000 (*GeNIMA*-O2000), on top of VMMC-O2000. The same protocol code runs on both WindowsNT and IRIX. Additionally, we extend *GeNIMA* to support a 64-bit address space. A number of architectural differences that arise as a result of faster communication, cc-NUMA nodes, and contention due to wider nodes. To address these issues we extend and optimize *GeNIMA*-O2000 as follows:

*Page invalidation (*mprotect*) operations:* *GeNIMA* uses *mprotect* calls to change the protection of pages and invalidating local copies of stale data. *mprotect* is a system call that allows user processes to change information in the process page table and can be expensive. Moreover, *mprotect* calls require invalidating TLBs of processors within each virtual node, and thus, the cost is affected by the system architecture. More information on the cost of *mprotects* will be presented in Section 4. *GeNIMA*-O2000 tries to minimize the number of *mprotects* by coalescing consecutive pages to a single region and changing its protection with a single call.

Data prefetching: We enhance data sharing performance by adding page prefetching to the page fault handler. When a process needs to fetch a new shared page, the subsequent N pages are also read into local memory. Empirically, we determined that $N=4$ offers the best prefetching performance on *GeNIMA*-O2000. Ideally, this approach can reduce both the number of `mprotect()` calls and the number of page protection faults by increasing demands on network bandwidth. The average cost of an *mprotect* call and the overhead of a page fault on the Origin2000 are relatively high: $\approx 150\mu s$ and $\approx 37\mu s$, respectively). Thus, prefetching not only takes advantage of the extra bandwidth in the system, but alleviates these costs as well. The actual prefetch savings are dependent on the applications' particular access pattern. In order to determine when prefetching is useful, "false" prefetches are tracked in a history buffer. A prefetch is marked as "false" when the additional pages that were fetched are not used. If a page has initiated a "false" prefetch in the past, the protocol determines that the current data access pattern is sufficiently non-sequential and disables prefetching for this page the next time it is needed.

Barrier synchronization: We enhance barriers in two ways. The first change involves serializing large sections of the barrier code to reduce *mprotect* contention. The second change introduces an ordering step prior to the processing of the page invalidations to reduce the number of *mprotect* calls. We use the *quick-sort* algorithm to order the updates by page number; groups of consecutive pages are then serviced together and the entire block is updated with a single *mprotect*.

Intra-node lock synchronization: The original *GeNIMA* protocol uses the *test-and-set* algorithm to implement inter-node locks. Although this approach works well for systems with small–scale SMP nodes, it is not adequate for systems with larger–scale, cc-NUMA nodes and leads to poor caching performance and increased inter-node traffic in *GeNIMA*-O2000. We have experimented with a number of lock implementations. Overall, ticket–based locks turn out to be the most efficient. For this reason we replace these locks with a variant of the *ticket-based* locking algorithm that generates far less invalidation traffic and offers FIFO servicing of lock requests to avoid potential starvation.

2.4   Applications Layer

We use the SPLASH-2 [15,9] application suite. A detailed classification and description of the application behavior for SVM systems with uniprocessor nodes is provided in [7]. 64-bit addressing and operating system limitations related to shared memory segments prevent some of the benchmarks from running at specific system configurations. We indicate these configurations in our results with 'N/A' entries. The application we use are: FFT, LU, Radix, Volrend, WaterSpatial, and SampleSort. We use both original versions of SPLASH-2 applications [7] and versions that have been restructured to improve their performance on SVM systems [9]. Finally, in this work we restructure FFT (FFTst) to stagger the transpose phase among processors within each node. This allows FFT to take advantage of the additional bandwidth available in the system.

3   Impact of Fast Interconnection Networks

In this Section we discuss the impact of faster interconnection networks on the performance of shared virtual memory. We use system configurations with 4 processors per node since most of the work so far with SVM on clusters has used Quad SMP nodes and we would like to place our results in context. This allows us to loosely compare our results with an actual cluster [8]. The two platforms cannot be compared directly due to a number of differences in the micro-processor and memory architectures they use, they offer invaluable insight in the needs of future clusters. A more detailed analysis of our results is presented in [5].

Figure 1 presents execution time breakdowns for each application for *GeNIMA*-Myrinet and *GeNIMA*-O2000 for 32– and 64–processors. Table 2 presents speedups and individual statistics for the applications run on *GeNIMA*-O2000. In particular, FFT is bandwidth–limited on the cluster. By modifying the original application to stagger the transpose phase, the speedup of FFTst is quadrupled with an impressive 23.8 on the 32-processor system.

Table 2. Parallel speedup of benchmarks running under *GeNIMA*-O2000.

	FFTst	LU	radixL	ssort	volrend	waterF
32 Proc (8x4)	23.8	N/A	1.9	6.7	17.8	9.2
64 Proc (16x4)	26.6	32.9	1.0	N/A	23.1	14.5

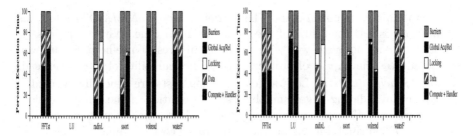

Fig. 1. Execution time breakdowns for each application. The leftmost graph provides breakdowns for a 32-processor system; the rightmost graph provides breakdowns for a 64-processor system. The left bar in each graph refers to *GeNIMA*-Myrinet, whereas the right bar refers to the *GeNIMA*-Origin2000. (FFT is the original version under *GeNIMA*-Myrinet and the staggered version under *GeNIMA*-O2000.)

Remote data wait: Overall, the faster communication layer and our optimizations provide significantly improved data wait performance. The direct remote read/write operations in *GeNIMA* eliminate protocol processing at the receive side (the page home) and make it easier for protocol performance to track improvements in interconnection network speed. In the majority of the applications data–wait time is reduced to at most 20% of the total execution time. Also, although a direct comparison is not possible, data wait time is substantially improved compared to the SAN cluster. Applications where prefetching is effective receive an additional benefit in reduced page-fault interrupt costs: anywhere from 30% to 80% of the total page faults are typically eliminated in prefetching, with the reduction in page faults resulting in a savings of 15% to 20% in total execution time on *GeNIMA*-O2000.

Table 3. Performance of *GeNIMA* on the SAN Cluster (Myrinet) and on the emulated system (O2000). The results are for for 32-processor systems (8 nodes, 4 processors/node). Percentages indicate percent of total execution time.

	Data Time		Barrier Time		Lock Time		*mprotect* Time		*diff* Time	
	Myrinet	O2000	Myrinet	O2000	Myrinet	O2000	Myrinet	O2000	Myrinet	O2000
FFTst	33.9%	18.3%	18.9%	17.9%	–	–	15.0%	6.1%	0.1%	0.6%
LU	N/A	N/A	N/A	N/A	–	–	N/A	N/A	N/A	N/A
radixL	29.9%	22.8%	30.9%	28.9%	3.5%	17.0%	16.7%	6.2%	18.4%	10.3%
ssort	N/A	N/A	N/A	N/A	–	–	N/A	N/A	N/A	N/A
volrend	1.1%	1.9%	16.1%	37.2%	–	–	0.5%	2.5%	0.3%	2.5%
waterF	20.6%	27.2%	16.7%	16.7%	0.1%	0.2%	14.6%	11.1%	0.2%	0.7%

Barrier Synchronization: The faster communication layer and our barrier–related optimizations result in moderately improved barrier synchronization performance for most of the applications. Protocol costs are the main overhead and faster interconnects are unlikely to benefit barrier costs in future clusters, unless they are accompanied by protocol, application, and/or operating system changes as well to improve *mprotect* costs. The barrier cost on *GeNIMA*-O2000 for most applications is less than 30% of the total execution time. With 64 processors, barrier cost does not scale well with the problem sizes we use here. Future work should address the impact of problem size on the scalability of these overheads.

Lock Synchronization: Lock synchronization costs are generally effected by the cost of invalidating pages and the resulting dilation of the critical sections. For this reason, although lock acquires and releases have lower overheads in *GeNIMA*-O2000 (in Table 1), the higher *mprotect* costs dominate (Tables 3, 4). Applications that rely heavily on lock synchronization such as radixL exhibit higher lock costs with *GeNIMA*-O2000. This change is due to radixL's small critical regions (several dozen loads and stores) and the larger locking overheads (under contention) on *GeNIMA*-O2000: more time is spent acquiring and releasing the locks than is actually spent inside the critical region. The result is that the application spends a comparatively small percentage of its time inside in the critical region, thus reducing the likelihood that acquires are issued while the lock is still local. Developing techniques to limit the effect of *mprotect* cost on lock synchronization is critical for further reductions in lock synchronization overheads.

4   Impact of Wide, CC-NUMA Nodes

In this section we attempt to answer the question of future performance on software shared memory clusters by demonstrating the performance of GeNIMA-O2000 on wide (8- and 16-processor) cc-NUMA nodes. A more detailed analysis of our results is presented in [5]. Table 5 summarizes the parallel speedups for the 4-, 8- and 16-processor configurations. Table 7 shows the major protocol costs as percentage of the total execution time at the 64–processor scale.

Local Data Placement: The NUMA effects of the nodes in *GeNIMA*-O2000 are negligible up to the 4-processor level. However, as wider nodes are introduced, processors in each node exhibit highly imbalanced compute times. Generally, applications that exhibit significant intra–node sharing of global data that are fetched from remote nodes, such as FFTst and waterF, incur highly imbalanced

Table 4. Performance of *GeNIMA* on the SAN Cluster (Myrinet) and on the emulated system (O2000). The results are for for 64-processor systems (16 nodes, 4 processors/node). Percentages indicate percent of total execution time.

	Data Time		Barrier Time		Lock Time		*mprotect* Time		*diff* Time	
	Myrinet	O2000	Myrinet	O2000	Myrinet	O2000	Myrinet	O2000	Myrinet	O2000
FFTst	42.2%	35.1%	17.1%	22.4%	–	–	6.8%	6.6%	0.4%	0.2%
LU	6.7%	3.2%	20.5%	34.8%	–	–	1.3%	1.4%	0.0%	0.3%
radixL	35.8%	14.7%	40.8%	32.3%	11.5%	35.2%	7.1%	10.2%	14.7%	4.4%
ssort	16.1%	4.1%	63.8%	38.9%	–	–	14.4%	2.1%	12.8%	0.3%
volrend	3.3%	2.6%	27.3%	56.1%	–	–	1.1%	7.0%	0.7%	0.8%
waterF	26.5%	28.4%	18.3%	23.7%	0.3%	1.0%	10.6%	11.2%	0.7%	0.8%

Table 5. Parallel speedups for 64-processor configurations with varying node widths.

	Parallel Speedup					
	32 Processors			**64 Processors**		
	4	8	16	4	8	16
FFTst	23.8	16.3	5.7	26.6	28.7	12.4
LU	N/A	18.5	N/A	32.9	32.5	N/A
radixL	1.9	1.8	N/A	1.0	1.2	N/A
volrend	17.8	17.7	16.1	23.1	24.8	17.6
waterF	9.2	4.9	1.3	14.5	8.5	2.7

Table 6. Average *mprotect* cost for 64-processor configurations with varying node widths.

	Avg. *mprotect* Cost		
	4	8	16
FFTst	$127.8\mu s$	$199.5\mu s$	$622.1\mu s$
LU	$55.4\mu s$	$43.6\mu s$	N/A
radixL	$325.8\mu s$	$296.4\mu s$	N/A
volrend	$404.2\mu s$	$402.5\mu s$	$569.4\mu s$
waterF	$87.1\mu s$	$163.4\mu s$	$674.9\mu s$

compute times. On the other hand, applications that either exhibit little intra-node sharing of global data, and/or have low data wait overheads overall, such as LU, volrend, and radixL, are not greatly affected. To explain this behavior we need to examine what happens when pages are fetched from remote nodes. Certain processors in each node fetch more shared pages than others. Due to the *bcopy* method used to fetch pages in VMMC-Origin, new pages are placed in the cache of the processor that performs the fetch operations. Thus, due to the NUMA node architecture, processors in each node exhibit varying local memory overheads depending on the page fetch patterns. An examination of data access patterns in Figure 2 shows that in FFTst processors with the lowest execution times also perform the highest number of remote data fetches. The processor that first fetches each page places data in its second–level cache, resulting in higher local memory access overheads for the rest of the processors in the same *GeNIMA*-O2000 node. Modifying FFTst, such that processors in each node compete less for fetching pages, reduces imbalances and improves average compute time from 45% to 60% of total execution time (Figure 2). However, processors in each node still exhibit varying memory overheads. Thus, dealing with NUMA effects in wide nodes is an important problem that future SVM protocol and possibly application design has to address.

Remote Data Wait: In contrast to local memory access overheads, remote data wait time is reduced across applications by up to 57%. Table 7 shows the

Table 7. Protocol overhead for 64-processor configurations with varying node widths. (Percentages indicate percent of total execution time.)

	Data Time			Barrier Time			Lock Time			*mprotect* Time		
	4	8	16	4	8	16	4	8	16	4	8	16
FFTst	35.1%	21.1%	17.8%	22.4%	28.7%	19.0%	–	–	–	6.6%	9.9%	12.6%
LU	3.2%	3.2%	N/A	34.8%	35.0%	N/A	–	–	–	1.4%	0.9%	N/A
radixL	14.7%	13.5%	N/A	32.3%	31.4%	N/A	35.2%	28.2%	N/A	10.2%	4.4%	N/A
volrend	2.8%	2.6%	1.8%	56.1%	52.8%	62.1%	–	–	–	7.0%	2.3%	1.0%
waterF	28.4%	17.6%	12.2%	23.7%	31.6%	44.8%	1.0%	0.2%	0.0%	11.2%	10.7%	11.1%

percentage of execution time associated with remote data fetches, and each of the other components of the protocol overhead. The number of remote fetches are reduced on 16-processor nodes by 19% on average over 4-processor nodes. This leads to an overall reduction in remote fetch times of 15% to 20% for all benchmarks, except LU where data wait time is small and accounts only for about 3% of the total execution time.

Lock Synchronization: Lock synchronization also benefits from wider nodes, achieving significant reductions in overhead with 8–processor nodes, and showing a moderate improvement at the 16-processor level. Local lock acquires and releases are very inexpensive in *GeNIMA* (equivalent to a few instructions). Wider nodes incur more local than remote acquires and all aspects of lock overhead are improved with 8-processor nodes.

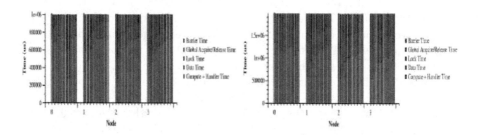

Fig. 2. Breakdown of protocol overhead in FFTst, before (left) and after (right) balancing.

Barrier Synchronization: In contrast to the gains observed in remote data access and lock synchronization, barrier performance is generally unchanged with 8-processor nodes. This is mainly due slightly higher intra– and inter–node imbalances in compute times. With wide nodes, barrier performance is dominated by higher imbalances and high *mprotect* costs, as explained next.

mprotect Costs: On the cluster *mprotect* calls typically cost between 40–80μs. In contrast, the cost of downgrading a page (e.g., moving from a *read-write* to an *invalid* state in a barrier or unlock) is between 250μs and 1ms in *GeNIMA-O2000*. The cost of *mprotect* on IRIX stems from four different sources: (i) broadcasting TLB invalidations to multiple processes in the same share group; (ii) changing the protection on large portions of the address space, necessitating the modification of multiple page table entries; (iii) changing the protection of a page to a state that differs from its neighbors, resulting in the breaking of the larger protection region into three smaller ones; and, (iv) locking contention in the kernel while executing multiple, unrelated processes (in our case, in *mprotect* code). Although we ensure that (i)–(iii) are minimized in *GeNIMA-O2000*, *mprotect* costs remain higher the comparable costs in the cluster. This leads us to suspect that (iv) is the reason for the additional overhead[1].

5   Related Work

The MGS system [16] examined clustering issues for SVM systems and in particular different partitioning schemes on Alewife, a hardware cache-coherent system. Unlike our work, the authors use a TreadMarks-like protocol and they find both synchronization and data movement across nodes to be a problem. The SVM protocol they use is tuned for traditional clusters with slow interconnection networks (they make use of interrupts and assume high overheads in communication initiation).

The authors in [11] take advantage of direct remote access capabilities of system area networks to address communication overheads in software shared memory protocols. This study is simulation-based and the performance characteristics of the interconnection network correspond to todays' state-of-the-art SANs. The focus is on protocol-level issues for extending the Cashmere protocol to clusters of SMPs and interconnection network with direct remote memory access capabilities. A subset of the features examined in the simulation studies was implemented on an actual cluster [14], which, however, corresponds more to state of the art clusters that can be built with today's technology.

The SoftFLASH system [4] provided a sequentially consistent software shared memory layer on top of 8–processor SMP nodes. They find that the cost for page invalidations within each node is very high. Another study examined the all-software home-based HLRC and the original Treadmarks protocols on a 64-processor Intel Paragon multiprocessor [17]. This study focused on the ability of a communication coprocessor to overlap protocol processing with useful computation in the two protocols but it also compared the protocols at this large scale. However, the architectural features and performance parameters of the Paragon system and its operating system are quite different from those of today's and future clusters. Also, the study used mostly simple kernels with little computational demand, and only two real applications (WaterNsquared and Raytrace).

The authors in [2] examine the effect of communication parameters on shared memory clusters. They use architectural simulation to investigate the impact of host overhead, bandwidth, network occupancy, and interrupt cost on end-system performance. However, they only examine systems with 4–way SMP nodes. Moreover, due to the infrastructure they use, they are limited to examining relatively small problem sizes.

Overall, while a lot of progress has been made, the impact of next generation SANs and wider nodes on shared memory clusters has not been adequately addressed. This work has addressed some of the related issues in this direction.

---

[1] To verify this, we perform a set of experiments that are somewhat intricate so we do not describe them fully. However, an effort was made to try and capture the desired system behavior.

6   Conclusions

In this paper we examine the implications of building software shared memory clusters with interconnection networks that offer latency and bandwidth comparable to hardware cache–coherent systems and wide, cc-NUMA nodes. We use an aggressive hardware cache-coherent system to investigate the impact of both of these architectural features on SVM performance. We first port a shared memory protocol that has been optimized for low-latency, high-bandwidth system area networks on a 64–processor Origin 2000. We then provide a number of optimizations that take advantage of the faster interconnection network.

Our results show that improvements in protocol data wait costs follow improvements in network speed, whereas improving synchronization requires further protocol and/or application work. The communication improvements and our protocol optimizations make communication–related costs less significant than on existing clusters. In addition, *diff* costs are very small in *GeNIMA-O2000*, typically less than 1% and no more than 5% of the total execution time. Barrier synchronization costs at the protocol level benefit from the faster interconnection network and the related optimizations but they do not scale well to the 64–processor level for some applications and the problems sizes we examine. At the 32–processor scale, barrier overheads are less than 30% of the execution time, whereas at the 64–processor scale these costs increase to to 40%. Lock synchronization costs are not able to benefit from the faster lock acquire and release times, due to the increased cost of *mprotect* calls. Finally, the most important remaining protocol overhead is the cost of *mprotect* calls.

When using wide nodes, although data wait time and lock synchronization overheads are reduced, compute time imbalances due to the NUMA node architecture and *mprotect* cost due to increased TLB invalidation overheads and OS contention result in either small application performance improvements with 8-processor nodes, or significant performance degradation with 16-processor nodes. In building future systems with wide, cc-NUMA nodes, SVM protocols and/or applications need to address issues in intra–node data access patterns and data placement. Moreover, these effects are important even with the relatively small-scale nodes we examine here. Future clusters could support cc-NUMA nodes with even wider nodes making these effects more significant.

Overall, software shared memory can benefit substantially from faster interconnection networks and can lead in building inexpensive shared memory systems for large classes of applications. However, current SVM protocols can only partially take advantage of faster interconnects and do not address the issues that arise when faster networks and wider nodes are used. Our work quantifies these effects and identifies the areas where more research is required for future SVM clusters.

Acknowledgments. We would like to thank Dongming Jiang for providing the modified versions of the SPLASH-2 applications and Bill Wichser for his help with using the SGI Origin 2000.

References

1. A. Bilas, C. Liao, and J. P. Singh. Accelerating shared virtual memory using commodity ni support to avoid asynchronous message handling. In *The 26th International Symposium on Computer Architecture*, May 1999.
2. A. Bilas and J. P. Singh. The effects of communication parameters on end performance of shared virtual memory clusters. In *Proceedings of Supercomputing 97, San Jose, CA*, November 1997.
3. C. Dubnicki, A. Bilas, Y. Chen, S. Damianakis, and K. Li. VMMC-2: efficient support for reliable, connection-oriented communication. In *Proceedings of Hot Interconnects*, Aug. 1997.
4. A. Erlichson, N. Nuckolls, G. Chesson, and J. Hennessy. SoftFLASH: analyzing the performance of clustered distributed virtual shared memory. In *The 7th International Conference on Architectural Support for Programming Languages and Operating Systems*, pages 210–220, Oct 1996.
5. C. Gibson and A. Bilas. Shared virtual memory clusters with next–generation interconnection networks and wide compute nodes. Technical Report TR-01-01-02, Department of Electrical and Computer Engineering, University of Toronto, Toronto, Ontario M5S3G4, Canada, 2001.
6. R. Grindley, T. Abdelrahman, S. Brown, S. Caranci, D. Devries, B. Gamsa, A. Grbic, M. Gusat, R. Ho, O. Krieger, G. Lemieux, K. Loveless, N. Manjikian, P. McHardy, S. Srblijic, M. Stumm, Z. Vranesic, and Z. Zilac. The NUMAchine Multiprocessor. In *The 2000 International Conference on Parallel Processing (ICPP2000)*, Toronto, Canada, Aug. 2000.
7. L. Iftode, J. P. Singh, and K. Li. Understanding application performance on shared virtual memory. In *Proceedings of the 23rd International Symposium on Computer Architecture (ISCA)*, May 1996.
8. D. Jiang, B. Cokelley, X. Yu, A. Bilas, and J. P. Singh. Applicaiton scaling under shared virtual memory on a cluster of smps. In *The 13th ACM International Conference on Supercomputing (ICS'99)*, June 1999.
9. D. Jiang, H. Shan, and J. P. Singh. Application restructuring and performance portability across shared virtual memory and hardware-coherent multiprocessors. In *Proceedings of the 6th ACM Symposium on Principles and Practice of Parallel Programming*, June 1997.
10. D. Jiang and J. P. Singh. Does application performance scale on cache-coherent multiprocessors: A snapshot. In *Proceedings of the 26th International Symposium on Computer Architecture (ISCA)*, May 1999.
11. L. I. Kontothanassis and M. L. Scott. Using memory-mapped network interfaces to improve t he performance of distributed shared memory. In *The 2nd IEEE Symposium on High-Performance Computer Architecture*, Feb. 1996.
12. J. P. Laudon and D. Lenoski. The SGI Origin2000: a scalable cc-numa server. In *Proceedings of the 24rd Annual International Symposium on Computer Architecture*, June 1997.
13. D. Lenoski, J. Laudon, K. Gharachorloo, A. Gupta, J. Hennessy, M. Horowitz, and M. Lam. Design of the Stanford DASH multiprocessor. Technical Report CSL-TR-89-403, Stanford University, December 1989.
14. R. Stets, S. Dwarkadas, N. Hardavellas, G. Hunt, L. Kontothanassis, S. Parthasarathy, and M. Scott. Cashmere-2L: Software Coherent Shared Memory on a Clustered Remote-Write Network. In *Proc. of the 16th ACM Symp. on Operating Systems Principles (SOSP-16)*, Oct. 1997.

15. S. Woo, M. Ohara, E. Torrie, J. P. Singh, and A. Gupta. Methodological considerations and characterization of the SPLASH-2 parallel application suite. In *Proceedings of the 23rd International Symposium on Computer Architecture (ISCA)*, May 1995.

16. D. Yeung, J. Kubiatowicz, and A. Agarwal. Multigrain shared memory. *ACM Transactions on Computer Systems*, 18(2):154–196, May 2000.

17. Y. Zhou, L. Iftode, and K. Li. Performance evaluation of two home-based lazy release consistency protocols for shared virtual memory systems. In *Proceedings of the Operating Systems Design and Implementation Symposium*, Oct. 1996.

Stream-Packing: Resource Allocation in Web Server Farms with a QoS Guarantee

Johara Shahabuddin[1], Abhay Chrungoo[2], Vishu Gupta[3], Sandeep Juneja[3], Sanjiv Kapoor[3], and Arun Kumar[1]

[1] IBM-India Research Lab, New Delhi, India, {sjohara, kkarun}@in.ibm.com
[2] IIT, Guwahati, cabhay@iitg.ernet.in
[3] IIT, Delhi, {vishu@cse,sandeepj@mech,skapoor@cse}.iitd.ernet.in

Abstract. Current web server farms have simple resource allocation models. One model used is to dedicate a server or a group of servers for each customer. Another model partitions physical servers into logical servers and assigns one to each customer. Yet another model allows customers to be active on multiple servers using load-balancing techniques. The ability to handle peak loads while minimizing cost of resources required on the farm is a subject of ongoing research.

We improve resource utilization through sharing. Customer load is expressed as a multidimensional probability distribution. Each customer is assigned to a server so as to minimize the total number of servers needed to host all the customers. We use the notion of complementarity of customers in simple heuristics for this stochastic vector-packing problem. The proposed method generates a resource allocation plan while guaranteeing a QoS to each customer. Simulation results justify our scheme.

1 Introduction

Web server farms offer managed clusters of servers to host web applications for multiple businesses. They allow businesses to focus on their core competencies while availing of technical skills, scalability, high availability of servers, and reduced system management costs due to economies of scale.

Resources on servers include CPU, memory, network bandwidth, and disk I/O, etc. The simplest approach used for hosting customers is to dedicate resources to each customer ([17], [18], [19]). A customer is given an integral number of servers, or, a physical server is partitioned into multiple isolated virtual servers for its customers. However, the peak-to-average ratios of customer load tend to be quite large ([2]). So, to handle peak loads the web server farm reserves costly resources that lie idle during average load conditions.

Another approach allows applications to reside on more than one shared server, so that each customer can be serviced from multiple servers. Load balancing techniques are used to spray the load evenly across servers ([6]). For complex applications, this may consume significant server resources. It also introduces the overhead of monitoring and interpreting the customers' load in

B. Monien, V.K. Prasanna, S. Vajapeyam (Eds.): HiPC 2001, LNCS 2228, pp. 182–191, 2001.
© Springer-Verlag Berlin Heidelberg 2001

real time. Where applications share servers, a surge in the traffic of one customer may adversely affect other customers. Given the drawbacks of the above schemes, systems which can improve resource utilization, while guaranteeing QoS to customers are a subject of ongoing research ([1]).

Access rates of web sites are observed to have periodicity on multiple time scales such as daily, weekly, and seasonal ([2], [3], [9]). Web sites experience peak loads at different times ([21]), such as sites accessed from opposite time zones. Again, a weather forecast application may be accessed mainly in the morning, while stock quotes may be heavily accessed in the evening. The proposed system makes use of customers' periodic access patterns to assign each customer to a server so as to minimize the total number of servers needed. We model this as a stochastic vector packing problem which is an NP-hard problem. We use simple off-line heuristics to give a resource allocation plan with QoS guarantees.

Related work has been done in the context of stochastic networks ([13], [14]). These are conservative approaches that use a single number to approximate burstiness, suitable for short-lived traffic (time scales of seconds). We use exact probability distributions, since we look at a different time scale. Also, by subdividing time intervals we improve on the notion of QoS in this work.

We believe that exploiting the temporal complementarity of customers' peak load using vector-packing is a novel approach for optimizing resource usage. Though the system we propose is in the context of web server farms, it is applicable more generally to other domains where resources may be divided into modifiable partitions. The implementation of our system would need OS-level support to partition resources and to measure their usage. Several such systems exist ([5], [11], [21]). We illustrate the potential of our system through simulations on a set of traces obtained from the Internet Traffic Archives [20].

2   Stream-Packing Approach

Customer Workload Characterization: Resource allocations on individual machines are modeled as independent variables in [4]. Resource allocation is optimized using periodically collected usage data. Their model is linear (like ours) but not stochastic. This simplification is required for ease of analysis. However, linearity does not hold for resources like disk bandwidth. Non-additivity also appears as the load increases especially where there are many customers. For example, at higher loads the total resource usage due to multiple processes may not be the sum of individual resource usage levels at lighter loads.

We model the customer resource requirement by means of a random vector having $d = r \times k$ dimensions, where there are r resources and k time slots (e.g., $k = 24$ if the day is divided into hour-long time slots). The resource requirement process is assumed to have the same distribution each day (or any other suitable period). The random variable X_{ij} denotes the requirement of customer i at any time t within the time interval corresponding to dimension j. Assuming stationarity, this random variable has a distribution that is independent of t. Stationary stochastic models for web traffic can be found in [9], [12], and [15].

Each customer i negotiates an agreement whereby he is always allocated a minimum resource requirement a_{ij} (at least 0), and can specify a maximum requirement of b_{ij} (less than or equal to the capacity of a server) for all $j = 1, ..., d$. Let Y_{ij} denote the capacity promised to i in j. Past access patterns are used to estimate the distribution of X_{ij} and hence Y_{ij} for $j = 1, 2, \ldots, d$.

$$Y_{ij} = a_{ij} \text{ if } X_{ij} < a_{ij}$$
$$= X_{ij} \text{ if } a_{ij} \le X_{ij} \le b_{ij}$$
$$= b_{ij} \text{ otherwise.}$$

We introduce the notion of α-*satisfiability* as a measure of QoS. A customer is said to be α-satisfied if he receives his promised capacity at least α proportion of time in each dimension. Let customers $C_1, ..., C_n$ be allocated to a server and let $Q = (Q_1, ..., Q_d)$ denote the server capacity. This allocation is α-**satisfiable** if for every $j \in \{1, ..., d\}$, and α (positive scalar constant less than 1),

$$P(Y_{1j} + Y_{2j} + \ldots + Y_{nj} \ge Q_j) \le 1 - \alpha. \tag{1}$$

This is a stronger guarantee than ensuring that during an entire 24-hour period a customer is given promised capacity α proportion of time (since a customer may fall short during his peak period). A notion of QoS that is based on usage history has been used in [13], [14]. However, it is for a single time-resource dimension. Unexpected surges are likely to cause a deterioration in the performance of such a QoS. Our scheme implicitly provisions for response time and throughput, since the notion of capacity of a server is derived from its performance under a given load [10].

Resource Allocation Problem: Given N customers, we wish to find the minimum number of servers needed to host them, subject to α-satisfiability as defined above. Each customer resides on one of the servers, sharing it with the other customers on it. This problem can be solved through a vector-packing approach, which is a generalization of bin-packing to multiple dimensions. Bin-packing is an integer programming problem where items of various sizes are to be packed into bins of a given capacity, so as to minimize the number of bins [8]. Multidimensional bin-packing packs volumes into a multi-dimensional space. However, the generalization we need is vector-packing which ensures that the jobs (customers) do not overlap in any dimension [7].

The assumption that a customer needs at most one server is not restrictive since larger sets of servers can be treated as a single one. To make reasonable comparisons between capacities of various resources and servers, we scale the measurement units appropriately. For example, if one resource has units in which a reasonable value is 1, whereas another has a reasonable value of 100, then we scale the units so that they match.

A part of each server's resources is partitioned among the customers on it, and the remaining is available as a common pool. The ith customer's partition has size $a_{ij}, j = 1, ..., d$. The partition protects against load surges from other customers on the same server.

Only the resources which are significant should be used for this scheme. Fewer resources and time intervals reduce computational complexity. A single abstract resource could model a system where there is a high correlation between load on multiple resources, or if there is a single bottleneck resource.

Stream-Packing Heuristic: Customers $\{C_1, ..., C_N\}$ are to be packed into servers. During the progress of the algorithm customers $\{C_n, ..., C_N\}$ are left to be packed, $n \leq N$, where N is the total number of customers. The ith customer is represented by a resource usage vector $Y_i = (Y_{i1}, ..., Y_{id})$, where each element is a random variable. Let $E(Y_i) = (E(Y_{i1}), ..., E(Y_{id}))$ denote its expected value. Let $E_{av}(Y_i)$ denote the mean usage over all dimensions: $E_{av}(Y_i) = \frac{\sum_{j=1}^{d} E(Y_{ij})}{d}$.

Servers are chosen to be packed in a sorted sequence such as decreasing order of mean capacity. At a given time, only one server is open for packing. Let $Q = (Q_1, ..., Q_d)$ be the capacity of this server. Let $B = (B_1, ..., B_d)$ denote the distribution of the resource utilized in the current server due to the customers C_B already in it. Let $E(B) = (E(B_1), ..., E(B_d))$ be the vector of expected values of the resource utilized in the dimensions of the current server.

If the ith customer is chosen as a suitable addition to the current server, the server's resource utilization vector is updated to the element-by-element convolution of Y_i with B, assuming these are independent. The new server distribution $B^{new} = (B_1 * Y_{i1}, ..., B_d * Y_{id}) = B * Y_i$, where $*$ represents the convolution operation. Let $C_\alpha \subset \{C_n, ..., C_N\}$ denote the set of customers that are α-satisfiable with the current server as in equation (1). The following heuristics are proposed for selecting the *most complementary customer* to add to the current server:

(i) *Roof Avoidance* : Let D be the dimension where the current server has the highest mean utilization: $E(B_D) = \max\{E(B_1), ..., E(B_d)\}$. Then the mth customer is said to be the most complementary customer if, $E(Y_{mD}) = \min\{E(Y_{iD})|C_i \in C_\alpha\}$, i.e., C_m is the α-satisfiable customer that has the smallest mean in the dimension D.

(ii) *Minimized Variance* : Let the empty space in the server be denoted by $e = (e_1, ..., e_d)$, where e_j is the maximum amount of resource that may be taken away from the jth dimension so that the current server stays α-satisfiable:

$$e_j = \max\{x | P(\sum_{\{i|C_i \in C_B\}} Y_{ij} \geq (Q_j - x)) \leq (1 - \alpha).\}$$

Denote the empty space in the current server if the ith customer was added to it by $e^i = (e_1^i, ..., e_d^i)$. Let e_{av}^i denote the mean empty space, $e_{av}^i = \frac{(e_1^i + ... + e_d^i)}{d}$. Then, variance of the empty space in the server if the ith customer was added is defined as: $\sigma_i = \sum_{j=1}^{d} \frac{(e_j^i - e_{av}^i)^2}{d}$. The mth customer is the most complementary customer if it gives rise to the least empty space variance, i.e., $\sigma_m = \min(\sigma_i | C_i \in C_\alpha)$. This variation attempts to equalize the utilization over all dimensions, thus preventing saturation in any one dimension.

(iii) *Maximized Minima* : We begin by finding C_m as in the Minimized Variance heuristic. Let C_v be a set of customers with empty space variance close to that of C_m: $C_v = \{C_i|((\sigma_i - \sigma_m) \leq V), (C_i \in C_\alpha)\}$, where V is a scalar constant. Define the maximum free space for the ith element of C_v as $f^i = \max_{\{j=1,...,d\}} e^i_j$. The most complementary customer C_m is updated to the customer in C_v with the minimum of this maximum free space, i.e., $f_m = \min\{f_i|C_i \in C_v\}$. The motivation here is to keep the variance low while also maximizing utilization.

(iv) *Largest Combination* : The set of customers C_v, that have a low empty space variance is found as in Maximized Minima. In this set we choose the customer with the minimum average empty space as the most complementary customer C_m, i.e., $e^m_{av} = \min\{e^i_{av}|C_i \in C_v\}$. The motivation here is to keep the variance low with also a high overall usage of the server.

Stream-Packing Algorithm:

> *let C be the set of customers, $\{C_1, ..., C_N\}$*
> *while ($C \neq \phi$)*
> > *open a new server B for packing*
> > *$C_B \leftarrow$ null customer set*
> > *let $C_\alpha \subset C$ be the α-satisfiable customer set for server B*
> > *while   ($C_\alpha \neq \phi$)*
> > > *$C_m \leftarrow$ most-complimentary-customer*
> > > *update the probability distribution vector $B \leftarrow B * Y_m$*
> > > *$C_B \leftarrow C_B + C_m$*
> > > *$C \leftarrow C - C_m$*
> > > *update $C_\alpha \subset C$ for new C_B*
> > *end*
> > *close B*
> *end*

Our algorithm finds its inspiration from one dimensional bin-packing ([8], [14]) together with the use of variance amongst the dimensions. In that context, off-line algorithms such as First-Fit Decreasing (FFD) and Best-Fit Decreasing (BFD) have been found effective with a worst case approximation bound of $11/9$. Our first approach is in the spirit of Best-Fit Decreasing. However the emphasis of the multi-dimensional algorithms defined above is to reduce variance and thus avoid over-utilization of any one single dimension. Our Maximized Minima approach buckets items into size classes as in Modified First-Fit Decreasing for the one-dimensional problem. But the criterion of the bucketing is the variance.

Handling of implementation issues related to addition, deletion of customers, change in requirements and high availability etc. are discussed in [16].

3   Experiments

We test the four stream-packing schemes discussed above for a single resource class, and servers of equal capacity $Q = 100$. We take the minimum and maximum requirements for customers as $a_{ij} = 0$ and $b_{ij} = 100$, respectively, for all $i = 1, \ldots, N$ and all $j = 1, \ldots, d$.

We compare stream-packing against a virtual server approach (see Introduction) in which the resource allocated is a percentile of the peak requirement. For instance, if α was 90% for our scheme then the natural comparison is against the virtual server scheme where the customer's allocation was 90% of his peak requirements. We chose the virtual server approach for the comparison because it encompasses the dedicated server approach. The packing is done using the next-fit decreasing heuristic.

Resource utilization data has been generated from web traffic traces as it is expected to be correlated. We obtained 16 traces, of which most were from the Internet Traffic Archives [20]. We generated multiple traces from these as follows. Variations in peak periods were simulated by time-shifting the web traces by a random interval (uniform over 0-24 hours). The trace arrival patterns were scaled to the desired average customer load level for each run. For each customer, 60 days of a trace was used. The first 30 days was used to determine the packing. The next 30 days were used for validation.

To generate the probability distributions, we totaled the number of hits arriving in each second. Each day was divided into 12 two-hour long time intervals to reduce the number of dimensions. The number of request arrivals per second X_{ij} takes integer values between 1 and 100. The average resource consumption over all dimensions for a customer is referred to as its average load. The number of servers required for the packing, was found out for all customers at a single average load level. This was done for average loads ranging from 2% to 30% of server capacity. The experiment was repeated for different QoS-levels (70 - 99.9%) and for different numbers of customers (20-500). In order to simulate a more realistic situation, we also tested the algorithms for customers with random average loads. Average customer load levels are sampled from a uniform distribution from 2% to 30% of server capacity.

4   Analysis of Results

All the stream-packing heuristics generate an allocation that require significantly fewer servers than the virtual server scheme. This is so across all QoS levels. For 95% α-satisfiability and 50 customers each of average load 2% of server capacity, our schemes require only 3 servers while the virtual server scheme requires 18 servers (Fig. 1(c)).

With increasing customer load, the improvement in aggregate resource requirement falls. The system experiences wasted space due to internal fragmentation at larger loads, and as a result more servers are needed. However, even at an average load 30% of server capacity a saving of about 50 servers is observed for 95% α-satisfiability and 500 customers (Fig. 1(d)).

Our schemes perform best for more realistic customers with random loads. The graph in Fig. 2(e) shows a linear (typical for all QoS levels) improvement in server requirement with number of customers for 95% α-satisfiability. Customers

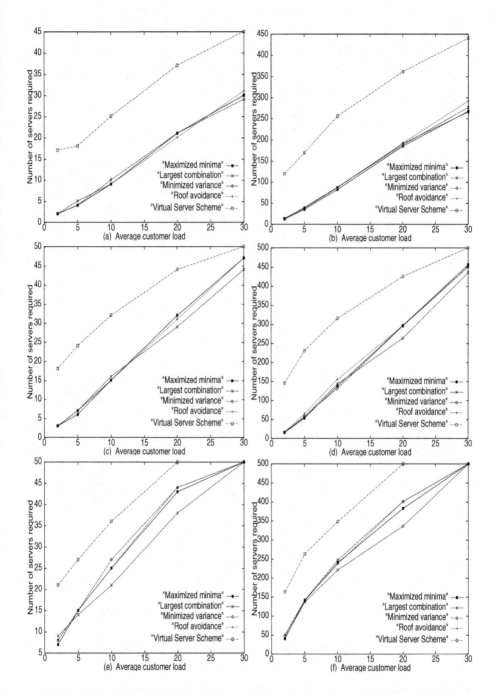

Fig. 1. Server requirements for different schemes against average customer load for 50 customers at (a) 80% QoS, (c) 95% QoS, (e) 99.9% QoS, and for 500 customers at (b) 80% QoS, (d) 95% QoS, (f) 99.9% QoS.

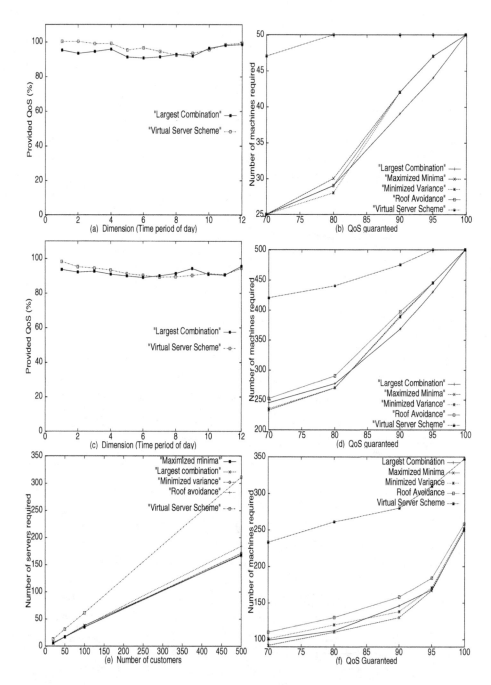

Fig. 2. (a)'Best day' and (c) 'worst day' for a random customer; Number of servers versus α (b) for 50 and (d) for 500 customers at 30% load, and (f) for 500 customers with random average loads; (e) Server requirement increases linearly with number of customers, for random average loads.

with smaller average loads fill up the servers effectively and result in better utilization.

The Largest Combination and Maximized Minima methods, require the least number of servers in general. The former performs better than the other three methods when used with all customers having the same average load.

There is a server cost to providing higher QoS (Fig. 2(b),(d),(f)). For 500 customers, average customer load of 30%, the number of servers needed was around 270 at 80% α as compared to 500 servers at 99.9% α. For 500 customers, our schemes needed about 120, 170, 240 servers for 80%, 95%, 99.9% α, respectively, for random average load. At 99.9% α and for the largest average customer loads (Fig. 1), no two customers were complementary and all five schemes needed as many servers as there were customers. (Recall that there is a nonzero probability of a customer requirement being equal to server capacity.)

In Fig. 2 (a),(c), the QoS provided to a randomly chosen server in the web cluster is plotted against time of day for 90% α-satisfiability, 100 customers with random average loads. For each hour of the day, the proportion of seconds when the capacity exceeded total demand is noted. The best QoS is plotted for the day on which average QoS is maximum (Fig. 2(a)). The worst QoS is analogously defined. Since all the customers are not equally affected by drops in QoS provided, the QoS provided to individual customers would be better than the total QoS provided by the server. The QoS delivered by our schemes is always more than required. Most of the time the virtual server scheme gives higher QoS (at the cost of more servers) than stream-packing schemes, except when the utilization is close to the peak.

5   Conclusion

We have proposed a resource allocation scheme that guarantees a probabilistic QoS to each customer. Stochastic vector packing was used to find complementary customers so as to reduce the number of servers needed. Simulation results showed reduction in total resource requirements, especially where there were customers of random loads to be packed. In future, this approach could be used as a basis for an on-line resource scheduling algorithm that 'knows' customer history, and allocates resources just-in-time by responding efficiently to actual resource usage levels.

References

1. Appleby, K., Fakhouri, S., Fong, L., Goldszmidt, G., Kalantar, M., Krishnakumar, S., Pazel, D.P., Pershing, J., Rochwerger, B.: Oceano - SLA Based Management of a Computing Utility. To appear in Proceedings of the IFIP/IEEE International Symposium on Integrated Network Management (2001)
2. Arlitt, M., Jin T.: Workload Characterization of the 1998 World Cup Website. HPL-1999-35(R.1), Internet Systems and Applications Laboratory, Hewlett Packard (1999) ({arlitt,tai}@hpl.hp.com)

3. Arlitt, M., Williamson, C.L.: Internet Web Servers: Workload Characterization and Performance Implications. ACM-SIGMETRICS, Philadelphia, Pennsylvania (1996) 631–645
4. Aron, M., Druschel, P., Zwaenepoel, W.: Cluster Reserves: A Mechanism for Resource Management in Cluster-based Network Servers. ACM-SIGMETRICS, Santa Clara, California, USA (2000) 90–101
5. Banga, G., Druschel, P., Mogul, J.C.: Resource Containers: A new facility for resource management in web server systems. Proceedings of the IIIrd Symposium on Operating Systems Design and Implementation (1999)
6. Cardellini V., Colajanni, M., Yu, P.S.: Dynamic load balancing on web server systems. Vol.3, No.3, IEEE Internet Computing (1999)
7. Chekuri, C., Khanna, S.: On Multi-Dimensional Packing Problems. Symposium On Discrete Algorithms (1999) 185–194
8. Coffman Jr., E.G., Garey, M.R., Johnson, D.S.: Approximation Algorithms for Bin Packing – a Survey. In: Hochbaum, D.S., (ed.): Approximation Algorithms for NP-Hard Problems. PWS Publishing Company, Boston (1995)
9. Crovella, M., Bestavros, A.: Self-similarity in World Wide Web traffic: Evidence and possible causes. ACM-SIGMETRICS, Philadelphia, PA (1996) 160–169
10. Dilley, J., Friedrich, R., Jin, T.: Measurement Tools and Modeling Techniques for Evaluating Web Server Performance. Hewlett-Packard Laboratories, Palo Alto, California, Tech Report HPL-96-161 (1996)
11. Govil, K., Teodosiu, D., Huang, Y., Rosenblum, M.: Cellular Disco: Resource Management Using Virtual Clusters on Shared-Memory Multiprocessors. 17th ACM Symposium on Operating Systems Principles (SOSP '99). In: Operating Systems Review 34(5) (1999) 154–169
12. Iyengar, A.K., Squillante, M.S., Zhang, L.: Analysis and Characterization of Large-Scale Web Server Access Patterns and Performance. The 8th World Wide Web Conference (1999)
13. Kelly, F.P.: Charging and Accounting for Bursty Connections. In: McKnight, L.W., Bailey, J.P. (eds.): Internet Economics. MIT Press, ISBN 0 262 13336 9, (1997) 253–278
14. Klienberg, J., Rabani, Y., Tardos, E.: Allocating Bandwidth for Bursty Connections. Proceedings of the 29th ACM Symposium on Theory of Computing (1997)
15. Leland, W.E., Taqqu, M.S., Willinger, W., Wilson, D.V.: On the self-similar nature of Ethernet traffic (extended version). IEEE/ACM Transactions on Networking 2, 1 (1994), 1–15
16. Shahabuddin, J., Chrungoo, A., Gupta, V., Juneja, S., Kapoor, S., Kumar, A.: Stream-Packing: Resource Allocation in Web Server Farms with a QoS Guarantee (extended version). Revised May 2001, available from the authors.
17. Netscape Hosting Services White Paper. (http://home.netscape.com/hosting/v4.0/whitepaper/, 5/2000)
18. Building a Scalable and Profitable Application Hosting Business, White Paper, Ensim Corporation. (http://www.ensim.com/Products/index.htm, 5/2000)
19. Hosting with Exchange 2000 Server, White Paper. (http://www.microsoft.com/exchange/techinfo/ASPplanning.htm/, 9/2000)
20. Internet Traffic Archives. (http://www.acm.org/sigcomm/ita/, 6/2000)
21. Sequent Technologies, Sequent's Application Region Manager. (http://www.sequent.com/dcsolutions/agile_wpl.html, 1/2000)

Weld: A Multithreading Technique Towards Latency-tolerant VLIW Processors

Emre Özer, Thomas M. Conte and Saurabh Sharma
{eozer,conte,ssharma2}.eos.ncsu.edu

Department of Electrical and Computer Engineering,
North Carolina State University, Raleigh, N.C., 27695.

Abstract. This paper presents a new architecture model, named *Weld*, for VLIW processors. Weld integrates multithreading support into a VLIW processor to hide run-time latency effects that cannot be determined by the compiler. It does this through a novel hardware technique called *operation welding* that merges operations from different threads to utilize the hardware resources more efficiently. Hardware contexts such as program counters and fetch units are duplicated to support multithreading. The experimental results show that the Weld architecture attains a maximum of 27% speedup as compared to a single-threaded VLIW architecture.

1. Introduction

Variable memory latencies in VLIW processors are one of the most crucial problems that VLIW processors face. A VLIW processor might have to stall for a large number of cycles until the memory system can return the data required by the processor. Multithreading is a technique that has been used to tolerate long latency instructions or run-time events such as cache misses, branch mispredictions, and exceptions. It has also been used to improve single program performance by spawning threads within a program, such as loop iterations or acyclic pieces of code. Weld is a new architecture model that uses compiler support for multithreading within a single program to combat variable latencies due to unpredictable run-time events in VLIW architectures.

There are two main goals that Weld aims to achieve: 1) better utilization of processor resources during unpredictable run-time events and 2) dynamic filling of issue slots that cannot be filled at compile time. Unpredictable events that cannot be detected at compile time, such as cache misses, may stall a VLIW processor for numerous cycles. The Weld architecture tolerates those wasted cycles by issuing operations from other active threads when the processor stalls for an unpredictable run-time event within the main thread. VLIW processors are also limited by the fact that they use a discrete scheduling window. VLIW compilers partition the program into several scheduling regions and each scheduling region is scheduled by applying different *instruction-level parallelism* (ILP) optimization techniques such as speculation, predication, loop unrolling, etc. The VLIW compiler cannot fill all schedule slots in every MultiOp [12][1] because the compiler cannot migrate operations from different scheduling regions. A hardware mechanism called the *operation welder* is introduced in this work to achieve our second goal. It merges operations

---

[1] A MultiOp is a group of instructions that can be potentially executed in parallel.

B. Monien, V.K. Prasanna, S. Vajapeyam (Eds.): HiPC 2001, LNCS 2228, pp. 192–203, 2001.
Springer-Verlag Berlin Heidelberg 2001

from different threads in the issue buffer during each cycle to eliminate empty issue slots, or NOPs. A scheduling region is a potential thread in our model. Executing operations from different scheduling regions and filling the issue slots from these regions simultaneously at run time can increase resource utilization and performance. The main objective of this study is to increase ILP using compiler-directed threads (scheduling regions) in a multithreaded VLIW architecture.

The remainder of this paper is organized as follows. Section 2 introduces the general Weld architecture. Section 3 explains the *bork insertion algorithm*, which defines how the compiler spawns new threads. Section 4 presents performance results and analysis. Section 5 discusses the related work in multithreading techniques. Finally, Section 6 concludes the paper and discusses future work.

2. The General Weld Architecture Model

The Weld model assumes that threads are generated from a single program by the compiler. During program execution there is a single main thread and potentially several speculative threads, but the main thread has the highest priority among all threads. The general Weld architecture is shown in Figure 1. Each thread has its own program counter, fetch unit and register file while all threads share the branch predictor and instruction and data caches. Weld consists of a basic 5-stage VLIW pipeline. The fetch stage fetches MultiOps from the Icache, and the decode/weld stage decodes and welds them together. The operation welder is

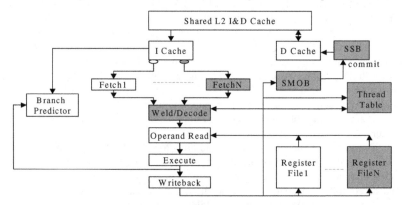

Figure 1. The general Weld architecture

integrated into the decode stage in the pipeline. The operand read stage reads operands into the buffer for each thread and sends them to the functional units. The execute stage executes operations, and the write-back stage writes the results into the register file and Dcache.

2.1. ISA Extension

A new instruction and some extensions to the ISA are required to support multithreading in a VLIW. A new instruction is needed to spawn threads. A branch and fork operation, or *bork*, is introduced to spawn new threads and create new hardware contexts for those threads. It has a target address, which is the address of the new, speculative thread.

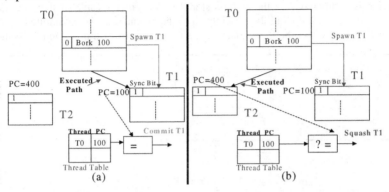

Figure 2. An example of synchronization of two threads

Two extra bits are added to each MultiOp in the ISA: *separability* and *synchronization* bits. A separability bit is necessary for each MultiOp to distinguish between separable and inseparable MultiOps. In a separable MultiOp, the individual operations that form the MultiOp can be issued across multiple cycles without violating dependencies between those operations. The reason for this classification is that there might be anti or output dependencies between operations in a MultiOp. Splitting such operations from the whole MultiOp may disrupt the correct execution of the program. The synchronization bit is added to each MultiOp in the ISA and is set in the first MultiOp of each thread at compile-time to help synchronize threads at run-time. With the combination of ISA and hardware support, threads can merge in a straightforward way as explained in Section 2.2.

2.2. Thread Creation and Synchronization

When a bork is executed, a new hardware context (register file, fetch unit and program counter) is assigned to the new thread if those resources are available. If there is not an available hardware context for the new thread, the bork behaves like a NOP. Once the new hardware context has been created, the register file of the ancestor thread is copied into the register file of the descendant thread, and the program counter of the descendant thread is initialized with the target address of the bork. Moreover, the target address is also stored in the Thread Table for thread synchronization. Speculative threads can then spawn other speculative threads, and so on. However, a thread (main or speculative) may spawn only one thread, i.e. the compiler guarantees that there is only one bork operation per executed path in order to reduce the complexity of thread synchronization. If there is a stall in one thread due to a cache miss, the other threads can still continue to fetch and execute. Even if there is no stall in any thread, the decode/weld stage can fill from multiple threads by taking advantage of any empty fields in the MultiOps.

An ancestor thread merges with its own descendant thread when the ancestor fetches the first MultiOp instruction in the descendant thread. The fetch unit checks if the synchronization bit of the fetched MultiOp is set. If set, the PC address of the instruction is compared with the address stored in the Thread Table. If they are the same, the descendant thread is correctly speculated and can be committed. The ancestor thread dies and the descendant thread takes over in case of a commit. If the

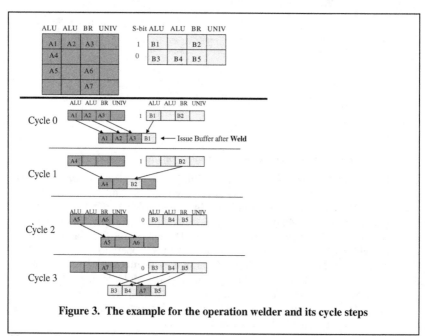

Figure 3. The example for the operation welder and its cycle steps

addresses do not match, then the descendant thread is incorrectly speculated. In this case, the descendant thread and its own descendant threads must be squashed. Also, the speculative stores in the Speculative Store Buffer (SSB) and the speculative loads in the Speculative Memory Operation Buffer (SMOB) are invalidated. Figure 2a shows an example of synchronization of two threads. Thread 0 (T0) spawns Thread 1 (T1) at address 100. Address 100 is also written in the Thread Table together with the thread id. If the control flow goes into T1, T0 fetches the first MultiOp, which is the MultiOp at address 100. The fetch unit for T0 checks if the synchronization bit is set. Then, the PC address 100 is compared with the address in the Thread Table, which is also 100. There is a match, therefore T1 can commit, which means T0 dies and T1 becomes the main thread. If the control flow goes into Thread 2, then T1 was misspeculated, as shown in Figure 2b. The PC address 400 is compared with the address 100 in the Thread Table. There is no match, so T1 must be squashed.

2.3. Operation Welder

The weld/decode stage in the pipeline takes MultiOps from each active thread, welds them together and decodes the welded MultiOp. The welding should be done before the operand read stage because the number of bits in an operation to be routed before decode is much less than the number of bits in the operand read stage. Therefore, the operation welder is integrated into the decode pipeline stage. It can send an operation to any functional unit as long as it is the correct resource to execute the operation. Each thread has a buffer called the *prepare-to-weld buffer* to hold a MultiOp. At each cycle, a MultiOp is sent from the fetch buffer to the weld/decode stage. Each operation has an empty/full bit that states whether an operation exists in the slot, and each MultiOp also has a separability bit. The operation welder consists of an array of multiplexers to forward the operations to the functional units. Control logic takes the empty/full and separability bits and sends control signals to the multiplexers. Each functional unit has a multiplexer in front of it. In the Weld architectural model, the main thread has the highest priority. Among the speculative threads, older threads have higher priority than younger ones. All operations in the main thread are always forwarded to the issue buffer. The empty slots in the issue buffer are filled from operations in the speculative threads. Some operations in a MultiOp of a speculative thread may be sent to the functional units while the remaining operations stay in the prepare-to-weld buffer. In this case, the fetch from this thread is stalled until the remaining operations in the buffer are issued.

The example in Figure 3 demonstrates how the operation welder works. There are two threads scheduled for a 4-issue VLIW machine. Two ALU units, one branch unit and one universal functional unit are used in the example architecture. The thread on the right shows the speculative thread and the separability bit for each of its MultiOps. At cycle 0, the first MultiOps from both threads are in the prepare-to-weld buffer. First, the whole MultiOp from A is forwarded into the issue buffer. Then a check is made if the separability bit of the MultiOp from B is set. Since it is set, the operations from B are separable. B1 is welded into slot 4 of the issue buffer. The remaining operations stay in the prepare-to-weld buffer of B. At cycle 1, the second MultiOp from A is forwarded into the issue buffer. B2 is welded into slot number 3 of the issue buffer. At cycle 2, thread A puts its third MultiOp and thread B puts its second MultiOp into their respective prepare-to-weld buffers. Again, thread A forwards all operations into the issue buffer. However, there can be no welding with thread B because its separability bit is 0. Therefore, all operations from this MultiOp must be sent to the issue buffer at the same time. Only the MultiOp from thread A is sent to the functional units. At cycle 3, the last MultiOp from thread A is put into the prepare-to-weld buffer and forwarded into the issue buffer. A check is made if all operations from B can be welded. Since there are three slots available in the issue buffer and they are the right resources, B3, B4 and B5 are all welded into the issue buffer.

2.4. Thread Table

The thread table (TT) keeps track of threads that have been spawned and merged. There is an entry for each active thread. Each entry keeps a thread id, register file id, and the borked PC address. The thread id is a unique number that identifies each active thread. This unique id can be obtained from a time-stamp counter. The time stamp represents the age of the threads. Each time a new thread is created, the counter is incremented by one and stamped into the thread and written into the TT. The time-stamp counter must be wide enough so as not to overflow. Otherwise, the time order of the threads would be disrupted. The time-stamp counter is reset to zero when there is only one active thread, which is the main thread. If the counter overflows while there are active threads, all of the speculative threads are squashed and the counter is reset to zero. The register file id is the identifier of the register file assigned to the thread. The time-stamp id is also attached to each operation in a thread to distinguish operations from different threads. When an operation completes, it searches the TT to find the proper register file id by comparing the attached time stamp with time stamps in the table. The TT can be designed as a shift register. When two threads merge, the ancestor thread dies and the descendant thread takes over. This involves deleting the ancestor thread's entry from the TT. Before removing it, its borked PC is copied into the borked PC field of its own ancestor thread, if there is any. Then the descendant thread's entry is shifted up and overwritten into the ancestor thread.

2.5. Speculative Memory Operation Buffer (SMOB) and Speculative Store Buffer (SSB)

Load operations from speculative threads are kept in a buffer called the *speculative memory operation buffer* (SMOB). The operating principle of the SMOB is similar to the ARB [14][15]. However, it is different from the ARB in the sense that only load operations are kept in the SMOB. It uses a single shared fully associative buffer to resolve run-time load/store violations. A SMOB entry contains the speculative load memory address, the speculative thread's time stamp, and a valid bit. An outstanding speculative/nonspeculative store memory address together with the time stamp is compared associatively in the SMOB for a conflict. A conflict can occur only when an outstanding store address matches a load address in a more speculative thread in time in the SMOB. The main thread store operations check all load operations in the SMOB. On the other hand, a speculative thread checks only the descendant speculative threads (i.e. by comparing time stamps) for a conflict. If there is a conflict, the speculative thread and all descendant speculative threads are squashed. The SMOB entry and the other entries with the same time stamp or larger are invalidated in the SMOB. The main thread re-executes all instructions in the speculative thread where a thread misspeculation has occurred.

Speculative store values are kept in the *speculative store buffer* (SSB) shared by all speculative threads. The main thread updates the data cache as soon as store operations are executed. Speculative threads write store values in the SSB until commit time. Each SSB entry contains a memory address, store value, a next link pointer, store write time and a valid bit. Each thread has a head pointer and a tail pointer to the SSB that denote the first and last store operations from that thread. Those pointers are saved in a table called the *pointer box*. The time stamp (i.e. thread id) of a store operation decides how to access the correct pointer set (i.e. the head and tail pointers). When a speculative store operation executes, a search is made to find an available entry in the SSB. If one exists, the entry is allocated and the memory address and store value are all written into the entry. The next link pointer of the previous store in the same thread pointed by the tail pointer is set to point to the current SSB entry. A speculative load operation in a thread can read the most recent store by reading the value with the latest store write time. When a store is written into the table, the store write time value is obtained from a counter that is incremented by one every time a new store is written. The counter is reset when there is no active speculative thread left. Speculative stores are written into the data cache, in order, as soon as the speculative thread is verified as a correct speculation since stores in a thread are executed in order. Those entries are removed at the time of a commit or a squash from the SSB by visiting all links starting from the thread's head pointer to the tail pointer. If the SSB becomes full, all speculative threads stall. The main thread continues and frees the SSB entries by committing threads.

3. Bork Insertion Algorithm

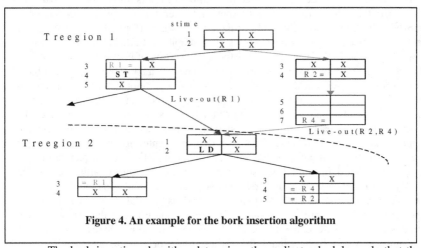

Figure 4. An example for the bork insertion algorithm

The bork insertion algorithm determines the earliest schedule cycle that the compiler can schedule a bork operation in the code and spawn a new thread. The bork insertion algorithm was implemented within the LEGO experimental compiler[10], which uses treegion scheduling [11] as a global, acyclic scheduling algorithm. Each node of a treegion is a basic block. Treegions are single entry, multiple exit regions and can be formed with or without profile information. The scheduler is capable of code motion and instruction speculation above dependent branches. Load and store operations are not allowed to be speculated during scheduling, but other operations can be hoisted up to any basic block within the treegion. Once treegions are formed, the scheduler schedules each treegion separately. After the program is scheduled, register allocation is performed and physical registers are assigned. Then, borks are inserted into the scheduled and register allocated code. Borks spawn speculative treegions and are inserted as early as possible in the code in an attempt to fully overlap treegions. True data dependencies between two treegions are considered when inserting borks. There is only one bork per path allowed in each treegion, enforcing the rule that each thread may spawn at most one speculative thread.

The algorithm scans through all treegions in the program. To consider the true data dependencies, it takes a treegion (*Tmain*) and computes the live-out set of operands from *Tmain* to each succeeding treegion, *Ts*. For each path in *Tmain*, if there are any register definitions of each live-out operand, the location and schedule time of those operands in that path are found. The completion times of all live-out operand definitions for the path are computed. The maximum completion time among all dependencies is found by taking the maximum of all completion times in the path. The earliest time to schedule a bork is determined by the maximum completion time. A schedule hole is sought to insert a bork into a cycle between the earliest cycle time and the last schedule time of the selected path in *Tmain*. The bork is inserted into the path in *Tmain*. The algorithm tries to insert a bork for every possible path in *Tmain* to *Ts*. When there is no path left, the next succeeding treegion is processed. This continues until all treegions in the source file are visited. More than one bork can appear on a path since a bork is inserted for each path in a treegion after all paths are visited. Elimination is needed to reduce the number of borks to one for each path. To accomplish this, the earliest bork is kept and later bork(s) in the schedule are removed. The elimination phase of redundant borks is performed after insertion of borks for each treegion. Figure 4 shows an example of bork insertion. The figure shows the schedule of a piece of code before the insertion of borks. Treegion 1 enters into Treegion 2 with two exits. Each rectangle represents a basic block. Within a basic block, a row represents a MultiOp that contains two operations. Only the operations that are relevant are shown in the figures. *X* denotes an operation not under consideration and empty slots are denoted by NOPs. Also, *stime* represents the scheduling time for each MultiOp. All operations are assumed to take one cycle and all functional units are universal. There are two paths entering into Treegion 2 from Treegion 1. On the left path, there is only one live-out (R1). The completion time of the operation that defines R1 is the sum of its stime and the operation latency, which is 4. Cycle 4 is the earliest cycle a possible bork can be scheduled on the left path. An available slot is found at cycle 4 in slot 2. A bork is scheduled in that slot. On the right path, R2 and R4 are the live-outs. The completion times of the operations that define R2 and R4 are computed similarly, which are 5 and 8 respectively. The maximum of 5 and 8 determines the earliest time for a bork, which is cycle 8. However, there is no cycle 8 on the right path. So, no bork is scheduled on this path.

4. Performance Evaluation

A trace-driven simulator was written to simulate multiple threads running at the same time. 2-way set associative 32KB L1 instruction and 32KB data caches, 512KB 2 way set associative L2 instruction and data cache, 16KB shared PAS branch predictor, 128-entry SMOB and 64-entry SSB, 1-cycle L1 cache hit time, 10-cycle L2 hit time, 30-cycle L2 miss time, 5-cycle thread squash penalty time are assumed in the simulations. ALU, BR, BORK, ST and floating-point ADD take one cycle. LD takes two cycles and finally floating-point multiplication and division take three cycles. The machine model used for the experiments is a 6-wide VLIW processor with 2 universal and 4 ALU/BR units, 128 integer and 128 floating-point registers. Universal units can execute any type of instructions and ALU/BR units can execute only ALU and branch instructions. The SPECint95 benchmark suite is used for all runs. 100 million instructions were executed using the training inputs. Multiple thread runs are compared to the baseline model with a single thread run of the same benchmark program. The same compiler optimizations are applied to produce codes for single thread and multiple thread models in the experiments.

Figure 5 shows speedup results for Weld models consisting of one to six threads. As shown in the graph, an average speedup of 23% is attained with two threads over all benchmarks. With three threads, it is 26%. After three threads, the speedup stabilizes at 27%. There is little change in speedup as the number of threads increases beyond three threads. This is because the number of penalty cycles increases with the number of threads. As the number of threads increases, the SMOB and SSB fill up more quickly and stall the speculative threads. Also, the chances for SMOB conflicts increase as the architecture keeps more active threads. This effect can be observed in 129.compress, 130.li, 147.vortex, 126.gcc and 124.m88ksim. In 132.ijpeg, the two-thread model gives the best performance. As the number of threads increases beyond two threads, the stalls due to SMOB conflicts take over and cancel out the benefit gained from multithreading. The same effect can be observed in 134.perl where the six-thread model is worse than the four- and five-thread models. This is because the number of SMOB conflicts in the four- and five-thread models is much less than the number of conflicts in the six-thread model. Therefore, the five-thread model can go ahead in time and finish earlier.

Figure 6 depicts speedup results of a model that uses the operation welder versus one that does not. Not using the welder implies that a speculative thread can only issue when the parent stalls. With two threads, the operation welder gives a 5% speedup over all the benchmarks. With three threads and beyond, the speedup is 8.5%. An increase in the number of threads increases the chances of welding operations from many speculative threads. However, in 132.ijpeg, the opposite effect can be observed because the number of SMOB stalls beyond two threads is so high that the benefit from welding is hidden behind the SMOB squashes.

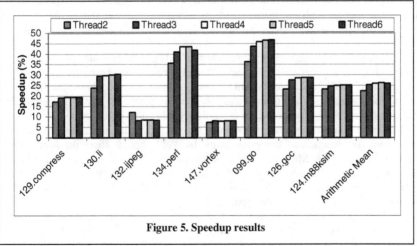

Figure 5. Speedup results

As seen from the experimental results, the models with three or more threads perform equally well. On the other hand, the two-thread model has performance of

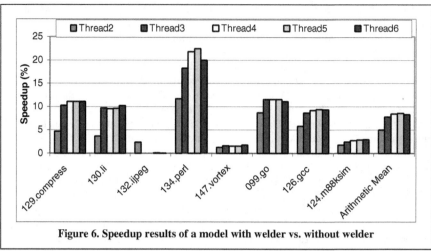

Figure 6. Speedup results of a model with welder vs. without welder

only 3-4% less in speedup than the other thread models, but with a much simpler register file organization, thread synchronization mechanism and less complicated SMOB and SSB.

5. Related Work

SPSM (Single-program Speculative Multithreading) [1] speculatively spawns multiple paths in a single program and simultaneously executes those paths or threads on a superscalar core. In SPSM there is a main thread that can spawn many speculative threads, whereas speculative threads can also spawn speculative threads in Weld. When the main thread merges with a speculative thread, the speculative thread's state merges with the main thread. At this point, the speculative thread dies and the main thread continues. SPSM is, however, for dynamic (superscalar) architectures.

Dynamic Multithreading Processors (DMT) [2] provide simultaneous multithreading on a superscalar core with threads created by the hardware from a single program. Each thread has its own program counter, rename tables, trace buffer, and load and store queues. All threads share the same register file, Icache, Dcache and branch predictor. DMT is proposed for dynamically scheduled processors.

MultiScalar Processors [3] consist of several processing units that have their own register file, Icache and functional units. In Weld, threads share the functional units and caches. Each MultiScalar processing unit is assigned a task, which is a contiguous region of the dynamic instruction sequence. Tasks are created statically by partitioning the control flow of the program. As in SPSM and DMT, MultiScalar is proposed for dynamically scheduled processors.

TME (Threaded Multiple Path Execution) [4] executes multiple alternate paths on a Simultaneous Multithreading (SMT) [5] superscalar processor. It uses free hardware contexts to assign paths of conditional branches. Speculative loads are allowed. In contrast, threads are created at compile time in Weld.

Prasadh [6] et al. proposed a multithreading technique in a statically scheduled RISC processor. Statically scheduled VLIWs from different threads are interleaved dynamically to utilize NOPs. If a NOP is encountered in a VLIW at run time, it is filled with an operation from another thread through a dynamic interleaver. The dynamic interleaver does not interleave instructions across all issue slots and therefore there is one-to-one mapping between functional units. Weld has neither of these limitations. Also, threads are the different benchmarks or programs in their experiments unlike multithreading from a single program in Weld.

In Processor Coupling [7][8], several threads are scheduled statically and interleaved into clusters at run time. A cluster consists of a set of functional units that share a register file. Operations from different threads compete for a functional unit within a cluster. Interleaving does not occur across all issue slots. The compiler inserts explicit *fork* and *forall* operations to partition code into several parallel threads. On the other hand, Weld allows operation migration at run time and speculative threads to be spawned.

XIMD [9] is a VLIW-like architecture that has multiple functional units and a large global register file similar to VLIW. Each functional unit has an instruction sequencer to fetch instructions. A program is partitioned into several threads by the compiler or a partitioning tool. The XIMD compiler takes each thread and schedules it separately. Those separately scheduled threads are merged statically to decrease static code density or to optimize for execution time. However, Weld merges threads at run time by taking advantage of dynamic events.

6. Conclusion and Future Work

In this paper, a new architecture model called *Weld* is proposed for VLIW processors. Weld exploits the ISA, compiler, and hardware to provide multithreading support. The compiler, through the insertion of bork instructions, creates multiple threads from a single program, which are acyclic regions of the control graph. At run time, threads are welded to fill in the holes by special hardware called the *operation welder*. The experimental results show that a maximum of 27% speedup using Weld is possible as compared to a single-threaded VLIW processor.

We will focus on the dual-thread Weld model for further research. The issues such as the sizes of SMOB and SSB on performance, variable memory latencies, and higher branch penalty in deeper pipelines will be studied for the dual-thread model. We are also working on a compiler model for Weld without the SMOB in order to avoid squashes due to load/store conflicts at the thread level.

Acknowledgements

This research was supported by generous hardware and cash donations from Sun Microsystems and Intel Corporation.

References

[1]   P. K. Dubey, K. O'Brien, K. M. O'Brien and C. Barton, "Single-Program Speculative Multithreading (SPSM) Architecture: Compiler-Assisted Fine-Grained Multithreading," in *Proc. Int'l Conf. Parallel Architecture and Compilation Techniques.* (Cyprus). June 1995.

[2]   H. Akkary and M. A. Driscoll, "A Dynamic Multithreading Processor," in *Proc. 31st Ann. Int'l Symp. Microarchitecture,* Nov. 1998.

[3]   G. S. Sohi, S. E. Breach and T. N. Vijaykumar, " Multiscalar Processors," in *Proc.22nd Ann. Int'l Symp. Computer Architecture.* (Italy). May 1995.

[4]   S. Wallace, B. Calder and D. M. Tullsen, "Threaded Multiple Path Execution," in *Proc. 25th Ann. Int'l Symp. Computer Architecture,* Barcelona, Spain, June 1998.

[5]   D. M. Tullsen, S. J. Eggers and H. M. Levy, "Simultaneous Multithreading: Maximizing On-chip Parallelism," in *Proc. 22nd Ann. Int'l Symp. Computer Architecture,* Italy, May 1995.

[6]   G. Prasadh and C. Wu, "A Benchmark Evaluation of a Multithreaded RISC Processor Architecture," in *Proc. of Int'l Conf. on Parallel Processing,* Aug. 1991.

[7]   S. W. Keckler and W. J. Dally, "Processor Coupling: Integrating Compile Time and Runtime Scheduling for Parallelism," in *Proc. 19th Ann. Int'l Symp. Computer Architecture,* Australia, May 1992.

[8]   M. Fillo, S. W. Keckler, W. J. Dally, N.P. Carter, A. Chang, Y. Gurevich and W.S. Lee, "The M-Machine Multicomputer," in *Proc.28th Ann. Int'l Symp. Microarchitecture,* Ann Arbor, MI, Dec. 1995.

[9]   A. Wolfe and J.P. Shen," A Variable Instruction Stream Extension to the VLIW Architecture," in *Proc. 4th Int'l Conf. on Architectural Support for Programming Languages and Operating Systems,* ACM Press, Apr. 1991.

[10] W. A. Havanki, "Treegion Scheduling for VLIW Processors", *Master's Thesis,* Dept. of Electrical and Computer Engineering, North Carolina State University, Raleigh, NC 27695-7911, July 1997.

[11] W. A. Havanki, S. Banerjia and T. M. Conte, "Treegion Scheduling for Wide-issue Processors", in *Proc. 4th Int'l Symp. High Performance Computer Architecture,* Las Vegas NV, Feb. 1998.

[12] B. R. Rau, "Dynamically Scheduled VLIW Processors," *Proc. 26th Ann. Int'l Symp. Microarchitecture,* Dec 1993.

[13] J. E. Smith and A. R. Pleszkun, "Implementing Precise Interrupts in Pipelined Processors", in *IEEE Trans. on Computers,* Vol. 37, NO. 5, May 1988.

[14] M. Franklin and G. S. Sohi, "The Expandable Split Window Paradigm for Exploiting Fine-grain Parallelism", *Proc. 19th Ann. Int'l Symp. Computer Architecture,* Gold Coast, Australia, May 1992.

[15] M. Franklin and G. S. Sohi, "ARB: A Hardware Mechanism for Dynamic Reordering of Memory References", in *IEEE Trans. on Computers,* May 1996.

Putting Data Value Predictors to Work in Fine-Grain Parallel Processors

Aneesh Aggarwal[1] and Manoj Franklin[2]

[1] Department of Electrical and Computer Engineering
[2] Department of Electrical and Computer Engineering and UMIACS
University of Maryland, College Park, MD 20742
{aneesh, manoj}@eng.umd.edu

Abstract. Recent work has shown that the hurdles imposed by data dependences on parallelism can be overcome to some extent with the use of data value prediction. This paper highlights how data value history is affected when implementing data value predictors in fine-grained parallel processors, wherein microarchitectural issues affect the recorded history. Simulation studies show that mispredictions increase and correct predictions decrease when the recorded history is not updated properly. The paper also investigates techniques for overcoming the effects of value history disruption. The investigated techniques rely on extrapolation of outdated history so as to make it up-to-date, and utilization of misprediction information to turn off predictions of subsequent instances of the mispredicted instruction. We evaluate the proposed techniques using a cycle-accurate simulator for a superscalar processor. Results from this study indicate that the extrapolation technique is indeed able to provide up-to-date history in most of the cases, and is able to recoup most of the ground lost due to microarchitectural effects. Utilization of misprediction information helps to further reduce the number of mispredictions, although in some cases it reduces the number of correct predictions also.

1 Introduction

A major hurdle to exploiting instruction-level parallelism is the presence of data dependences, which delays the execution of instructions. Recent studies [1] [2] [3] [7] [6] [8] [9] [12] have shown that it is possible to overcome this hurdle to some extent with the use of *data value prediction*. That is, the result of an instruction is predicted based on the past behavior of previous instances of the instruction, and passed on to subsequent instructions that use the result. Later, when the actual operands of the instruction become available, the instruction is executed, and the correct result is compared with the earlier prediction. If the values match, then all of the instructions in the active window that used the predicted value are informed of the match. If the values do not match, then the correct result is forwarded to the instructions that require this value, and those instructions are re-executed.

Although the concepts behind data value prediction are similar to those behind control value prediction, there are some fundamental differences between

B. Monien, V.K. Prasanna, S. Vajapeyam (Eds.): HiPC 2001, LNCS 2228, pp. 204–213, 2001.
© Springer-Verlag Berlin Heidelberg 2001

the two concerning implementation in a real processor. In both cases, a prediction may not be validated before making several additional predictions. If the subsequent predictions are performed before doing the history update, then the prediction accuracy is likely to be poor (because of using outdated history). This problem can be overcome in control flow speculation by speculatively updating the history [5]. Whenever a control misprediction is detected, the processor squashes all of the incorrect instructions, including the incorrect branches. The speculative updates done for those branches are setback. In the case of data value prediction, there are two major differences:

- When a misprediction is detected, subsequent instructions need not be squashed; instead those that used the erroneous value can be re-executed. Thus, the instructions that used erroneous speculative history for their prediction continue to hold on to their earlier prediction, and do not get a fresh chance to predict.
- Data value predictors are usually equipped with a confidence estimator to reduce the number of mispredictions [2] [12]. When an instruction' s result is not predicted because of poor confidence, it becomes difficult to update the history speculatively.

To exploit parallelism, modern processors typically perform speculative execution with a large instruction window. Such a processor will often fetch several instances of an instruction before executing any of them. Because data value prediction is done in the fetch part of the pipeline, the results of several instances of an instruction may be predicted before any of them have been validated. Moreover, if the history is updated only after validation, then many predictions will have to be based on obsolete history.

This paper investigates the topic of adapting a data value predictor for application in a real processor. It analyzes different scenarios that lead to incorrect history, and investigates techniques to perform accurate data value prediction in processors. These techniques deal with reconstructing the correct value history pattern (using "extrapolation" techniques), and utilization of misprediction information to turn off predictions of subsequent instances of a mispredicted instruction. These schemes are shown to offset most of the microarchitectural effects on data value history.

The rest of this paper is organized as follows. Section 2 illustrates how value history is impacted when multiple instances of an instruction are present in the active window. Section 3 evaluates existing data value predictors in a dynamically scheduled superscalar processor. Section 4 discusses techniques to alleviate the microarchitectural impact on value history. Section 5 presents an experimental evaluation of the proposed schemes. Section 6 presents the conclusions.

2   Impact of Multiple Instances on Value Prediction

2.1   Motivation

Many of the earlier studies on data value prediction describe and evaluate the prediction schemes in the context of an ISA-level simulator. A hallmark of such evaluations is that each value prediction is validated immediately after performing the prediction. The history recorded by the predictor is also immediately updated with the correct outcome, before performing further predictions. Thus, the history information recorded by the data value predictor depends only on the program being executed and the set of inputs with which the program is being executed (barring the effects of exceptional events such as context switches). In other words, microarchitectural issues do not play any role in deciding the history recorded in the data value predictor. The location of a dynamic instruction within the dynamic program order and the set of inputs with which the program was being run, completely determined the instruction's context.

To implement data value prediction in a real processor, several additional issues need to be resolved. This is because *the microarchitecture plays a major role in determining the history recorded in the data value predictor*. For example, the results of several instances of an instruction may get predicted before any of these instances gets executed and validated. This results in the use of outdated history for performing predictions. A similar effect is shown in [1] for load address predictions, and in [4] for branch predictions in a multithreaded processor. In particular, microarchitectural impact can result in the following 4 types of incorrect data value history (or their combinations) to be recorded:

- **Discontinuous history:** the recorded history doesn't include some updates.
- **Outdated history:** the recorded history doesn't contain the latest updates.
- **Scrambled history:** the recorded history is updated in an incorrect order.
- **Inaccurate history:** the recorded history contains erroneous updates.

When the processor microarchitecture influences the history recorded in the data value predictor, it is not obvious if the recorded history is capable of predicting future behavior accurately, because several uncertainties are introduced. Below, we discuss how specific microarchitectural issues impact the nature of history available to a data value predictor.

2.2   Microarchitectural Issues Impacting Data Value History

When multiple instances of an instruction are simultaneously active in a processor, it is likely that many of the younger instances among them were fetched before the older ones are executed and validated. The history used for predicting the younger instances may, therefore, not be up-to-date. The extent to which a recorded history is obsolete depends on the specifics of the microarchitecture.

Value History Update Time. An issue that has a bearing on value history is the time at which the recorded history is updated with an instruction outcome. We can think of at least three different times for updating the history:

- instruction fetch time (value prediction time)
- instruction validation time
- instruction commit time

If the update is performed at prediction time, as discussed for branch prediction [5], then the update is speculative in nature. A motivation for performing such an update is to provide the latest value history for subsequent predictions. It is important to notice the difference between branch prediction and data value prediction, in this respect. In the case of branch prediction, when the processor detects an incorrect prediction, it recovers by squashing all subsequent instructions. This results in undoing the speculative updates corresponding to the squashed branches, thereby restoring a "clean slate" that is devoid of any debilitative effects of the wrong predictions. Thus, in the branch prediction case, the effect of a misprediction does not linger in the recorded history, and branches belonging to the correct execution path always see up-to-date and correct history when making predictions. By contrast, in the case of data value prediction, recovery actions arising from mispredictions can be done just by re-executing the affected instructions (without squashing them). Herein lies the big difference! Subsequent instructions that have already used the speculatively updated *incorrect history* for their predictions are very likely to have been mispredicted, but are neither squashed nor re-predicted. Instead, their misprediction is detected only when they are executed with correct operands.

The second option is to perform updates at instruction validation time. A motivation for this is that the update is always correct. However, in processors exploiting parallelism, instructions are very likely to be validated out-of-order, and so updation at validation time will lead to *scrambled history* being recorded in the predictor, which is likely to affect the prediction accuracy.

The third option is to perform the update at instruction commit time. A motivation for this is that the updates correspond only to the correct predictions, and are also done in the correct order. However, as far as predictions are concerned, the available history is *outdated*.

It is important to note that the data value prediction schemes proposed so far in the literature record history on a per-address basis. The recorded history of a per-address predictor is affected only when multiple instances of the same instruction are present in the active window.

Confidence Estimation. If a prediction is supplied for every prediction-eligible instruction, irrespective of the confidence in the prediction, then the number of mispredictions can be quite high (30-50%). Because a misprediction can be costly, it is highly desirable to use confidence mechanisms to pare down the number of mispredictions. When a confidence mechanism is used, most of the mispredictions and a few correct predictions end up being not predicted. Thus, about 20-40% of the prediction-eligible instructions are not predicted. When a prediction is not supplied for a prediction-eligible instruction, it becomes difficult to update the recorded history speculatively. If the updates corresponding to these low-confidence instances are omitted from the speculative updates, then the recorded history becomes *discontinuous*. If, on the other hand, an update is performed with the low-confidence predicted value, then many of those updates will be wrong, resulting in *inaccurate history* to be recorded.

3   Evaluation of Existing Data Value Predictors

The previous section identified microarchitectural issues that affect data value history and data value prediction in a fine-grain parallel processor. In this section, we evaluate how these issues impact the performance of existing predictors.

3.1   Experimental Setup

The experiments are conducted using a cycle-accurate simulator that accepts programs compiled for a MIPS-based DECstation and simulates their execution, keeping track of relevant information on a cycle-by-cycle basis. The simulated processor is a 5-stage out-of-order execution pipeline. Value predictions are done in the fetch stage of the pipeline. Prediction validations are done after executing the instruction and confirming the correctness of the operands.

For benchmarks, we use integer programs from the SPEC '95 benchmark suite. All programs were compiled using the MIPS C compiler with the optimization flags distributed in the SPEC benchmark makefiles. The metrics we use are: (i) *Percentage of instructions correctly predicted*, (ii) *Percentage of instructions mispredicted*, and (iii) *Percentage of instructions not predicted*. All metrics are expressed as percentage of the total number of single register result-producing instructions.

Default Parameters for the Study: The benchmarks are simulated up to 100 million instructions. The default instruction window size is 64. For the data value predictor, we use a hybrid of stride and 2-level predictors [12]. The stride predictor adds the current stride to the last value to get the predicted value. The Value History Table (VHT) of the predictor has 8K entries, and is 2-way set-associative. The default pattern size for the 2-level predictor part is 8, and the number of values stored per VHT entry is 4. Thus, the PHT has 64K entries. The PHT counters saturate at 11. The threshold counter value used for confidence estimation is 6. When updating a PHT entry, the counter corresponding to the correct result is incremented by 3, and the others are decremented by 1.

3.2   Multiple Instances in the Active Window

As the window size increases, the number of instances of the same static instruction in the window is likely to increase. This means that more and more instructions see an out-of-date history, leading to more mispredictions and fewer correct predictions. Figure 1 gives the average distribution of the number of instances of the same static instruction present in the window, for 3 different window sizes. For each benchmark, the figure presents three bars, for window sizes of 64, 128, and 256, respectively. Each bar can have up to 5 parts: the bottom-most part denotes instructions with a single instance in the window, the one above it depicts instructions with 2 instances, and so on. As can be seen in the figure, as the window size increases the number of multiple instances increases, and the percentage of instructions that have just one instance in the window decreases.

This means that, as the window size is increased, the microarchitectural impact on value history is likely to increase.

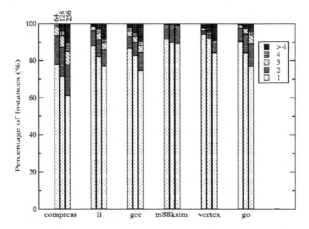

Fig. 1. Percentage distribution of instances of the same static instruction in a super-scalar processor

3.3   Impact of Outdated History on Prediction Accuracy

Next we present the impact on prediction accuracy due to outdated history. Figure 2 presents these results. In the figure, two bars are presented for each benchmark, corresponding to perfect history (PH) and commit-time updated history (CUH). Each bar has three parts: the lowest part represents the percentage of incorrect predictions, the middle part represents the percentage of non-predictions, and the top part represents the percentage of correct predictions. The results show that `compress95` and `li` suffer the maximum increase in misprediction percentage—9.6% and 9%, respectively. The average increase in percentage mispredictions, when going from PH to CUH, is 4.5%. The average drop in percentage of correct predictions is 5.5%. Thus, *outdated history* caused by commit-time history update, has a double-barreled impact of higher mispredictions and lower correct predictions.

3.4   Impact of Prediction-to-Validation Latency on Prediction Accuracy

Pipeline depths have been steadily increasing over the years. But this may have a degrading impact on the performance of value predictors. Because prediction and validation are typically done at the opposite ends of the pipeline, deeper pipelines increase the prediction-to-validation latency. As this latency increases, the value history becomes more out-of-date. We experimented with three different prediction-to-validation latencies: default pipeline, default + 4 cycles, and default + 8 cycles. These experiments showed that the number of correct predictions decreases and the number of mispredictions increases by a small amount when the latency between predictions and validations increases.

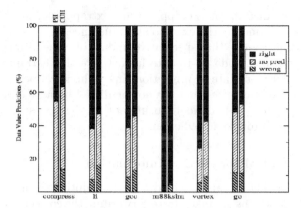

Fig. 2. Data Value Prediction Statistics for a Hybrid Predictor with Perfect History (PH) and with Commit-time Updated History (CUH)

4   Techniques to Reduce Microarchitectural Impact

The previous sections highlighted the impact of microarchitectural features on data value history and the accuracy of data value prediction. This section focuses on reducing this impact. This discussion also focuses on a hybrid of stride and 2-level predictors.

4.1   Extrapolation of Outdated History

The extrapolation technique attempts to reconstruct the value history that would be available, had the processor been updating history in program order and in a timely manner. In this scheme, the value history is updated only at instruction commit time. However, a count (called *extrapolation distance*) is kept for each instruction currently mapped to the data value predictor. This count keeps track of the number of instances of the corresponding instruction in the instruction window. When a prediction-eligible instruction is fetched, the history corresponding to the current extrapolation distance is used to make the prediction. The extrapolation technique is particularly useful for predicting the instances of instructions that belong to a loop.

For the stride predictor, extrapolation can be easily achieved by multiplying the recorded stride by the extrapolation distance and adding this product to the last committed result for that static instruction, similar to what is proposed in [1]. The implementation of extrapolation is a little bit more complex for the 2-level predictor. One option is to keep multiple PHTs (Pattern History Tables) as done for multiple-branch prediction in [10], one for each possible extrapolation distance. Whenever a predictable instruction is committed, the VHT (Value History Table) is updated as before. In addition, each PHT is updated, taking into consideration the extrapolation distance it models. Another option is to have a single PHT, which records history based on zero extrapolation distance, and determine the prediction for the appropriate extrapolation distance, by using the PHT as many times as the extrapolation distance.

It is important to note that speculative update of value history at prediction time also strives to provide up-to-date history like this extrapolation scheme. However, speculative update at prediction time is handicapped by the presence of many instruction instances that are not predicted because of low confidence. As mentioned earlier, it is difficult to update the history speculatively if the predictor does not provide a prediction.

4.2   Feedback of Misprediction Information

Data value mispredictions are detected while attempting to validate instruction results. Recovery actions are taken by re-forwarding the correct result to succeeding data-dependent instructions. The occurrence of a misprediction after a series of correct predictions for an instruction is due to a deviation from the prevailing steady-state condition. In all likelihood, such a misprediction is followed by a series of mispredictions for subsequent instances of the same static instruction. However, in the schemes discussed so far, this deviation from steady-state behavior is not reflected in the recorded history until the first mispredicted instance among them is committed. Although the extrapolation technique attempts to make the outdated history up-to-date, the extrapolation part is not considering such deviations from steady-state behavior. This is because extrapolation works on the premise that the prevailing state-of-affairs will continue to remain so.

One way to deal with this problem is to keep a `Misprediction Bit` for each entry in the predictor table, and set this bit to 1 when a misprediction has occurred for a particular instruction. As long as this bit is set, no further predictions are made for subsequent instances of that instruction. Once the history (recorded at commit time) reflects a return to a steady state, this bit can be reset, allowing subsequent instances of this instruction to be predicted. If the mispredictions of a static instruction occur in bunches, then this technique will help reduce the number of mispredictions by converting them to non-predicted instructions. If the mispredictions do not appear in bunches, but are scattered throughout the program, then this technique will hurt performance by reducing the number of correct predictions!

5   Experimental Evaluation

This section presents an experimental evaluation of the two techniques discussed in Section 4 for overcoming the impact of incorrect updates and out-of-order updates to value history. These experiments are conducted using the same simulator and setup that were used to obtain the results in Section 3. Figure 3 presents these results. There are 5 histogram bars for each benchmark, corresponding to commit-time updated history (CUH), extrapolation of commit-time updated history with single PHT (E-CUH-SP), extrapolation of commit-time updated history with multiple PHTs (E-CUH-MP), use of misprediction bit (E-CUH-MP-MB), and perfect history (PH).

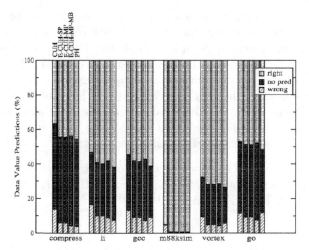

Fig. 3. Data Value Prediction Statistics for Hybrid Predictor when implementing Different History Update Schemes

For the hybrid predictor, extrapolation with a single PHT and with multiple PHTs give roughly the same performance. Both schemes are able to provide substantial improvement over commit-time updated history (CUH). In particular, the percentage of mispredictions decreases, and the percentage of correct predictions increases for all the benchmark programs. For **go**, the number of mispredictions with these schemes is lower than that with perfect history. This surprising result confirms that for some instructions, the history of previous instances alone is not a good indicator of future behavior [11]. The utilization of the misprediction bit (E-CUH-MP-MB scheme) reduces the percentage of mispredictions considerably, but at the expense of a slight reduction in the percentage of correct predictions.

6   Summary and Conclusions

We have studied the performance of data value prediction in a speculative processor. We observed that the accuracy of the recorded value history is dependent on when the history is updated with instruction results. Simulation-based experimental studies showed that this impact on history increases the number of mispredictions and decreases the number of correct predictions. This means that microarchitectural considerations such as history update time has to be taken into consideration while designing data value predictors for real processors.

In order to minimize the affects of mispredicted instructions and non-predicted instructions, we perform history updates at instruction commit time (which results in outdated history). We explored an extrapolation technique to reduce the impact of outdated history. This technique starts with the outdated, but correct, history and extrapolates the history by an appropriate amount before making any prediction. We also explored the potential of utilizing misprediction information to improve the confidence estimation mechanism.

Simulation-based experimental studies showed that the extrapolation technique is able to recoup most of the ground lost due to microarchitectural effects. Utilization of misprediction information helps to further reduce the number of mispredictions; however, in some cases it decreases the number of correct predictions. In the future, as the effective window size increases, the microarchitectural issues impacting data value prediction will become more and more important, necessitating the use of techniques such as the ones presented in this paper.

Acknowledgements. This work was supported by the US National Science Foundation (NSF) through a CAREER Award (MIP 9702569). We thank the reviewers for their helpful comments.

References

1. M. Bekerman, S. Jourdan, R. Ronen, G. Kirshenboim, L. Rappoport, A. Yoaz, and U. Weiser, "Correlated Load-Address Predictors," *Proc. 26th International Symposium on Computer Architecture*, 1999.
2. B. Calder, G. Reinman, and D. M. Tullsen, "Selective Value Prediction," *Proc. 26th International Symposium on Computer Architecture*, 1999.
3. F. Gabbay and A. Mendelson, "Using Value Prediction to Increase the Power of Speculative Execution Hardware," *ACM Transactions on Computer Systems*, Vol. 16, No. 3, pp. 234-270, August 1998.
4. J. Gummaraju and M. Franklin, "Branch Prediction in Multi-Threaded Processors," *Proc. International Conference on Parallel Architectures and Compilation Techniques (PACT)*, 2000.
5. E. Hao, P-Y. Chang, and Y. N. Patt, "The Effect of Speculatively Updating Branch History on Branch Prediction Accuracy, Revisited," *Proc. 27th International Symposium on Microarchitecture (MICRO-27)*, 1994.
6. M. H. Lipasti and J. P. Shen, "Exceeding the Dataflow Limit via Value Prediction," *Proceedings of the 29th International Symposium on Microarchitecture (MICRO-29)*, pp. 226-237, 1996.
7. J. González and A. González, "The Potential of Data Value Speculation to Boost ILP," *Proc. ACM International Conference on Supercomputing*, 1998.
8. T. Nakra, R. Gupta, and M. L. Soffa, "Global Context-based Value Prediction," *Proc. 5th International Symposium on High Performance Computer Architecture (HPCA-5)*, pp. 4-12, 1999.
9. Y. Sazeides and J. E. Smith, "Modeling Program Predictability," *Proc. 25th International Symposium on Computer Architecture (ISCA)*, 1998.
10. A. Seznec, S. Jourdan, P. Sainrat, P. Michaud, "Multiple-Block Ahead Branch Predictors," *Proc. 7th International Conference on Architectural Support for Programming Languages and Operating Systems (ASPLOS VII)*, 1996.
11. R. Thomas and M. Franklin, "Using Dataflow Based Context for Accurate Value Prediction," *Proc. International Conference on Parallel Architectures and Compilation Techniques (PACT)*, 2001.
12. K. Wang and M. Franklin, "Highly Accurate Data Value Prediction using Hybrid Predictors," *Proceedings of the 30th International Symposium on Microarchitecture (MICRO-30)*, pp. 281-290, 1997.

Confidence Estimation for Branch Prediction Reversal

Juan L. Aragón[1], José González[1], José M. García[1] and Antonio González[2]

[1]Dpto. Ingeniería y Tecnología de Computadores
Universidad de Murcia, 30071 Murcia (Spain)
{jlaragon,joseg,jmgarcia}@ditec.um.es
[2]Departament d'Arquitectura de Computadors
Universitat Politècnica de Catalunya, 08034 Barcelona (Spain)
antonio@ac.upc.es

Abstract. Branch prediction reversal has been proved to be an effective alternative approach to dropping misprediction rates by means of adding a Confidence Estimator to a correlating branch predictor. This paper presents a *Branch Prediction Reversal Unit* (*BPRU*) especially oriented to enhance correlating branch predictors, such as the *gshare* and the Alpha 21264 metapredictor. The novelty of this proposal lies on the inclusion of data values in the confidence estimation process. Confidence metrics show that the *BPRU* can correctly tag 43% of branch mispredictions as low confident predictions, whereas the *SBI* (a previously proposed estimator) just detects 26%. Using the *BPRU* to reverse the *gshare* branch predictions leads to misprediction reductions of 15% for the SPECint2000 (up to 27% for some applications). Furthermore, the *BPRU+gshare* predictor reduces the misprediction rate of the *SBI+gshare* by an average factor of 10%. Performance evaluation of the *BPRU* in a superscalar processor obtains speedups of up to 9%. Similar results are obtained when the *BPRU* is combined with the Alpha 21264 branch predictor.

1 Introduction

One of the main characteristics of each new processor generation is an increase in the complexity and in the size of the hardware devoted to predict branches. Two main factors motivate the necessity of accurate branch predictors:

In order to exploit more ILP, processors inspect an increasing number of instructions in each cycle. In such processors, the fetch engine must provide instructions from the correct path at a high rate. To meet these requirements, complex instruction caches and powerful branch prediction engines are needed.

Reducing the clock cycle has been traditionally used to improve the performance of processors. On a given technology, fewer gates per pipeline stage result in higher frequencies, leading to longer pipelines. This is a challenge for superscalar processors since the branch misprediction penalty is increased.

Branch predictor accuracy can be improved by augmenting its size or complexity. However, traditional misprediction rate figures show that, reached a certain size, the misprediction rate is scarcely reduced. *Confidence Estimation* has become an interesting approach to increasing prediction accuracy by means of assessing the quality of

B. Monien, V.K. Prasanna, S. Vajapeyam (Eds.): HiPC 2001, LNCS 2228, pp. 214–223, 2001.
© Springer-Verlag Berlin Heidelberg 2001

branch predictions [8][10]. The confidence estimator generates a binary signal indicating whether the branch prediction can be considered high or low confidence.

This paper presents a confidence estimator used to reverse the low confidence branch predictions (*Branch Prediction Reversal Unit*). The novelty of this estimator is the inclusion of data values to assign confidence to branch predictions, attempting to use different information from that employed by the branch predictor (usually branch history and branch PC). This can be very efficient in different scenarios such as branches that close loops, list traversals or pathological *if-then-else* structures, where the involved branches may not be correlated with previous history but may be correlated with data values. The main contributions of this paper are:

> We propose a generic *BPRU* that can be used in conjunction with any branch predictor. The *gshare* [14] and the 21264 branch predictor [11] are used as test cases.
>
> The *BPRU* is compared with a previously proposed confidence estimator, the *SBI* [13]. We first use the metrics proposed in that work and show that including data values as additional information to assign confidence, leads to better confidence estimation. We also compare both estimators for branch prediction reversal.
>
> We evaluate the impact of adding the *BPRU* to traditional branch predictors in a superscalar processor, and compare it with the impact of adding the *SBI*.

The rest of this paper is organized as follows. Section 2 presents the motivation of our proposal and related work. The proposed *BPRU* is described in Section 3. Section 4 analyzes the accuracy of the *BPRU* in terms of misprediction rate, as well as its performance in terms of IPC. Finally, Section 5 summarizes the conclusions of this work.

2 Related Work and Background

Different branch prediction strategies have been presented to improve prediction accuracy. Most of previous proposals correlate the behavior of a branch with its own history [16][17], the history of previous branches [14], or the path [15]. These initial approaches were later extended with anti-aliasing techniques [3][12] and hybrid predictors [4][14]. Data values have been also incorporated in the branch prediction process. Heil *et al* [9] use the history of data values to access branch prediction tables. Value prediction was directly used to compute the branch outcome in [7], by predicting the inputs of branches and compare instructions and speculatively compute the branch outcome in some dedicated functional units.

Assigning confidence to branch predictions appeared as an orthogonal way of improving accuracy. Jacobsen *et al* [10] propose different schemes to calculate the likelihood of a branch misprediction. They suggest that confidence estimation can be applied to improve prediction accuracy by means of prediction reversal, although no particular implementations are presented. Grunwald *et al* [8] propose different confidence estimators and Manne *et al* [13] evaluate their usefulness to *Selective Branch Inversion*. In [1], we present a branch prediction confidence estimator especially tuned to improve the *Branch Predictor through Value Prediction* proposed in [7]. This confidence estimator relies on the fact that many value mispredictions are concentrated on a small range of data values. Detecting such critical values and reversing those branch predictions made according with them improves the *BPVP* accuracy.

The work we present here differs from previous proposals in the following: a) branch confidence estimators presented in [8][13] are based on branch history and branch PC, whereas our confidence estimator is also based on data values; and b) with respect to our initial proposal [1], we extend here its functionality to make it suitable for any branch predictor, instead of just focusing on improving the *BPVP*.

In order to compare different confidence estimators, Grunwald *et al* [8] defined four metrics: *Sensitivity* (SENS) represents the fraction of correct branch predictions identified as high confidence; *Predictive Value of a Positive Test* (PVP) identifies the probability that a high confidence estimation is correct; *Specificity* (SPEC) is the fraction of incorrect predictions identified as low confidence; and *Predictive Value of a Negative Test* (PVN) is the fraction of low confidence branches that are finally mispredicted. For the application of branch reversal, we are interested in the last two metrics: SPEC and PVN, obtaining a significant potential only if both metrics are high.

3 Branch Prediction Reversal Mechanism

3.1. Quantitative Analysis of the Branch Reversal Benefit

We have first performed an off-line analysis in order to gain some insight into the processor parameters that provide a better knowledge of branch mispredictions. We have examined the following parameters:

1. The predicted value of the branch input.
2. The predicted value of the branch input and the branch PC.
3. The predicted value of the branch input and the PC of the branch input producer.
4. The predicted value of the branch input, the PC of the branch input producer and the recent path followed to reach the branch.
5. The predicted value of the branch input, the PC of the branch input producer and the recent history of branch outcomes.

We have run some benchmarks from the SPECint2000 using a modified version of the *sim-safe* simulator [2]. The number of branch hits and misses of a baseline branch predictor have been measured for all the above combinations of processor state parameters, assuming unbounded storage resources. For those parameter instances whose number of branch mispredictions is greater than the number of hits, the prediction is reversed. Thus, a new misprediction rate is obtained, showing the potential of reversing the branch prediction when considering this *a priori* information.

Fig. 1 shows the new misprediction rate obtained for *bzip2*, *eon*, *mcf* and *twolf* for the five evaluated scenarios. We have used the *Stride Value Predictor* with an unrealistic size of 512 KB trying to isolate the potential of our proposal from the performance of the value predictor. The underlying branch predictor is a 32 KB *gshare*. It can be observed that using the predicted value of the branch input together with the PC of the branch input producer and the history of branch outcomes is the best approach. These results indicate the potential of branch reversal but are not an upper-bound since they have been obtained assuming that each instance of the chosen parameters can be reversed either always or never. In practice, the proposed mechanism will be able to dynamically reverse the prediction only for a certain part of the execution.

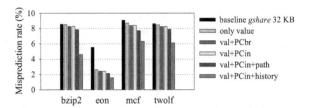

Fig. 1. Potential misprediction rate for a 32 KB *gshare*.

3.2. Branch Prediction Reversal Unit (*BPRU*)

This section presents the implementation of the *Branch Prediction Reversal Unit (BPRU)*. This unit could be included in any branch predictor, although in this work we have used the *BPRU* along with the *gshare* and the alpha 21264 branch predictors.

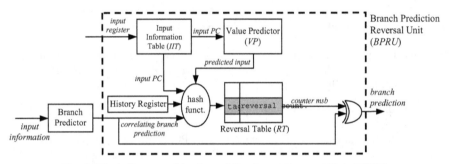

Fig. 2. Block diagram of the *Branch Prediction Reversal Unit (BPRU)*.

Fig. 2 depicts a diagram of the *BPRU*. It is composed of an *Input Information Table (IIT)*, a *Value Predictor (VP)*, a *History Register (HR)* and a *Reversal Table (RT)*. The *IIT* is a simplified version of the one proposed in [7][1]. The *Value Predictor* can be any known predictor, although in this work we have assumed the *Stride Value Predictor (STP)* because of its effectiveness [5][7]. Therefore, the *VP* provides either the predicted branch input value for load/arithmetic/logical instructions or the predicted difference between both inputs for compare instructions. The *History Register* collects the outcomes of recent executed branches. Finally, the confidence for each conditional branch is estimated by the *Reversal Table (RT)*.

Each entry of the *RT* stores an up-down saturating counter that determines whether the prediction must be reversed, and a tag. The *RT* is accessed when the branch is predicted, by hashing some processor state information. According to the analysis of the previous section, the most effective approach to reversing branch predictions is to

---

[1] The *IIT* has the same number of entries as total logical registers. Each *IIT* entry only has one field: the PC of the latest instruction that had this register as destination. During decoding, load/arithmetic/logical instructions store their PC in the corresponding entry of the *IIT*, whereas compare instructions first access the *IIT* to obtain the PC of the producers of both inputs, and then store a *xor*-hashed version of both PCs.

correlate with the predicted value, the PC of the branch input producer and the history of recent branch outcomes. In addition, the prediction bit of the baseline predictor is used to index the *RT*, since it is useful for reducing interference in the confidence estimator [7][13]. The most significant bit of the counter of the *RT* entry indicates whether the branch prediction must be reversed. Once the actual branch outcome is computed, the *RT* entry is updated, incrementing the counter if the prediction was incorrect, and decreasing the counter otherwise.

Conflicts in the *RT* are one of the major problems that may limit the accuracy of the *BPRU*. Experimental results showed that a tagged *RT* provides higher accuracy than a non-tagged *RT* of the same total storage, despite the space occupied by tags. Besides, the replacement policy of the *RT* has to be carefully selected. In our case, it gives priority to entries with lower values in their counters.

Since the *BPRU* has to perform three table accesses (*IIT*, *VP* and *RT*) to provide the prediction, its latency may be higher than the latency of the correlating branch predictor. Thus, during the fetch stage the correlating predictor provides its prediction as the initial one. Later, if the *BPRU* changes the initial prediction, the fetched instructions after the branch are flushed and the fetch is redirected to the new PC.

4 Experimental Results

This section analyzes the performance of the proposed *BPRU* engine when it is integrated into the *gshare* and the Alpha 21264 [11][2] branch predictors. For comparison purposes we have also implemented the *SBI* confidence estimator [13]. All experiments compare predictors with the same total size, including the space occupied by tags, counters and any other required storage. We have considered the whole SPE-Cint2000 suite for the evaluation of our proposed mechanism. All benchmarks were compiled with maximum optimizations by the Compaq Alpha compiler, and they were run using the *SimpleScalar/Alpha* v3.0 tool set [2]. We have reduced the input data set of the SPEC2000 suite while trying to keep a complete execution for every benchmark, as showed in Table 1.

Table 1. Benchmark characteristics.

Benchmark	Input Set	Total # dyn. Instr. of Input Set (Mill.)	Total # simulated Instr. (Mill.)	# skipped Instr (Mill.)	# dyn.cond. branch (Mill.)
bzip2	input source 1	2069	500	500	43
crafty	test (modified)	437	437	-	38
eon	kajiya image	454	454	-	29
gap	test (modified)	565	500	50	56
gcc	test (modified)	567	500	50	62
gzip	input.log 1	593	500	50	52
mcf	test	259	259	-	31
parser	test (modified)	784	500	200	64
twolf	test	258	258	-	21
vortex	test (modified)	605	500	50	51
vpr	test	692	500	100	45

---

[2] The Alpha 21264 processor uses a branch predictor composed by a metapredictor that chooses between the predictions made by a global and a local predictor.

4.1. Confidence Estimator Evaluation

Before analyzing the effect of our *BPRU* on branch prediction accuracy, we evaluate its usefulness as a confidence estimator. The total size of the *BPRU+gshare* is 36 KB (16 KB for the *RT*, 16 KB for the *gshare* and 4 KB for the value predictor) whereas the total size of the *SBI+gshare* is 32 KB (16 KB for the confidence estimator and 16 KB for the *gshare*). Table 2 shows that for both estimators, SENS and PVP metrics are quite similar and very high, which means that correct branch predictions are correctly estimated by both schemes. PVN, is also very similar for both schemes, i.e., 70% of branches estimated as "low confidence" finally miss the prediction. This metric must be greater than 50% in order to a get a positive inversion benefit. Finally, we can observe that the *BPRU* estimator provides a better accuracy for the SPEC metric. On average, 43% of incorrect predictions are identified as low confidence by our estimator, whereas just 26% of them are identified as low confidence by the *SBI*. This metric gives an insight about the "quality" of the confidence estimation: the *BPRU* can detect branch mispredictions much better than the *SBI*.

Table 2. Confidence estimation metrics for the *BPRU+gshare* and the *SBI+gshare*.

Benchmark	BPRU+gshare				SBI+gshare			
	Spec	PVN	Sens	PVP	Spec	PVN	Sens	PVP
bzip2	53,0%	65,9%	97,7%	96,1%	24,1%	62,3%	98,8%	93,9%
crafty	45,7%	65,4%	98,3%	96,3%	28,1%	77,4%	99,4%	95,2%
eon	19,6%	81,3%	99,7%	95,6%	18,2%	61,2%	99,3%	95,5%
gap	44,7%	74,1%	99,4%	97,9%	25,1%	76,3%	99,7%	97,2%
gcc	47,9%	80,3%	99,3%	96,9%	39,0%	82,3%	99,5%	96,4%
gzip	35,1%	62,9%	99,2%	97,5%	18,3%	65,3%	99,6%	96,9%
mcf	36,2%	62,7%	98,3%	95,1%	17,9%	57,8%	99,0%	93,8%
parser	35,7%	67,6%	98,9%	95,9%	29,5%	73,2%	99,3%	95,6%
twolf	41,7%	66,6%	97,8%	94,0%	20,7%	67,0%	98,9%	92,1%
vortex	71,4%	89,5%	99,9%	99,6%	53,0%	89,3%	99,9%	99,4%
vpr	39,0%	55,5%	97,6%	95,4%	16,6%	53,7%	98,9%	93,9%
AVERAGE	42,7%	70,2%	98,7%	96,4%	26,4%	69,6%	99,3%	95,4%

4.2. Results for Immediate Updates

The first set of experiments update prediction tables immediately, in order to evaluate the potential of the selective reversal mechanism when it is isolated from other aspects of the microarchitecture (using the *sim-safe* simulator). The *RT* is implemented as an 8-way set-associative table using 13 bits for tags and 3 bits for the reversal counters. A *RT* index of $k+1$ bits is the result of *XORing*: k bits from the PC of the branch input producer, 4 bits of history, k-4 bits from the predicted value and finally, the *gshare* prediction bit as the most significant bit.

Fig. 3 shows the misprediction rates for the *BPRU+gshare*, *SBI+gshare* and *gshare* predictors for five selected SPECint2000 programs (those with the highest misprediction rate) and the arithmetic mean of the whole SPECint2000. Based on experimental results, the *BPRU+gshare* allocates 1/9 of the size for the value predictor, 4/9 for the *RT* and 4/9 for the *gshare*. The *SBI+gshare* allocates 1/2 for the confidence estimator and 1/2 for the *gshare*.

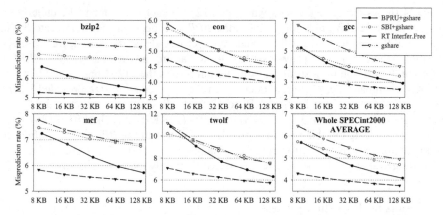

Fig. 3. Branch misprediction rates for the *BPRU+gshare*, *SBI+gshare* and *gshare* predictors.

The *BPRU+gshare* predictor significantly reduces the misprediction rate of the *gshare* of the same total capacity for all benchmarks and all evaluated sizes. For predictors of around 32 KB, the misprediction rate of the *gshare* is reduced by an average of 15% (up to 27% for *gcc*). It is important to note that the *BPRU+gshare* with a total size of 18 KB has the same misprediction rate (5.1%) as the *gshare* of 64 KB.

Comparing the *BPRU* and the *SBI* schemes, the *BPRU+gshare* outperforms the *SBI+gshare* for all benchmarks for sizes greater than 8 KB. On average, the *BPRU+gshare* with a total size of 36 KB reduces 10% the misprediction rate of the *SBI+gshare* of 32 KB (up to 18% for *bzip2*). Summarizing, for the whole SPECint2000 suite, the *BPRU+gshare* reduces the misprediction rate by a factor that ranges from 12% (8 KB) to 18% (128 KB) with respect to the *gshare*, and from 5% (16KB) to 13% (128 KB) with respect to the *SBI+gshare*.

Finally, Fig. 3 shows that the interference-free *RT* provides great improvements for all benchmarks. For instance, in the *twolf* application, the misprediction rate of an 8 KB *gshare* is reduced by a factor of 37%. This shows the potential of the *BPRU* and an opportunity for improvement by using better indexing schemes to access the *RT*.

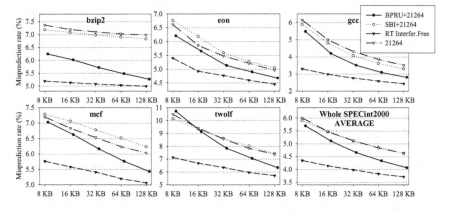

Fig. 4. Branch misprediction rates for the *BPRU+21264*, *SBI+21264* and *21264* predictors.

Fig. 4 shows the same analysis when the 21264 branch predictor is the underlying predictor. For 32 KB predictors, adding the *BPRU* reduces misprediction rate by an average of 9% (up to 19% for *bzip2* and *gcc*). We can also observe that, for these applications, the *SBI* hardly provides benefits. Regarding predictor sizes, the *BPRU+ 21264* of 36 KB obtains a similar misprediction rate as the 21264 branch predictor of 128 KB. Summarizing, average reductions in the misprediction rate provided by the *BPRU+21264* range from 5.1% (8 KB) to 12.1% (128 KB) with respect to the 21264 branch predictor, and from 3.5% (8 KB) to 12.0% (128 KB) over the *SBI+21264*.

Results presented in this Section show the usefulness of the *BPRU* as a confidence estimator for branch prediction reversal, demonstrating that adding the *BPRU* to a correlating branch predictor is more effective than simply increasing its size.

4.3. Results for Realistic Updates

This Section presents an evaluation of the proposed *BPRU* engine in a dynamically-scheduled superscalar processor. Table 3 shows the configuration of the simulated architecture. In addition, the original *SimpleScalar* simulator pipeline has been lengthened to 20 stages, following the pipeline scheme of the Pentium 4 processor [6].

Table 3. Configuration of the simulated processor.

Fetch engine	Up to 8 instructions/cycle, allowing 2 taken branches. 8 cycles misprediction penalty.
Execution engine	Issues up to 8 instructions/cycle, 128-entries reorder buffer, 64-entries loads/store queue.
Functional Units	8 integer alu, 2 integer mult, 2 memports, 8 FP alu, 1 FP mult.
L1 Instruct.-cache	128 KB, 2-way set associative, 32 bytes/line, 1 cycle hit latency.
L1 Data-cache	128 KB, 2-way set associative, 32 bytes/line, 1 cycle hit latency.
L2 unified cache	512 KB, 4-way set associative, 32 bytes/line, 6 cycles hit latency, 18 cycles miss latency.
Memory	8 bytes/line, virtual memory 4 KB pages, 30 cycles TLB miss.

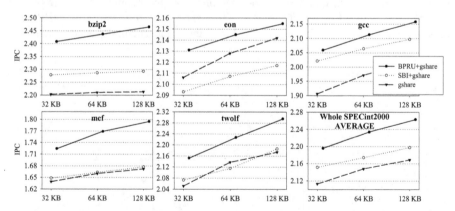

Fig. 5. IPC obtained by the *BPRU+gshare*, *SBI+gshare* and *gshare* branch predictors.

Fig. 5 shows the IPC obtained by the simulated processor when it incorporates the *gshare*, *SBI+gshare* and *BPRU+gshare*. The latency of both the *gshare* and the

SBI+gshare is considered to be one cycle, whereas the latency considered for the *BPRU+gshare* is 3 cycles. For 32 KB predictors, and in spite of its higher latency, the *BPRU+gshare* provides performance improvements over the *gshare* of 9% and 8% for *bzip2* and *gcc* respectively. As the size of the predictors grows, performance improvements are also increased, up to 11% for *bzip2* (128 KB). Comparing both confidence estimators, the *BPRU+gshare* outperforms the *SBI+gshare* for all programs and predictor sizes, obtaining speedups of up to 6% for 32 KB predictors.

Fig. 6 presents performance results using the underlying 21264 branch predictor. For 32 KB predictors, the *BPRU+21264* outperforms the 21264 branch predictor by 7% and 5% for *bzip2* and *gcc*. Comparing both confidence estimators, the *BPRU+ 21264* obtains speedups of up to 6% over the *SBI+21264*, for 32 KB predictors. These speedups are similar to those obtained when the *gshare* is the baseline predictor.

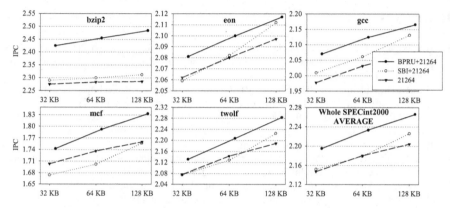

Fig. 6. IPC obtained by the *BPRU+21264, SBI+21264* and 21264 branch predictors.

5 Conclusions

In this work, we have presented a new branch prediction confidence estimator, the *Branch Prediction Reversal Unit (BPRU)*. The key characteristic of this estimator is the inclusion of predicted data values in the process of assigning confidence to the predictions. We have shown that the accuracy of conventional correlating predictors such as the *gshare* and the Alpha 21264 branch predictor can be effectively correlated with the predicted data value, the PC of the producers of the branch input and recent branch history. We have first analyzed the *BPRU* as a confidence estimator, comparing it with an already proposed scheme, the *SBI*. Confidence metrics show that the *BPRU* is able to label as "low confidence" 43% of branch mispredictions, conversely to the *SBI*, which just assigns low confidence to 26% of branch mispredictions.

We have also evaluated the *BPRU* as a branch prediction reversal mechanism for the *gshare* and the 21264 branch predictors. Misprediction reductions are quite similar for both correlating predictors. Results for immediate updates of the prediction tables show misprediction reductions of 15% on average for the whole SPECint2000 suite (up to 27% for some applications) when 32 KB predictors are considered. Comparing the *BPRU* and the *SBI*, we show that, for 32 KB predictors, the *BPRU+ gshare* reduces the misprediction rate of the *SBI+gshare* by an average of 10%. Finally, performance evaluation of the *BPRU* in a superscalar processor show IPC improvements of up to 9% over both the *gshare* and the *SBI+gshare*, using 32 KB predictors. To conclude, the *BPRU* is a cost-effective way of improving branch prediction accuracy, being more effective than simply increasing the size of the predictor.

Acknowledgements

This work has been supported by the project TIC2000-1151-C07-03 of the *Plan Nacional de I+D+I* and by the Spanish CICYT under grant TIC98-0511.

References

1. Aragón, J.L., González, J., García, J.M., González, A.: Selective Branch Prediction Reversal by Correlating with Data Values and Control Flow. In: Proceedings of the Int. Conf. on Computer Design. (2001)
2. Burger, D., Austin, T.M.: The SimpleScalar Tool Set, Version 2.0. Technical Report #1342, University of Wisconsing-Madison, Computer Sciences Department. (1997)
3. Chang, P.Y., Evers, M., Patt, Y.N.: Improving Branch Prediction Accuracy by Reducing Pattern History Table Interference. In: Proceedings of the Int. Conf. on Parallel Architectures and Compilation Techniques. (1996)
4. Chang, P.Y., Hao, E., Patt., Y.N.: Alternative implementations of Hybrid Branch Predictors. In: Proceedings of the Int. Symp. on Microarchitecture. (1995)
5. Gabbay, F., Mendelson, A.: Speculative Execution Based on Value Prediction. Technical Report #1080, Technion, Electrical Engineering Department. (1996)
6. Glaskowsky, P.N.: Pentium 4 (Partially) Previewed. Microprocessor Report, Microdesign Resources. (August 2000)
7. González, J., González, A.: Control-Flow Speculation through Value Prediction for Superscalar Processors. In: Proc. of the Int. Conf. on Parallel Arch. and Comp. Tech. (1999)
8. Grunwald, D., Klauser, A., Manne, S., Pleszkun, A.: Confidence Estimation for Speculation Control. In: Proceedings of the Int. Symp. on Computer Architecture. (1998)
9. Heil, T.H., Smith, Z., Smith, J.E.: Improving Branch Predictors by Correlating on Data Values. In: Proceedings of the Int. Symp. on Microarchitecture. (1999)
10. Jacobsen, E., Rotenberg, E., Smith, J.E.: Assigning Confidence to Conditional Branch Predictions. In: Proceedings of the Int. Symp. on Microarchitecture. (1996)
11. Kessler, R.E., McLellan, E.J., Webb, D.A.: The Alpha 21264 Microprocessor Architecture. In: Proceedings of the Int. Conf. on Computer Design. (1998)
12. Lee, C.C., Chen, I.C.K., Mudge, T.N.: The Bi-Mode Branch Predictor. In: Proceedings of the Int. Symp. on Microarchitecture. (1996)
13. Manne, S., Klauser, A., Grunwald, D.: Branch Prediction using Selective Branch Inversion. In: Proc. of the Int. Conf. on Parallel Architectures and Compilation Techniques. (1999)
14. McFarling, S.: Combining Branch Predictors. Technical Report #TN-36, Digital Western Research Lab. (1993)
15. Nair, R.: Dynamic Path-Based Branch Correlation. In: Proceedings of the Int. Symp. on Microarchitecture. (1995)
16. Smith, J.E.: A Study of Branch Prediction Strategies. In: Proceedings of the Int. Symp. on Computer Architecture. (1981)
17. Yeh, T.Y., Patt, Y.N.: A Comparison of Dynamic Branch Predictors that Use Two Levels of Branch History. In: Proceedings of the Int. Symp. on Computer Architecture. (1993)

Retargetable Program Profiling Using High Level Processor Models

Rajiv Ravindran and Rajat Moona

Department of Computer Science & Engineering
Indian Institute of Technology
Kanpur 208016, India {rajiva, moona}@cse.iitk.ac.in

Abstract. Program profiling helps in characterizing program behavior for a target architecture. We have implemented a retargetable simulation driven code profiler from a high-level processor description language, *Sim-nML*. A programming interface has been provided for building customized profilers. The retargetability makes the profiling tool independent of the target instruction set.

1 Introduction

During the design of embedded systems, need is felt to automatically generate processor and application development tools like assemblers, disassemblers, compilers, instruction-set simulators etc. Automated generation of such tools yields faster turnaround time with lower costs for system design and simplifies the process of incorporating design changes. We have developed a retargetable environment in which processors are modeled at a high level of abstraction using the *Sim-nML* [3] specification language.

We use *Sim-nML* to describe the instruction set architecture of the processor from which various tools to aid processor design are automatically generated. The overall goal of the *Sim-nML* based project is to model a complete system environment with a processor core specified using *Sim-nML* for automatic architecture exploration. As part of the integrated development environment, we have developed a retargetable functional simulator [5], cache simulator [6], assembler, disassembler [13] and a compiler back-end generator. Work on a cycle accurate timing simulator is in progress from the *Sim-nML uses* [14] model. In this paper, we describe a mechanism for code analysis. Program analysis tools are extremely useful for understanding program behavior. Computer architects could employ them to analyze program performance on new architectures. It can be used to characterize instruction usage, branch behavior etc. Critical and time consuming portions of the code can be identified for optimizations. Compilers could use the profile information to guide its optimization pass.

Program profiling has always been instruction-set specific. Profiling tools instrument the input binary at specific points to sample the program behavior. Other techniques employed include hardware counters which are built into the processor. But most of these techniques are tied to a particular architecture. In the embedded world, the designer has to iterate over multiple design options to decide on the best target architecture for a given application, within a short period of time, before actual processor design. Hence, there is a need for a more generic model.

B. Monien, V.K. Prasanna, S. Vajapeyam (Eds.): HiPC 2001, LNCS 2228, pp. 224–233, 2001.
© Springer-Verlag Berlin Heidelberg 2001

We have tried to build a mechanism that provides architects and software developers a means to implement various profiling policies in a retargetable manner. A retargetable instruction set simulator provides the platform for code profiling. A retargetable functional simulator generator - *Fsimg* [5], generates a processor specific functional simulator from the processor model written in *Sim-nML* for a given program. A high-level processor modeling language makes our design retargetable. In our approach, the profiling of code is accomplished by instrumenting the functional simulator, for the given program on the target architecture, at chosen points. For example, to count the basic blocks traversed at run time, a counter could be placed at the end of each basic block. Similarly, to analyze branch behavior, routines could be added after conditional branch instructions. We view the program as a set of procedures each containing a collection of basic blocks, each of which is further composed of processor instructions. A user defined procedure for profiling the application program can be inserted before or after an instruction, a basic block or a procedure. We have provided an application programming interface (API) using which the user can insert his function calls at chosen points in the input binary. This technique is inspired from ATOM [7] which is a framework for building a wide range of customized program analysis tools. Through this approach, the user can construct a custom profiling tool. The retargetable program profiler is a step towards our original goal of complete system simulation consisting of a processor simulator, cache simulator etc. from *Sim-nML*.

Our code profiling strategy consists of two phases. In the first phase, the retargetable functional simulator generator *Fsimg*, takes as input the *Sim-nML* processor model describing the target architecture and the program binary. The API-calls helps the user define his routines and the points in the program binary to be instrumented. The granularity of instrumentation can be at the procedural, basic block or at instruction boundaries. *Fsimg* then generates an instruction set simulator -*Fsim*, specific to the given program and target processor instruction set. The instrumentation routines are linked with the simulator with user provided calls added at specified points in *Fsim*. In the second phase, the user runs the functional simulator which executes the instrumented code while simulating the input program. Thus the program profile is generated.

Our target through this paper has been two fold. Firstly, as an extension of the integrated development environment we have implemented a code profiler. Unlike earlier works on profiling, we have attempted to develop a retargetable profiler from a high-level processor description. Secondly, the API-calls provides an infrastructure to create a customized profiling environment through user specified functions.

The rest of the paper is organized as follows. In section 2, we list the related work. In section 3, we give an overview of *Sim-nML*. We describe the design of the code profiler in section 4. We explain the API with some examples of how they could be used for different kinds of profiling. Finally in section 5, we present some sample simulation and profiling results for simple programs and conclude.

2   Related Work

Performance modeling of a system is a growing area and a lot of research has been pursued in this area. We briefly review some of them here.

ATOM [7] provides a framework for providing customized program analysis tools. It instruments the input program through a programmable interface. **ATOM** uses no simulation or interpretation.

Pixie [8] is a utility that allows one to trace, profile or generate dynamic statistics for any program that runs on a MIPS processor. It works by annotating executable object code with additional instructions that collect the dynamic information during run time.

QPT [9] [10] is a profiler and tracing system. It rewrites a program's executable file (a.out) by inserting code to record the execution frequency or sequence of every basic block or control-flow edge.

EEL [11] (Executable Editing Library) is a C++ library that provides abstractions which allow a tool to analyze and modify executable programs without being concerned with particular instruction sets, executable file formats, or consequences of deleting existing code and adding foreign code.

EXPRESSION [1] is similar to *Sim-nML* and is used in architecture exploration. It has been used for automatic generation of compiler/simulator toolkit.

LISA [2] is a machine description language for generation of bit and cycle accurate models for DSP processors based on behavioral operation description.

UNIX tools **prof** and **gprof** record a statistical sample.

All of these tools are architecture specific or implement a specific set of profiling techniques (except ATOM). We try to improve upon them through retargetability and customization.

3   Sim-nML

Sim-nML[3][4] is a direct extension of *nML*[15] machine description formalism. It includes several features that are useful for performance simulation that are not present in *nML*.

Sim-nML is targeted for describing wide range of processor architectures including CISC, RISC and VLIW architectures at the instruction set level hiding its implementation details. Some of the target architectures described using *Sim-nML* include PowerPC, ARM, UltraSparc, MIPS-IV, 8085, Motorola 68HC11, ADSP2101. The instruction set is described in a hierarchical manner. The semantic actions of the instructions are captured as fragments of code spread all over the instruction tree. *Sim-nML* specifications are described using an attribute grammar. There is a fixed start symbol called *instruction*, and two types of productions, *and-rule* and *or-rule*.

There are certain fixed attributes defined for *and-rules* that capture various aspects of the instruction set. The *syntax* attribute captures the textual assembly language syntax of the instructions. The *image* attribute captures the binary image of the instructions. The *action* attribute captures the semantics of the instructions. The *uses* attribute captures the resource usage model and is used for timing simulation. The resource usage model captures the micro-architectural features of the processor and is used in cycle accurate timing simulation. The details of the uses model is given in [14].

The following illustration is a specification for a simple processor with four instructions – *add, sub, bim* (branch immediate) and *bin* (branch indirect).

The processor has two addressing modes – immediate and register indirect. The *mode* rule is associated with a *value* attribute, 'n' in case of *IMM* and 'R[n]' in case of *REG_IND*, where 'n' is a constant provided in the instruction encoding. The four instructions are hierarchically described. The branch immediate (*bim*) instruction modifies the PC with an immediate branch offset. The branch indirect (*bin*) takes the branch address in the specified register and copies it to the PC. The *add* instruction adds two registers and puts the results in the first register. The *sub* instruction subtracts two registers and puts the result in the first register.

```
let REGS = 5
type word = int ( 16 )

reg R [ 2**REGS, word ]
reg PC [ 1 , word ]

resource bu, alu

mode IMM ( n : card ( 12 ) ) = n
  syntax = format( "%d", n )
  image  = format( "%12b", n )

mode REG_IND ( n : card ( 5 ) ) = R [ n ]
  syntax = format( "r%d", n )
  image  = format( "%5b", n )

op instruction ( x : instr_action )
  uses   = x.uses
  syntax = x.syntax
  image  = x.image
  action = {
            PC = PC + 2;
            x.action;
          }
op instr_action = branch_inst | arithmetic_inst

op branch_inst = bim | bin

op bim ( d : IMM )
  uses = bu #1
  syntax = format( "bim %s", d.syntax )
  image  = format( "1000%s", d.image )
  action = { PC = PC + (d << 4); }
```

```
op bin ( r : REG_IND )
  uses = bu #1
  syntax = format( "bin %s", r.syntax )
  image  = format( "10010000000%s", r.image )
  action = { PC = r; }

op arithmetic_inst = add | sub

op add ( r1 : REG_IND, r2 : REG_IND )
  uses = alu #1
  syntax = format( "add %s %s", r1.syntax, r2.syntax )
  image  = format( "101000%s%s", r1.image, r2.image )
  action = { r1 = r1 + r2; }

op sub ( r1 : REG_IND, r2 : REG_IND )
  uses = alu #1
  syntax = format( "sub %s %s", r1.syntax, r2.syntax )
  image  = format( "101100%s%s", r1.image, r2.image )
  action = { r1 = r1 − r2; }
```

Retargetable tools flatten out the hierarchical description to enumerate out the complete instruction set and its associated attribute definitions. Different tools, depending on their needs, use different attributes. For example, an assembler uses the definition of attributes *syntax* and *image*. A detailed description of *Sim-nML* can be found in [3][4] (*http://www.cse.iitk.ac.in/sim-nml*).

4   Code Profiling

In this section, we give an overview of the functional simulator, discuss the design and implementation of the code profiler, the API and how they are incorporated into the functional simulator.

4.1   Overview of Functional Simulation Process

Fsimg [5] initially flattens the hierarchy in *image* and *action* attributes of the *Sim-nML* description. The *action* attribute captures the semantics of the instruction-set. For each machine instruction, *Fsimg* emits a corresponding unique *C* function. This function is obtained by translating flattened *action* attribute definition to *C*. *Fsimg* then reads the program binary and for every matched input instruction image, generates a call to the corresponding function defining that instruction. All these calls to functions are captured in a table of instruction function pointers with each entry pointing to the function corresponding to the respective instruction. Along with this table, *Fsimg* also generates data structures for memory, registers and a driving routine for simulation. Thus, it generates a list of function calls corresponding to all instructions in the input program.

The driver routine of the functional simulator simulates the program by calling these functions sequentially until the program terminates.

4.2 Instrumentation

The table of instruction function pointers generated by the functional simulator provide a convenient means for code instrumentation. To create a customized profiling tool, the user defines a set of routines which are to be executed at chosen points in the program. Calls to these user defined routines are inserted into the table of instruction function pointers between function calls at specified instrumentation points which represent the instruction boundaries. A set of predefined routines - an application programming interface (API) is provided which allows the user to add his procedure calls before or after instructions. A set of *Basicblock analysis* routines are provided for profiling at the level of procedures, basic blocks and instructions within basic blocks.

The profiler generation consists of the following steps.

– *Fsimg* analyzes the input program for basic blocks. *Fsimg* is provided with the set of conditional and unconditional control flow instructions in the *Sim-nML* hierarchy. Since *Sim-nML* is a hierarchical description, if the hierarchy allows, we could provide the top level branch node instead. The actual branch instruction is then enumerated from this. Once a list of branch instructions are enumerated, we split the input instruction stream at procedure boundaries. For a given procedure, the basic block boundaries are marked just after every branch instruction. The branch target address can be calculated from the *action* corresponding to the branch instructions.
– User adds his routines through the instrumentation-API within a predefined *Instrument* function.
– During generation of the functional simulator, the API calls are used to instrument the application program at appropriate places in the generated simulator.
– User runs the simulator which executes the instrumented code.

Application Programming Interface - API For code instrumentation, we list some important instrumentation-API below.

– **AddCallFuncbyName(iname, func, position)**: This function can be used to instrument the application program before/after (*position*) the specific instruction (*iname*) with the user defined routine (*func*).
– **AddCallFunc(inst, func, pos)**: This function can be used to instrument the application program at specific addresses i.e, whenever an instruction is fetched from the address *inst*.
– **AddTrailerFunc(func)**: The user can add any routines (*func*) to be executed after simulation. *Fsimg* adds these routines after the simulation engine. They can be used to collect the final statistics, dump profiling information etc.
– **GetFirstProc, GetNextProc, GetFirstBlock, GetNextBlock, GetLastInst**: These procedures are used at the procedure and basic block level. They can be used as iterators over all procedures and basic blocks in the given program.

For more details refer to [12].

Implementation   The instrumentation routines are added in 3 files - *instrument.c*, *bblockanal.c*, *userfuncs.c*. *Instrument.c* contains a call to a predefined routine *Instrument* in which the user adds the API calls.

For example, to count the occurrences of *add* instructions executed in the program, we specify the following

```
void Instrument()
{
    AddCallFuncbyName("add", INSTR_TYPE, "addcounter", AFTER);
    AddTrailerFunc("printaddcnt");
}
```

The file *userfuncs.c* contains the user defined routines. The function *addcounter* could be defined as follows:

```
long addcnt = 0;
void addcounter()
{
        addcnt++;
}
```

where *addcnt* is a global counter. *AddTrailerFunc* is used to add the user function *printaddcnt* at the end of simulation which could be defined as follows

```
void printaddcnt()
{
        printf("%d", addcnt);
}
```

The file *bblockanal.c* contains the instrumentation routines associated with basic block related analysis. It contains a call to a predefined routine *BasicblockAnal*, in which the user adds the API calls for basic block related profiling.

For example, to count the number of basic blocks that are traversed during program execution, we specify the following

```
void BasicblockAnal()
{
    Proc *p;
    Block *b;
    Inst inst;
    for (p = GetFirstProc(); p; p = GetNextProc(p))
      for (b = GetFirstBlock(p); b; b = GetNextBlock(b))
        inst = GetLastInst(b);
        AddCallFunc(inst, "countbb" AFTER);
      AddTrailerFunc("printbb");
}
```

The user defined function *countbb* is added *after* the last instruction in each basic block. The user might want to call different functions at the same address boundary. Multiple user defined instructions can be engineered at address boundaries by calling *AddCallFunc* with different function names at the same instruction address.

5   Results

Five benchmarks program were written in C (Table 1) and compiled for *PowerPC603 Sim-nML* processor description.

Program	Description
mmul.c	Integer matrix multiplication. This program initializes two integer matrices of 100x100 size and multiplies them.
bsort.c	Bubble sort. This program initializes an array of 1500 integers in descending order and sorts them to ascending order using bubble sort algorithm.
qs.c	Quick sort. This program initializes array of 1,00,000 integers in descending order and sorts them to ascending order using quick sort algorithm.
fmmul.c	Matrix multiplication of floating-point numbers. Initializes and multiplies two floating point matrices of size 100x100.
nqueen.c	This program finds all possible ways that N queens can be placed on an NxN chess board so that the queens cannot capture one another. Here N is taken as 12.

Table 1. Benchmark Programs

Table 2 gives the total number of dynamically executed instructions during the simulation.

Program	Total No. of Instructions
mmul.c	91,531,966
bsort.c	60,759,034
qs.c	80,773,862
fmmul.c	92,131,966
nqueen.c	204,916,928

Table 2. Total number of instructions simulated for test programs.

We have implemented a simple profiling tool which counts the number of basic blocks that are traversed at run time. At the same time, the number of *PowerPC603*

addi instructions executed is found. The code instrumentation technique used is the one specified in section 4.2.

The profiling output is given in table 3.

Program	Total No. of basic block traversed	Total No: of addi instructions executed
mmul.c	2081207	1030305
bsort.c	4506008	2253005
qs.c	7315513	242144
fmmul.c	2081207	1110305
nqueen.c	40030204	60766515

Table 3. Profiling output for test programs.

The functional simulator performance without any profiling code is shown in table 4

Program	Total Time in Seconds	Instructions per second
mmul.c	62	1,476,322
bsort.c	106	573,198
qs.c	109	741,044
fmmul.c	64	1,439,549
nqueen.c	225	910,741

Table 4. Performance Results of the functional simulator

We compare the simulation slow down from different profiling techniques (those in table 3) in table 5.

Program	Slowdown factor
mmul.c	1.01x
bsort.c	1x
qs.c	1.01x
fmmul.c	1x
nqueen.c	1.01x

Table 5. Performance results of profiling of test programs.

6   Conclusion

In this paper, we presented a simulation driven program profiler. The profiler is generic in the following ways. Firstly, we have used a retargetable functional simulator generated from a high-level processor description language, *Sim-nML*. Secondly, through a programming interface, we have provided a mechanism for implementing customized profilers. Thus, we do not tie the profiler to a particular instruction-set or to specific profiling techniques.

References

1. Ashok Halambi, Peter Grun , Vijay Ganesh, Asheesh Khare, Nikil Dutt, Alex Nicolau: *EXPRESSION: A Language for Architecture Exploration through Compiler/Simulator Retargetability*. Proceedings of the Conference on Design, Automation and Test in Europe, Munich, Germany, March 1999
2. S. Pees, A. Hoffmann, V. Zivojnovic, H. Meyr: *LISA - Machine Description Language for Cycle-Accurate Models of Programmable DSP Architectures*. Proceedings of the 36th Design Automation Conference, New Orleans, June 1999
3. V. Rajesh, Rajat Moona: *Processor Modeling for Hardware Software Co-Design*. Proceedings of the 12th International Conference on VLSI Design, Goa, India, January,1999
4. Rajat Moona: *Processor Models for Retargetable Tools*. Proceedings of Rapid Systems Prototyping 2000 (IEEE), Paris, June, 2000
5. Subhash Chandra and Rajat Moona: *Retargetable Functional Simulator Using High Level Processor Models*. Proceedings of the 13th International Conference on VLSI Design, Calcutta, India., January, 2000
6. Rajiv Ravindran and Rajat Moona: *Retargetable Cache Simulation Using High Level Processor Models*. Proceedings of the 6th Australasian Computer Systems Architecture Conference, Gold Coast, Australia, January, 2001
7. Amitabh Srivastava and David Wall: *ATOM: A System for Building Customized Analysis Tools*. Proceedings of the SIGPLAN '94 Conference of Programming Language Design and Implementation, June, 1994, 196-205
8. Michael D. Smith: *Tracing with Pixie*. Memo from Center for Integrated Systems, Stanford Univ., April, 1991
9. James R. Larus: *Efficient Program Tracing*. IEEE Computer, May, 1993, 26(5):52-61
10. James R. Larus, Thomas Ball: *Rewriting Executable Files to Measure Program Behavior*. Software: Practice and Experience, Feb, 1994, 24(2):197-218
11. James R. Larus, Eric Schnarr: *EEL: Machine-Independent Executable Editing*. SIGPLAN Conference on Programming Language Design and Implementation (PLDI), June 1995
12. Rajiv Ravindran: *Retargetable Profiling Tools and their Application in Cache Simulation and Code Instrumentation*. Masters thesis report, Dept. of Computer Science and Engg., IIT Kanpur, India, Dec 1999. http://www.cse.iitk.ac.in/research/mtech1998/9811116.html
13. Nihal Chand Jain: *Disassembler Using High Level Processor Models*. Masters thesis report, Dept. of Computer Science and Engg., IIT Kanpur, India, January 1998. http://www.cse.iitk.ac.in/research/mtech1997/9711113.html
14. Anand Shukla, Arvind Saraf: *A Formalism for Processor Description*. Bachelors thesis report, Dept. of Computer Science and Engg., IIT Kanpur, India, May 2001.
15. M. Freerick: *The nML Machine Description Formalism*. http://www.cs.tu-berlin.de/~mfx/dvi_docs/nml_2.dvi.gz, 1993

Towards Automatic Synthesis of High-Performance Codes for Electronic Structure Calculations: Data Locality Optimization

D. Cociorva[1], J. Wilkins[1], G. Baumgartner[2], P. Sadayappan[2], J. Ramanujam[3], M. Nooijen[4], D. Bernholdt[5], and R. Harrison[6]

[1] Physics Dept., Ohio State Univ., USA. {cociorva,wilkins}@pacific.mps.ohio-state.edu
[2] CIS Department, Ohio State University, USA. {gb,saday}@cis.ohio-state.edu
[3] ECE Department, Louisiana State University, USA. jxr@ee.lsu.edu
[4] Chemistry Department, Princeton University, USA. Nooijen@Princeton.edu
[5] Oak Ridge National Laboratory, USA. bernholdte@ornl.gov
[6] Pacific Northwest National Laboratory, USA. Robert.Harrison@pnl.gov

Abstract. The goal of our project is the development of a program synthesis system to facilitate the development of high-performance parallel programs for a class of computations encountered in computational chemistry and computational physics. These computations are expressible as a set of tensor contractions and arise in electronic structure calculations. This paper provides an overview of a planned synthesis system that will take as input a high-level specification of the computation and generate high-performance parallel code for a number of target architectures. We focus on an approach to performing data locality optimization in this context. Preliminary experimental results on an SGI Origin 2000 are encouraging and demonstrate that the approach is effective.

1 Introduction

The development of high-performance parallel programs for scientific applications is usually very time consuming. The time to develop an efficient parallel program for a computational model can be a primary limiting factor in the rate of progress of the science. Therefore, approaches to automated synthesis of high-performance parallel programs are very attractive. In general, automatic synthesis of parallel programs is not feasible. However, for specific domains, a synthesis approach is feasible, as is being demonstrated, e.g., by the SPIRAL project [35] for the domain of signal processing.

Our long term goal is to develop a program synthesis system to facilitate the development of high-performance parallel programs for a class of scientific computations encountered in computational chemistry and computational physics. The domain of our focus is electronic structure calculations, as exemplified by coupled cluster methods, where many computationally intensive components are expressible as a set of tensor contractions. We plan to develop a synthesis system that can generate efficient parallel code for a number of target architectures from an input specification expressed in a high-level notation. In this paper, we provide an overview of the planned synthesis system, and focus on an optimization approach for one of the components of the synthesis system that addresses data locality optimization.

The computational structures that we address arise in scientific application domains that are extremely compute-intensive and consume significant computer resources at national supercomputer centers. These computational forms arise in some computational physics codes modeling electronic properties of semiconductors and metals [2, 8, 28], and in computational chemistry codes such as ACES II, GAMESS, Gaussian, NWChem, PSI, and MOLPRO. In particular, they comprise the bulk of the computation with the coupled cluster approach to the accurate description of the electronic structure

B. Monien, V.K. Prasanna, S. Vajapeyam (Eds.): HiPC 2001, LNCS 2228, pp. 237–248, 2001.
© Springer-Verlag Berlin Heidelberg 2001

of atoms and molecules [21, 23]. Computational approaches to modeling the structure and interactions of molecules, the electronic and optical properties of molecules, the heats and rates of chemical reactions, etc., are crucial to the understanding of chemical processes in real-world systems.

The paper is organized as follows. In the next section, we elaborate on the computational context of interest and the pertinent optimization issues. Sec. 3 provides an overview of the synthesis system, identifying its components. Sec. 4 focuses on data locality optimization and presents a new approach and algorithm for effective tiling in this context. Sec. 5 presents experimental performance data on the application of the new algorithm. Related work is covered in Sec. 6, and conclusions are provided in Sec. 7.

2   The Computational Context

In the class of computations considered, the final result to be computed can be expressed using a collection of multi-dimensional summations of the product of several input arrays. Due to commutativity, associativity, and distributivity, there are many different ways to compute the final result, and they could differ widely in the number of floating point operations required. Consider the following expression:

$$S(a, b, i, j) = \sum_{c,d,e,f,k,l} A(a, c, i, k) \times B(b, e, f, l) \times C(d, f, j, k) \times D(c, d, e, l)$$

If this expression is directly translated to code (with ten nested loops, for indices $a - l$), the total number of arithmetic operations required will be $4 \times N^{10}$ if the range of each index $a - l$ is N. Instead, the same expression can be rewritten by use of associative and distributive laws as the following:

$$S(a, b, i, j) = \sum_{c,k} \left(\sum_{d,f} \left(\sum_{e,l} B(b, e, f, l) \times D(c, d, e, l) \right) \times C(d, f, j, k) \right) \times A(a, c, i, k)$$

This corresponds to the formula sequence shown in Fig. 1(a) and can be directly translated into code as shown in Fig. 1(b). This form only requires $6 \times N^6$ operations. However, additional space is required to store temporary arrays $T1$ and $T2$.

Generalizing from the above example, we can express multi-dimensional integrals of products of several input arrays as a sequence of formulae. Each formula produces some intermediate array and the last formula gives the final result. A formula is either: **(i)** a multiplication formula of the form: $Tr(\ldots) = X(\ldots) \times Y(\ldots)$, or **(ii)** a summation formula of the form: $Tr(\ldots) = \sum_i X(\ldots)$, where the terms on the right hand side represent input arrays or intermediate arrays produced by a previously defined formula. Let IX, IY and ITr be the sets of indices in $X(\ldots)$, $Y(\ldots)$ and $Tr(\ldots)$, respectively. For a formula to be well-formed, every index in $X(\ldots)$ and $Y(\ldots)$, except the summation index in the second form, must appear in $Tr(\ldots)$. Thus $IX \cup IY \subseteq ITr$ for any multiplication formula, and $IX - \{i\} \subseteq ITr$ for any summation formula. Such a sequence of formulae fully specifies the multiplications and additions to be performed in computing the final result.

The problem of determining an operation-optimal sequence that is equivalent to the original expression has been previously addressed by us. We have shown that the problem is NP-complete and have developed a pruning search procedure that is very efficient in practice [17, 19, 18].

An issue of great significance for this computational context is the management of memory required to store the elements of the various arrays involved. Often, some of

$T1\ (\ b,c,d,f)$

$$= \sum_{e,l} B(b,e,f,l) \times D(c,d,e,l)$$

$T2\ (\ b,c,j,k)$

$$= \sum_{d,f} T1(b,c,d,f) \times C(d,f,j,k)$$

$S\ (\ a,b,i,j)$

$$= \sum_{c,k} T2(b,c,j,k) \times A(a,c,i,k)$$

(a) Formula sequence

```
read B(*,*,*,*)
read D(*,*,*,*)
T1(*,*,*,*) = 0
FOR b = 1, Nb
  FOR c = 1, Nc
    FOR d = 1, Nd
      FOR e = 1, Ne
        FOR f = 1, Nf
          FOR l = 1, Nl
            T1(b,c,d,f)
               += B(b,e,f,l) * D(c,d,e,l)
END FOR l, f, e, d, c, b
read C(*,*,*,*)
T2(*,*,*,*) = 0
FOR b = 1, Nb
  FOR c = 1, Nc
    FOR d = 1, Nd
      FOR f = 1, Nf
        FOR j = 1, Nj
          FOR k = 1, Nk
            T2(b,c,j,k)
               += T1(b,c,d,f) * C(d,f,j,k)
END FOR k, j, f, d, c, b
read A(*,*,*,*)
S(*,*,*,*) = 0
FOR a = 1, Na
  FOR b = 1, Nb
    FOR c = 1, Nc
      FOR i = 1, Ni
        FOR j = 1, Nj
          FOR k = 1, Nk
            S(a,b,i,j)
               += T2(b,c,j,k) * A(a,c,i,k)
END FOR k, j, i, c, b, a
write S(*,*,*,*)
```

(b) Direct implementation (unfused code)

Fig. 1. A sequence of formulae and the corresponding unfused code.

the input, output, and intermediate temporary arrays are too large to fit into the available physical memory. Therefore, the computation must be structured to operate on memory resident blocks of the arrays that are suitably moved between disk and main memory as needed. Similarly, effective use of cache requires that appropriate blocking or tiling of the computation is performed, whereby data reuse in cache is facilitated by operating on the arrays in blocks.

If all arrays were sufficiently small, the computation could be simply expressed as shown in Fig. 1(b). Here, all elements of the input arrays A, B, C, and D are first read in, the three sets of perfectly nested loops perform the needed computations, and the result array S is output. However, if any of the arrays is too large to fit in memory, the computation must be restructured to process the arrays in blocks or "slices." If, instead of fully creating an intermediate array (like T1) before using it, a portion of the array could be created and used before other elements of the array are created, the space required for the array could be significantly reduced. Similarly, even for input arrays, instead of reading in the entire array, blocks of the array could be read in and used before other blocks are brought in. A systematic approach to explore ways of reducing the memory requirement for the computation is to view it in terms of potential loop fusions. Loop fusion merges loop nests with common outer loops into larger imperfectly nested loops. When one loop nest produces an intermediate array that is consumed by another loop nest, fusing the two loop nests allows the dimension corresponding to the fused loop to be eliminated in the array. This results in a smaller intermediate array and thus reduces the memory requirements. For the example considered, the application of fusion is illustrated in Fig. 2(a). In this case, most arrays can be reduced to scalars without changing the number of arithmetic operations.

For a computation consisting of a number of nested loops, there will generally be a number of fusion choices that are not all mutually compatible. This is because different

```
FOR b = 1, Nb
 Sf(*,*,*) = 0
 FOR c = 1, Nc
  T2f(*,*) = 0
  FOR d = 1, Nd
   FOR f = 1, Nf
    T1f = 0
    FOR e = 1, Ne
     FOR l = 1, Nl
      Bf = read B(b,e,f,l)
      Df = read D(c,d,e,l)
      T1f += Bf * Df
    END FOR l, e
    FOR j = 1, Nj
     FOR k = 1, Nk
      Cf = read C(d,f,j,k)
      T2f(j,k) += T1f * Cf
   END FOR k, j, f, d
   FOR a = 1, Na
    FOR i = 1, Ni
     FOR j = 1, Nj
      FOR k = 1, Nk
       Af = read A(a,c,i,k)
       Sf(a,i,j) += T2f(j,k) * Af
   END FOR k, j, i, a, c
   write S(a,b,i,j) = Sf(a,i,j)
END FOR b
```

(a) Memory-reduced (fused) version

```
FOR b = 1, Nb
 Sf(*,*,*) = 0
 FOR cT = 1, Nc, Tc
  T2f(*,*,*) = 0
  FOR d = 1, Nd
   FOR f = 1, Nf
    T1f(*) = 0
    FOR e = 1, Ne
     FOR l = 1, Nl
      Bf = read B(b,e,f,l)
      Df(*) = read_tile D(cT,d,e,l)
      FOR cI = 1, Tc
       T1f(cI) += Bf * Df(cI)
    END FOR cI, l, e
    FOR j = 1, Nj
     FOR k = 1, Nk
      Cf = read C(d,f,j,k)
      FOR cI = 1, Tc
       T2f(cI,j,k) += T1f(cI) * Cf
   END FOR cI, k, j, f, d
   FOR a = 1, Na
    FOR i = 1, Ni
     FOR j = 1, Nj
      FOR k = 1, Nk
       Af(*) = read_tile A(a,cT,i,k)
       FOR cI = 1, Tc
        Sf(a,i,j) += T2f(cI,j,k) * Af(cI)
   END FOR cI, k, j, i, a, cT
   write S(a,b,i,j) = Sf(a,i,j)
END FOR b
```

(b) Tiling of the memory-reduced version

Fig. 2. Pseudocodes for (a) the memory-reduced (fused) solution, and (b) a tiled example of the same solution. In (b), the loop over c is split up in a tiling loop c_T, and an intra-tile loop c_I. The "tile size" is denoted by T_c. The procedure read_tile reads T_c elements from a four-dimensional array on the disk (A or D) into a one-dimensional memory array of size T_c (A_f or D_f).

fusion choices could require different loops to be made the outermost. In prior work, we addressed the problem of finding the choice of fusions for a given formula sequence that minimized the total space required for all arrays after fusion [15, 16, 14].

Having provided information about the computational context and some of the optimization issues in this context, we now provide an overview of the overall synthesis framework before focusing on the specific data locality optimization problem that we address in this paper.

3   Overview of the Synthesis System

Fig. 3 shows the components of the planned synthesis system. A brief description of the components follows:

Algebraic Transformations: It takes high-level input from the user in the form of tensor expressions (essentially sum-of-products array expressions) and synthesizes an output computation sequence. The input is expressed in terms of multidimensional summations of the product of terms, where each term is either an array, an elemental function with each arguments being an expression index, or an expression involving only element-wise operations on compatible arrays. The Algebraic Transformations module uses the properties of commutativity and associativity of addition and multiplication and the distributivity of multiplication over addition. It searches for all possible ways of applying these properties to an input sum-of-products expression, and determines a combination that results in an equivalent form of the computation with minimal operation cost. The application of the distributive law to factor a term out of a summation implies the need to use a temporary array to hold an intermediate result.

Memory Minimization: The operation-minimal computation sequence synthesized by the Algebraic Transformation module might require an excessive amount of memory due to the large arrays involved. The Memory Minimization module attempts to perform

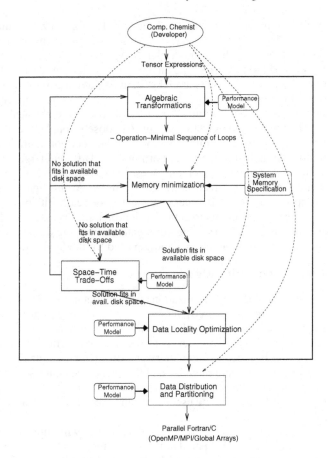

Fig. 3. The Planned Synthesis System

loop fusion transformations to reduce the memory requirements. This is done without any change to the number of arithmetic operations.

Space-Time Transformation: If the Memory Minimization module is unable to reduce memory requirements of the computation sequence below the available disk capacity on the system, the computation will be infeasible. This module seeks a trade-off that reduces memory requirements to acceptable levels while minimizing the computational penalty. If no such transformation is found, feedback is provided to the Memory Minimization module, causing it to seek a different solution. If the Space-Time Transformation module is successful in bringing down the memory requirement below the disk capacity, the Data Locality Optimization module is invoked. A framework for modeling space-time trade-offs and deriving transformations is currently under study.

Data Locality Optimization: If the space requirement exceeds physical memory capacity, portions of the arrays must be moved between disk and main memory as needed, in a way that maximizes reuse of elements in memory. The same considerations are involved in effectively minimizing cache misses — blocks of data must be moved between physical memory and the limited space available in the cache. In this paper, we focus on this step of the synthesis process. Given an imperfectly nested loop generated by the Memory Minimization module, the Data Locality Optimization module is responsible for generating an appropriately blocked form of the loops to maximize data reuse in the different levels of the memory hierarchy.

Data Distribution and Partitioning: The final step is to determine how best to partition the arrays among the processors of a parallel system. We assume a data-parallel model, where each operation in the operation sequence is distributed across the entire parallel machine. The arrays are to be disjointly partitioned between the physical memories of the processors. This model allows us to decouple (or loosely couple) the parallelization considerations from the operation minimization and memory considerations. The output of this module will be parallel code in Fortran or C. Different target programming paradigms can be easily supported, including message-passing with MPI and paradigms with global shared-memory abstractions, such as Global Arrays and OpenMP. Even with the shared-space paradigms, in order to achieve good scalability on highly parallel computer systems, careful attention to data distribution issues is essential. Thus the underlying abstraction used in determining good data partitioning decisions remains the same, independent of the programming paradigm used for the final code.

4   Data Locality Optimization Algorithm

We now address the data locality optimization problem that arises in this synthesis context. Given a memory-reduced (fused) version of the code, the goal of the algorithm is to find the appropriate blocking of the loops in order to maximize data reuse. The algorithm can be applied at different levels of the memory hierarchy, for example, to minimize data transfer between main memory and disk (I/O minimization), or to minimize data transfer between main memory and the cache (cache misses). In the rest of the paper, we focus mostly on the cache management problem. For the I/O optimization problem, the same approach could be used, replacing the cache size CS by the physical memory size MS.

There are two sources of data reuse: a) temporal reuse, with multiple references to the same memory location, and b) spatial reuse, with references to neighboring memory locations on the same cache line. To simplify the treatment in the rest of the paper, the cache line size is implicitly assumed to be one. In practice, tile sizes are determined under this assumption, and then the tile sizes corresponding to loop indices that index the fastest-varying dimension of any array are increased, if necessary, to equal the cache line size. In addition, other tiles may ned to be sized down slightly so that the total cache capacity is not exceeded.

We introduce a memory access cost model (*Cost*), that estimates the number of cache misses, as a function of tile sizes and loop bounds. For each loop — each node in the parse tree representation — we count the number *Accesses* of distinct array elements accessed in its scope. If this number is smaller than the number of elements that fit into the cache, then *Cost* = *Accesses*. Otherwise, it means that the elements in the cache are not reused from one loop iteration to the next, and the cost is obtained by multiplying the cost of the inner loop(s) — child node(s) in the parse tree — by the loop range.

To illustrate the cost model, we consider the fused code presented in Figure 2(a). The corresponding parse tree in Figure 4(a) shows the nesting structure of the loops. For the subtree rooted at node l, the number of elements accessed in the l loop is N_l for the arrays B and D, and 1 for the array $T1$, for a total of $2N_l + 1$ accesses. The computation of the number of accesses at the next node in the parse tree e (next loop nesting level) depends on the relative size of $2N_l + 1$ with respect to the cache size CS. If $2N_l + 1 > CS$, there can be no reuse of the array elements from one iteration of e to

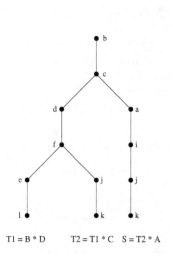

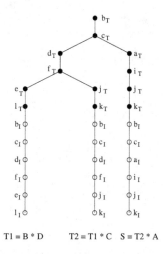

$T1 = B * D$ $T2 = T1 * C$ $S = T2 * A$

$T1 = B * D$ $T2 = T1 * C$ $S = T2 * A$

(a): No tiling: A node in the tree represents a loop nest; a parent-child pair represents an outer loop (parent node), and an inner loop (child node).

(b): With tiling: Each loop (node) in (a) is split into a tiling/intra-tile pair of loops (nodes). The intra-tile loops are then moved by fission and permutation operations toward the bottom of the tree.The tiling loops are represented by black nodes in the figure, while the intra-tile loops are represented by empty nodes.

Fig. 4. Parse trees for the fused loop structure shown in Figure 2(a).

the next. Therefore, the number of accesses for the subtree rooted at e is $N_e(2N_l + 1)$. However, if $2N_l + 1 < CS$, there is the possibility of data reuse in the e loop. The number of elements accessed in the e loop is $N_l * N_e$ for the arrays B and D, and 1 for the array $T1$ (the same element of $T1$ is repeatedly accessed), for a total of $2*N_l*N_e+1$ accesses. The new cost is again compared to CS, and then the next node in the parse tree is considered.

In practice, the problem has two additional aspects: the parse tree has branches (a parent node with multiple children nodes, corresponding to an outer loop with several adjacent inner loops), and each node in the tree is split into a parent-child pair, corresponding to a tiling loop node, and an intra-tile loop node. Figures 4(a) and 4(b) present the parse trees for the same computation, performed without and with loop tiling, respectively. For a given N-node untiled parse tree, the data locality optimization algorithm proceeds as follows: first, each node is split into a tiling/intra-tile pair. Subsequently, the resulting $2N$-node parse tree is transformed by loop permutation and intra-tile loop fission into an equivalent parse tree with the property that any tiling loop is exterior to any intra-tile loop (Figure 4(b)). Then, we pick starting values for the N tile sizes, thus fixing the loop ranges for the $2N$ nodes in the parse tree (if the original range of a loop is N_i, choosing a tile size T_i for the intra-tile loop also fixes the range N_i/T_i of the tiling loop).

We thus obtain a $2N$-node parse tree with well-defined loop ranges. Using a recursive top-down procedure, we compute the memory access cost for the computation (Figure 5) by traversing the corresponding nesting tree. Each node is associated with a loop index *LoopIndex*, a boolean value *Exceed* that keeps track of the number of distinct accesses in the loop scope in relation to the cache size CS, a list of arrays *ArrayList* accessed in its scope, and a memory access cost *Cost*. The core of the algorithm is the

```
Node: {
    Index LoopIndex
    boolean Exceed
    int Cost
    Node[] Children
    Array[] ArrayList
}

Array: {
    Index[] Indices
    int Accesses
}


boolean ContainIndex (Array A, Index LoopIndex) ≡
{T if LoopIndex ∈ A.Indices; F otherwise}

InsertArrayList (Node X, Node Y)
    foreach Array A ∈ Y.ArrayList
        if (A ∈ X.ArrayList) then
            X.ArrayList = X.ArrayList ∪ A
```

```
ComputeCost (Node X):
    X.ArrayList = NULL
    X.Cost = 0
    X.Exceed = F
    foreach Node Y ∈ X.Children
        ComputeCost (Y)
        InsertArrayList (X, Y)
        if (Y.Exceed) then X.Exceed = T
    if (X.Exceed) then
        foreach Node Y ∈ X.Children
            X.Cost += Y.Cost * X.Index.Range
        foreach Array A ∈ X.ArrayList
            A.Accesses = A.Accesses * X.Index.Range
    else
        int NewCost = 0
        foreach Array A ∈ X.ArrayList
            if (ContainIndex (A, X.LoopIndex)) then
                NewCost += A.Accesses * X.Index.Range
            else NewCost += A.Accesses
        if (NewCost < CacheSize) then
            X.Exceed = F
            X.Cost = NewCost
            foreach Array A ∈ X.ArrayList
                if (ContainIndex (A, X.LoopIndex)) then
                    A.Accesses *= X.Index.Range
        else
            X.Exceed = T
            foreach Array A ∈ X.ArrayList
                A.Accesses *= X.Index.Range
                X.Cost += A.Accesses
```

Fig. 5. Procedure **ComputeCost** for computing the memory access cost by a top-down recursive traversal of the parse tree.

procedure **ComputeCost**, presented in a pseudo-code format in Figure 5. **Compute-Cost** (Node X) computes the memory access cost of a sub-tree rooted at X, using the cost model based on the counting of distinct array elements accessed in the scope of the loop. The procedure outlines the computation of $X.Cost$ for the general case of a branched sub-tree with any number of children nodes.

Using this cost model, we arrive at a total memory access cost for the $2N$-node parse tree for given tile sizes. The procedure is then repeated for different sets of tile sizes, and new costs computed. In the end the lowest cost is chosen, thus determining the optimal tile sizes for the parse tree. We define our tile size search space in the following way: if N_i is a loop range, we use a tile size starting from $T_i = 1$ (no tiling), and successively increasing T_i by doubling it until it reaches N_i. This ensures a slow (logarithmic) growth of the search space with increasing array dimension for large N_i. If N_i is small enough, an exhaustive search can instead be performed.

5   Experimental Results

In this section, we present results of an experimental performance evaluation of the effectiveness of the data locality optimization algorithm developed in this paper. The algorithm from Section 4 was used to tile the code shown in Figure 1(b). Measurements were made on a single processor of a Silicon Graphics Origin 2000 system consisting of 32 300MHz IP31 processors and 16GB of main memory. Each processor has a MIPS R12000 CPU and a MIPS R12010 floating point unit, as well as 64KB on-chip caches (32KB data cache and 32KB instruction cache), and a secondary, off-chip 8MB unified data/instruction cache. The tile size selection algorithm presented earlier assumes a single cache. It can be extended in a straightforward fashion to multi-level caches, by multi-level tiling. However, in order to simplify measurement and presentation of experimental data, we chose to apply the algorithm only to the secondary cache, in order to minimize the number of secondary cache misses. For each computation, we determined the number of misses using the hardware counters on the Origin 2000. Three alternatives were compared:

x	Memory requirement			Performance (MFLOPs)			Cache misses		
	TA	FUS	UNF	TA	FUS	UNF	TA	FUS	UNF
0.2	8MB	30KB	32MB	484	88	475	7.2×10^5	5.2×10^6	3.2×10^6
0.4	8MB	0.2MB	0.5GB	470	101	451	1.4×10^7	8.3×10^7	5.3×10^7
0.5	8MB	0.5MB	1.2GB	491	80	460	2.9×10^7	1.8×10^9	1.4×10^8
0.6	8MB	0.8MB	2.6GB	477	97	407	6.8×10^7	3.3×10^9	4.2×10^8
0.7	8MB	1.3MB	4.8GB	482	95	N/A	1.0×10^8	7.0×10^9	N/A
0.8	8MB	1.9MB	8.2GB	466	83	N/A	1.9×10^8	7.9×10^9	N/A
1	16MB	3.7MB	20GB	481	92	N/A	4.8×10^8	2.9×10^{10}	N/A
1.2	48MB	6.5MB	41GB	472	98	N/A	1.1×10^9	4.8×10^{10}	N/A
1.4	80MB	10MB	77GB	486	85	N/A	1.8×10^9	8.5×10^{10}	N/A
1.6	120MB	15MB	131GB	483	82	N/A	3.5×10^9	1.6×10^{11}	N/A
1.8	192MB	22MB	211GB	465	77	N/A	5.1×10^9	3.2×10^{11}	N/A
2	240MB	29MB	318GB	476	79	N/A	8.1×10^9	5.0×10^{11}	N/A

Table 1. Performance data for fusion, plus tiling algorithm (TA), compared with performance data for fused alone (FUS), and unfused loops (UNF).

- UNF: no explicit fusion or tiling
- FUS: use of fusion alone, to reduce memory requirements
- TA: use of fusion, followed by tiling using algorithm presented in Section 4

We chose various array sizes for the problem to test the algorithm for a range of calculations typical in size for computational chemistry codes: $N_i = N_j = N_k = N_l = 40x$, $N_a = N_b = N_c = N_d = 300x$, and $N_e = N_f = 70x$, for x running from 0.2 to 2 (Table 1). For larger x, some arrays (e.g., $T1$) are so large that the the system's virtual memory limit is exceeded, so that loop fusion is necessary to bring the total memory requirements under the limit.

The codes were all compiled with the highest optimization level of the SGI Origin 2000 FORTRAN compiler (-O3). The performance data was generated over multiple runs, and average values are reported. Standard deviations are typically around 10MFLOPs. The experiments were run on a time-shared system; so some interference with other processes running at the same time on the machine was inevitable, and its effects are especially pronounced for the larger x tests.

Table 1 shows the memory requirement, measured performance, and the number of secondary cache misses generated by the three alternatives. The main observations from the experiment are:

- The total memory requirement is minimized by fusion of the loops over b, c, d, and f (FUS), bringing all four-dimensional arrays (e.g., $T2$) down to at most two explicit dimensions. However, the memory requirement of fusion plus the tiling algorithm (TA) is not much higher, since the tiling loops are fused, and the arrays are reduced to much smaller "tile" sizes. The UNF version has significantly higher memory requirements since no fusion has been applied to reduce temporary memory requirements. As x is increased, the UNF version requires more memory than the per-process virtual memory limit on the system.
- The maximally fused version (FUS) has the lowest memory requirement, but incurs a severe performance penalty due to the constraints imposed on the resulting loops that prevents effective tiling and exploitation of temporal reuse of some of the arrays, which leads to a higher number of cache misses, as shown in Table 1.
- The TA and UNF versions show comparable performance for smaller x. The SGI compiler is quite effective in tiling perfectly nested loops such as the sequence of

three matrix-matrix products present in the UNF version. The performance using the BLAS library routine DGEMM was found to be the same as that of the UNF version with a sequence of three nested loops corresponding to the three matrix products.

6 Related Work

Much work has been done on improving locality and parallelism through loop fusion. Kennedy and co-workers [11] have developed algorithms for modeling the degree of data sharing and for fusing a collection of loops to improve locality and parallelism. Singhai and McKinley [29] examined the effects of loop fusion on data locality and parallelism together. Although this problem is NP-hard, they were able to find optimal solutions in restricted cases and heuristic solutions for the general case. Gao et al. [6] studied the contraction of arrays into scalars through loop fusion as a means to reduce array access overhead. Their study is motivated by data locality enhancement and not memory reduction. Also, they only considered fusions of conformable loop nests, i.e., loop nests that contain exactly the same set of loops.

However, the work addressed in this paper considers a different use of loop fusion, which is to reduce array sizes and memory usage of automatically synthesized code containing nested loop structures. Traditional compiler research has not addressed this use of loop fusion because this problem does not arise with manually-produced programs. Recently, we investigated the problem of finding optimal loop fusion transformations for minimization of intermediate arrays in the context of the class of loops considered here [15]. To the best of our knowledge, the combination of loop tiling for data locality enhancement and loop fusion for memory reduction has not previously been considered.

Memory access cost can be reduced through loop transformations such as loop tiling, loop fusion, and loop reordering. Although considerable research on loop transformations for locality has been reported in the literature [22, 24, 33], issues concerning the need to use loop fusion and loop tiling in an integrated manner for locality and memory usage optimization have not been considered. Wolf et al. [34] consider the integrated treatment of fusion and tiling only from the point of view of enhancing locality and do not consider the impact of the amount of required memory; the memory requirement is a key issue for the problems considered in this paper. Loop tiling for enhancing data locality has been studied extensively [27, 33, 30], and analytic models of the impact of tiling on locality have been developed [7, 20, 25]. Recently, a data-centric version of tiling called data shackling has been developed [12, 13] (together with more recent work by Ahmed et al. [1]) which allows a cleaner treatment of locality enhancement in imperfectly nested loops.

The approach undertaken in this project bears similarities to some projects in other domains, such as the SPIRAL project which is aimed at the design of a system to generate efficient libraries for digital signal processing algorithms [35]. SPIRAL generates efficient implementations of algorithms expressed in a domain-specific language called SPL by a systematic search through the space of possible implementations. Several factors such as the lack of a need to perform space-time trade-offs renders the task faced by efforts such as SPIRAL and FFTW [5] less complex than what computational chemists face. Other efforts in automatically generating efficient implementations of programs include the telescoping languages project [10], the ATLAS [32] project for deriving efficient implementation of BLAS routines, and the PHIPAC [3] and TUNE [31] projects.

Recently, using a very different approach, we considered the data locality optimization problem arising in this synthesis context [4]. In that work, we developed an integrated approach to fusion and tiling transformations for the class of loops addressed. However, that algorithm was only applicable when the sum-of-products expression satisfied certain constraints on the relationship between the array indices in the expression. The algorithm developed in this paper does not impose any of the restrictions assumed in [4]. It takes a very different approach to effective tiling — first perform fusion to minimize memory requirements, followed by a combination of loop fission, tiling and array expansion transformations to maximize data reuse.

7   Conclusion

This paper has described a project on developing a program synthesis system to facilitate the development of high-performance parallel programs for a class of computations encountered in computational chemistry and computational physics. These computations are expressible as a set of tensor contractions and arise in electronic structure calculations. The paper has provided an overview of the planned synthesis system and has presented a new optimization approach that can serve as the basis for a key component of the system for performing data locality optimizations. Preliminary results are very encouraging and show that the approach is effective.

Acknowledgments

We would like to thank the Ohio Supercomputer Center (OSC) for the use of their computing facilities, and the National Science Foundation for partial support through grants DMR-9520319, CCR-0073800, and NSF Young Investigator Award CCR-9457768.

References

1. N. Ahmed, N. Mateev, and K. Pingali. Synthesizing transformations for locality enhancement of imperfectly-nested loops. *ACM Intl. Conf. on Supercomputing*, 2000.
2. W. Aulbur. *Parallel Implementation of Quasiparticle Calculations of Semiconductors and Insulators*, Ph.D. Dissertation, Ohio State University, Columbus, OH, October 1996.
3. J. Bilmes, K. Asanovic, C. Chin, and J. Demmel. Optimizing matrix multiply using PHiPAC. In *Proc. ACM International Conference on Supercomputing*, pp. 340–347, 1997.
4. D. Cociorva, J. Wilkins, C.-C. Lam, G. Baumgartner, P. Sadayappan, and J. Ramanujam. Loop optimization for a class of memory-constrained computations. In *Proc. 15th ACM International Conference on Supercomputing*, pp. 500–509, Sorrento, Italy, June 2001.
5. M. Frigo and S. Johnson. FFTW: An adaptive software architecture for the FFT. In *Proc. ICASSP 98*, Volume 3, pages 1381–1384, 1998, http://www.fftw.org.
6. G. Gao, R. Olsen, V. Sarkar and R. Thekkath. Collective Loop Fusion for Array Contraction. *Proc. 5th LCPC Workshop* New Haven, CT, Aug. 1992.
7. S. Ghosh, M. Martonosi and S. Malik. Precise Miss Analysis for Program Transformations with Caches of Arbitrary Associativity. *8th ACM Intl. Conf. on Architectural Support for Programming Languages and Operating Systems*, San Jose, CA, Oct. 1998.
8. M. S. Hybertsen and S. G. Louie. Electronic correlation in semiconductors and insulators: band gaps and quasiparticle energies. *Phys. Rev. B*, 34:5390, 1986.
9. J. Johnson, R. Johnson, D. Rodriguez, and R. Tolimieri. A methodology for designing, modifying, and implementing Fourier transform algorithms on various architectures. *Circuits, Systems and Signal Processing*, 9(4):449–500, 1990.
10. K. Kennedy et. al., Telescoping Languages: A Strategy for Automatic Generation of Scientific Problem-Solving Systems from Annotated Libraries. To appear in *Journal of Parallel and Distributed Computing*, 2001.

11. K. Kennedy. Fast greedy weighted fusion. *ACM Intl. Conf. on Supercomputing*, May 2000.
12. I. Kodukula, N. Ahmed, and K. Pingali. Data-centric multi-level blocking. In *Proc. SIGPLAN Conf. Programming Language Design and Implementation*, June 1997.
13. I. Kodukula, K. Pingali, R. Cox, and D. Maydan. An experimental evaluation of tiling and shackling for memory hierarchy management. In *Proc. ACM International Conference on Supercomputing (ICS 99)*, Rhodes, Greece, June 1999.
14. C. Lam. *Performance Optimization of a Class of Loops Implementing Multi-Dimensional Integrals*, Ph.D. Dissertation, The Ohio State University, Columbus, OH, August 1999.
15. C. Lam, D. Cociorva, G. Baumgartner and P. Sadayappan. Optimization of Memory Usage and Communication Requirements for a Class of Loops Implementing Multi-Dimensional Integrals. *Proc. 12th LCPC Workshop* San Diego, CA, Aug. 1999.
16. C. Lam, D. Cociorva, G. Baumgartner, and P. Sadayappan. Memory-optimal evaluation of expression trees involving large objects. In *Proc. Intl. Conf. on High Perf. Comp.*, Dec. 1999.
17. C. Lam, P. Sadayappan, and R. Wenger. Optimal reordering and mapping of a class of nested-loops for parallel execution. In *9th LCPC Workshop*, San Jose, Aug. 1996.
18. C. Lam, P. Sadayappan and R. Wenger. On Optimizing a Class of Multi-Dimensional Loops with Reductions for Parallel Execution. *Par. Proc. Lett.*, (7) 2, pp. 157–168, 1997.
19. C. Lam, P. Sadayappan and R. Wenger. Optimization of a Class of Multi-Dimensional Integrals on Parallel Machines. *Proc. of Eighth SIAM Conf. on Parallel Processing for Scientific Computing*, Minneapolis, MN, March 1997.
20. M. S. Lam, E. E. Rothberg, and M. E. Wolf. The cache performance and optimizations of blocked algorithms. In *Proc. of Fourth Intl. Conf. on Architectural Support for Programming Languages and Operating Systems*, April 1991.
21. T. J. Lee and G. E. Scuseria. Achieving chemical accuracy with coupled cluster theory. In S. R. Langhoff (Ed.), *Quantum Mechanical Electronic Structure Calculations with Chemical Accuracy*, pp. 47–109, Kluwer Academic, 1997.
22. W. Li. Compiler cache optimizations for banded matrix problems. In *International Conference on Supercomputing*, Barcelona, Spain, July 1995.
23. J. M. L. Martin. In P. v. R. Schleyer, P. R. Schreiner, N. L. Allinger, T. Clark, J. Gasteiger, P. Kollman, H. F. Schaefer III (Eds.), *Encyclopedia of Computational Chemistry*. Wiley & Sons, Berne (Switzerland). Vol. 1, pp. 115–128, 1998.
24. K. S. McKinley, S. Carr and C.-W. Tseng. Improving Data Locality with Loop Transformations. *ACM TOPLAS*, 18(4):424–453, July 1996.
25. N. Mitchell, K. Högstedt, L. Carter, and J. Ferrante. Quantifying the multi-level nature of tiling interactions. *Intl. Journal of Parallel Programming*, 26(6):641–670, June 1998.
26. G. Rivera and C.-W. Tseng. Data Transformations for Eliminating Conflict Misses. *ACM SIGPLAN PLDI*, June 1998.
27. G. Rivera and C.-W. Tseng. Eliminating Conflict Misses for High Performance Architectures. *Proc. of 1998 Intl. Conf. on Supercomputing*, July 1998.
28. H. N. Rojas, R. W. Godby, and R. J. Needs. Space-time method for Ab-initio calculations of self-energies and dielectric response functions of solids. *Phys. Rev. Lett.*, 74:1827, 1995.
29. S. Singhai and K. S. McKinley. A Parameterized Loop Fusion Algorithm for Improving Parallelism and Cache Locality. *The Computer Journal*, 40(6):340–355, 1997.
30. Y. Song and Z. Li. New Tiling Techniques to Improve Cache Temporal Locality. *ACM SIGPLAN PLDI*, May 1999.
31. M. Thottethodi, S. Chatterjee, and A. Lebeck. Tuning Strassen's matrix multiplication for memory hierarchies. In *Proc. Supercomputing '98*, Nov. 1998.
32. R. Whaley and J. Dongarra. Automatically Tuned Linear Algebra Software (ATLAS). In *Proc. Supercomputing '98*, Nov. 1998.
33. M. E. Wolf and M. S. Lam. A Data Locality Algorithm. *ACM SIGPLAN PLDI*, June 1991.
34. M. E. Wolf, D. E. Maydan, and D. J. Chen. Combining loop transformations considering caches and scheduling. In *Proceedings of the 29th Annual International Symposium on Microarchitecture*, pages 274–286, Paris, France, December 2–4, 1996.
35. J. Xiong, D. Padua, and J. Johnson. SPL: A language and compiler for DSP algorithms. *ACM SIGPLAN PLDI*, June 2001.

Block Asynchronous I/O: A Flexible Infrastructure for User-Level Filesystems

Muthian Sivathanu, Venkateshwaran Venkataramani, and
Remzi H. Arpaci-Dusseau

University of Wisconsin-Madison

Abstract. Block Asynchronous I/O (BAIO) is a mechanism that strives
to eliminate the kernel abstraction of a filesystem. In-kernel filesystems
serve all applications with a generic set of policies, do not take advantage
of application-level knowledge, and consequently deliver sub-optimal performance to a majority of applications. BAIO is a low-level disk access
mechanism that solves this problem by exporting the filesystem component of the kernel to the application level, thereby facilitating construction of customized user-level filesystems. The role of the kernel is
restricted to regulating access to disk by multiple processes, keeping track
of ownership information, and enforcing protection boundaries. All other
policies, including physical layout of data on disk and the caching and
prefetching of data, are implemented at application-level, in a manner
that best suits the specific requirements of the application.

1 Introduction

Generality and performance have long been at odds in operating systems [1].
Often designed as general-purpose facilities, operating systems must support a
wide class of potential applications, and the result is that they offer reasonable
performance for a majority of programs, while providing truly excellent performance for only a very few.

Recent work in application-specific operating systems such as the Spin [3] and
Vino [8] has addressed this inherent limitation by restructuring the operating
system to be *extensible*. Applications running on top of such systems can tailor
policies of the OS so as to make better-informed decisions, and thus can improve
performance by orders of magnitude in certain scenarios. However, these efforts
all mandate that an entirely new operating system structure be accepted and
widely deployed in order for such functionality to become available, which may
not be realistic [2].

Of particular importance to a large class of important applications, including
web servers, file servers, and databases, is the performance of the file system. To
achieve optimal disk I/O performance, it is imperative that applications have
full control over all aspects of the layout and management of data on disk.

B. Monien, V.K. Prasanna, S. Vajapeyam (Eds.): HiPC 2001, LNCS 2228, pp. 249–261, 2001.
© Springer-Verlag Berlin Heidelberg 2001

Applications often have a complete understanding of the semantics of the data they manage, and hence are the ideal candidates to dictate decisions on how the data is to be laid out on disk, what data is to be cached, and when prefetching should occur.

In this paper, we describe Block Asynchronous I/O (BAIO), an in-kernel substrate that enables direct access to disk. User-level library-based file systems can be built on top of BAIO, and thus can deliver high performance to applications by exploiting application-specific knowledge. The role of the kernel is limited to regulating access to the disk by multiple processes and imposing protection boundaries. BAIO differs from the large body of work on extensible operating systems in that it only requires small modifications to a stock Linux kernel.

There are three keys to the implementation of BAIO: an asynchronous interface, limited but powerful control over disk placement, and a flexible caching scheme. Asynchrony enables the construction of highly efficient I/O-bound applications without the overhead of threads, can be used to overlap I/O latency with other operations, and can increase the effectiveness of disk scheduling. Control over layout allows applications to place related data items near one another, and thus minimize disk seek and rotational overheads. Flexible caching allows applications to place data likely to be accessed by many applications or across many runs in a shared global cache, and to turn off caching (or cache in the application itself) data that is likely important only to a particular run of the application. In combination, these three features combine to allow applications or user-level filesystems built on top of BAIO to extract a high level of performance from the underlying disk system, by specializing their file layout and access to best suit their needs.

The rest of this paper is organized as follows. In Section 2, we discuss conventional disk I/O methods and their disadvantages. In Section 3, we introduce BAIO and explain its advantages as compared to conventional methods, and in Section 4, we discuss implementation details. Finally, in Section 5, we provide a performance evaluation, discuss related work in Section 6, and conclude in Section 7.

2   Conventional I/O

There are predominantly two ways in which disk I/O is conventionally performed. One is through an in-kernel filesystem, and the other is through the raw-disk interface provided by most UNIX systems. Both these methods suffer from major limitations. As discussed above, filesystems suffer due to their generality, as they are designed to be "reasonable" for a majority of applications, rather than being "optimal" for a specific one. For example, filesystems commonly have a read-ahead scheme where they prefetch a certain number of blocks that are contiguous to the requested block. This policy fails to consider the fact that the access patterns of applications need not necessarily be sequential, and in such a case, the additional prefetch that the filesystem does is unnecessary work that only degrades performance. I/O-intensive applications such as web servers and databases suffer greatly from this generality, as these applications have highly-specialized requirements [9].

To accommodate specialized applications like those mentioned above, many UNIX systems provide a raw-disk interface, allowing applications to gain direct control over a logical disk partition. Though this eliminates many of the problems due the generality of a filesystem, it has its own set of limitations. First, access to raw disk is limited to privileged applications, so non-privileged applications must still use the filesystem. Second, the granularity of access is a whole device partition, and hence multiple applications cannot co-exist within the same partition. Third, there is no notion of ownership within a partition.

Thus filesystems and the raw disk are two extremes, one in which the kernel provides a full I/O service, and the other in which it does not even provide the basic OS service of multiplexing. A mechanism that leaves policy decisions to the user level but still manages ownership and protection boundaries across applications would be clearly more effective and flexible than either of these schemes.

A limitation that both the filesystem interface and the raw-disk interface have in common is the fact that these interfaces are essentially synchronous, i.e., an application that initiates an I/O request is blocked until completion of the I/O. Considering that disks perform several orders of magnitude slower than the CPU, a blocking I/O mechanism incurs a great performance penalty. Though the impact of a synchronous mechanism may not be drastic in a heavily multiprogrammed workload where the overall throughput of the system is more important rather than that of individual applications, the impact is quite significant in dedicated systems like web-servers and database systems. For example, it is not reasonable to impose that a web-server application be blocked on I/O to fetch a page from its disk, thereby disabling the server from accepting or servicing subsequent connections in the meantime. What would clearly be more desirable, is a mechanism where the disk I/O and processing could be overlapped. Currently, this is being done by employing multiple threads to service different I/O requests, but this model of multiple threads scales very poorly and the overhead of context-switching and management of threads very soon becomes a performance bottleneck. A better solution would be an interface that exposes the inherent asynchrony of disks to applications. With such an interface, the process issuing an I/O request is returned control immediately and is notified by the kernel on completion of the I/O. An asynchronous interface will also facilitate application-specific prefetching of data from disk.

3   Block Asynchronous I/O

3.1   Overview

Block Asynchronous I/O (BAIO) is a solution aimed at addressing the above-mentioned limitations in performing disk I/O through conventional methods. BAIO is an in-kernel infrastructure that provides a direct, protected interface to the disk. With this mechanism, applications can specify which disk blocks to read or write, and the kernel only takes care of assigning capabilities to disk blocks and ensuring protected access by the applications. All other policies with regard to organization and management of data on disk are left to the user applications. In effect, applications that utilize BAIO have the power and flexibility to create their own filesystems customized to their peculiarities and specific requirements.

With BAIO, applications also have precise control over how data is laid out on disk. Since the BAIO interface is at the disk-block level, applications can place closely related data in physically contiguous disk locations, thereby ensuring that access to such data incurs minimal seek and rotational overhead.

The interface that BAIO provides is asynchronous, which makes it flexible and efficient, because applications can overlap disk I/O with useful processing without incurring the additional overhead of multiple threads of control. Moreover, this enables applications to implement customized prefetching policies. Applications can decide, based on the semantic knowledge available to them, which data blocks are accessed together. This notion of "semantic contiguity" of data is more accurate than the "physical contiguity" that generic filesystems try to exploit. Because applications know the exact location of their data, they can prefetch those blocks that contain semantically contiguous data, which have a very strong likelihood of being accessed given knowledge of application access patterns. In doing so, the performance degradation due to the *ad hoc* prefetching techniques of the kernel is avoided. Note that the benefit of customized prefetching is difficult to realize even with the raw-disk interface, because of its synchronous nature.

Applications can also decide the extent of caching required and the exact disk blocks to be cached. In-kernel filesystems utilize a global buffer cache and treat all data blocks equally (likely caching those that have been accessed recently), whether or not those blocks are likely to be accessed again. With BAIO, applications or user-level libraries can decide which blocks to cache based on specific knowledge of access patterns. The benefits of application controlled caching have already been studied [2], and an exhaustive discussion on the various benefits is beyond the scope of this paper. By providing for application-specific caching, BAIO allows applications to take advantage of these benefits.

Though BAIO provides a direct interface to the disk to applications, multiple applications can co-exist and share a single logical device, as opposed to the raw-disk interface where the unit of ownership is a whole device. The kernel keeps track of fine-grained ownership and capability information for different sets of disk blocks and ensures that protection boundaries are not violated – applications can only access those blocks that they are authorized to access.

3.2   Basic Operation

An important aspect of the BAIO interface is the notion of a "disk segment." We define a disk segment as a sequence of physically contiguous disk blocks. The application sees a disk segment as a logically separate disk with a starting block number of zero. The logical block numbers are not the same as the actual physical disk block numbers, but blocks that are contiguous with respect to logical block numbers are guaranteed to be physically contiguous. We arrived at this interface because we felt that applications are more concerned with ensuring that data accessed together are physically contiguous on disk and do not generally need to decide or know about the exact physical position in which data is stored. Thus, regardless of where a disk segment is mapped onto the physical disk, blocks in a disk segment are guaranteed to be physically contiguous, and therefore any decision that the application makes based on logical contiguity of disk blocks will remain valid in the physical layout also.

An application makes a request to the kernel for a disk segment of a specified size, upon which the kernel looks for a physically contiguous set of free disk blocks of the desired size, and subject to quota and other limitations, creates a capability for the relevant user to the disk segment. Subsequently, the application can open the disk segment for use, whereupon it is granted the capability to that disk segment by the kernel. This capability is valid for the lifetime of the application, and the capability allows the application to perform I/O on the disk segment directly by specifying the exact blocks to read or write. Each read or write request must also include the capability that is granted during open. A capability is akin to a file descriptor in UNIX for normal files. As the BAIO read/write interface is non-blocking, the process regains control immediately after issuing a request, and is notified upon completion of the I/O.

Multiple per-block read and write calls can be merged into a single BAIO operation, which has two primary advantages. First, the disk driver can create a better schedule of disk requests, minimizing seeks and rotational overhead. The more requests the driver has, the better job of scheduling it is likely to do (note that the asynchronous interface BAIO provides also helps in this regard). Second, system call overhead is reduced, as merging reduces the effective number of system calls an application invokes. Thus, the application can control the number of user-kernel crossings, in a manner similar to but more flexible than the readv() and writev() system calls.

By giving applications total control over a disk segment, BAIO creates the potential for every disk segment to have its own filesystem, tailor-made to the specialized requirements of the application that manages it. Traditional filesystems can be implemented on top of BAIO for the large class of applications that do not require the sophistication of an interface like BAIO. Such applications can simply link with the standard filesystem implementation and operate as before. Another possible approach to support backwards compatibility is to deploy file-server applications each implementing one standard filesystem. Applications can connect to whichever file server they are interested in, and read and write calls can be transparently redirected through stubs to the appropriate file servers. This will permit existing applications to be easily adapted to run on top of BAIO, since it will only involve relinking of the object binary with the appropriate user-level filesystem implementation.

3.3   Structure of a BAIO Device

In the kernel, BAIO keeps track of information pertaining to various disk segments, including the mapping of a disk segment to the physical disk, ownership attributes for disk segments, and so forth. This information is held in a per-disk-segment structure called an inode. The inode also contains the name of the disk segment within itself. This is in contrast to the UNIX style of having names and their inode numbers in directories, and was necessitated because we chose to have a flat namespace for naming disk segments, in-order to provide flexibility to the user-level filesystem to implement its own naming scheme, rather than impose a fixed hierarchical naming scheme. Inodes are hashed by the disk segment name. The kernel also maintains global information on the free blocks available for future allocation.

The first block in a BAIO device contains the superblock structure which maintains information on the total number of blocks in the device, the block-size of the device, the number of inodes on disk and pointers to the start of free list blocks, inode blocks and data blocks. Free lists are maintained in the form of bitmaps, with a free list hint indicating which position in the free list to start looking for a contiguous stream of free blocks. Considering that disk segments are meant to be reasonably large (spanning several megabytes), this organization of free block information is reasonable, especially in light of the fact that creation of new disk segments is a rare event compared to regular open, read, or write operations.

In most cases, the free-block hint maintained enables fast location of a free chunk. Again this decision on maintenance of free block information is not as crucial in our model as it is in normal filesystems because, in the latter, free blocks have to be found every time a write is made, and therefore sub-optimal tracking of free block information will slow down all writes. In BAIO, the free list is accessed only at the time of disk segment creation and once a segment is created; applications just read or write into pre-allocated disk blocks and hence excessive optimization in this area was not warranted. The number of inodes has been fixed at 1/1000th of the total number of data blocks in the system. Clearly, the average length of segments should be large enough to amortize this fixed overhead.

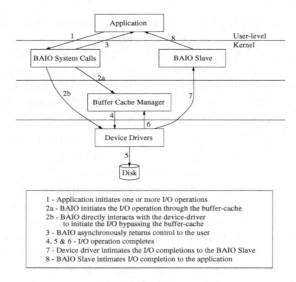

Fig. 1. Control flow during a BAIO operation.

3.4   Asynchrony and Caching

The BAIO architecture achieves asynchrony by queueing I/O requests at the device and returning control to the application immediately. A BAIO service daemon takes care of intimating completion of I/O to the application. The service daemon enters the kernel through a specific entry-point and never returns. Its sole duty is to look through the request queue to find which I/O operations have completed and notify the processes on whose behalf the I/O was performed.

Since this daemon is also in-charge of copying data to the address space of the application in the event of a read operation, it needs to have access to the virtual address space of the application. There are two methods in which this could be implemented. The first method is to use in-kernel shared-memory objects to share a memory region between the application and the service daemon. In this model, there will be a single BAIO service daemon for the entire system, managing the I/O requests of all processes. The drawback of this model is that considerable overhead is incurred at the service daemon in binding to shared-memory objects and the single service daemon can become a performance bottleneck. Moreover, this would require that the application allocate a shared-memory object and attach its data buffer to the shared-memory region prior to invoking the BAIO system call. Since there is a system-wide limit on the number of shared memory objects that can exist, this naturally places a limitation on the number of BAIO requests that can be pending simultaneously. However, this method does have certain advantages in that the single service daemon has a complete view of all pending requests and hence can potentially make better scheduling decisions.

The other method, which we have chosen for our model, is to have the service daemon share the application's virtual address space. This means that each process has its own BAIO service thread and this thread can directly write into the application's address space. This is more scalable than the first alternative because now the service thread is only in-charge of BAIO requests initiated by a single process and the overhead of attaching to shared memory objects on every I/O completion is eliminated.

The overall control flow during an I/O operation through BAIO is depicted in Figure 1. There are two kinds of interfaces that BAIO provides to applications. In the first method, the I/O takes place through the global buffer cache of the kernel, which allows for sharing of caching across applications, at the cost of generality. The advantage of this method is that multiple applications will be able to share the global kernel buffer cache, which may not be possible with application-level caching alone. This method benefits a scenario in which multiple applications access the same set of disk blocks, as otherwise, each application will try to cache it separately, wasting memory. However, this method has the drawback of introducing some generality into the caching scheme for disk blocks, and could lead to double-buffering.

In the second method, I/O operations bypass the buffer cache of the kernel entirely. In this method, the application directly interacts with the device driver and does not use the buffer cache. This is similar to the raw interface provided in UNIX systems. The option of whether or not to use the buffer cache can be specified on a per-operation basis, thereby providing for a fine-grained choice by the application. Blocks which the application thinks will be shared by other processes can be made to go through the buffer cache while others may be read/written directly. For example, if there is a large read to be made, the application may wish to ensure that this read does not flush the current contents of the buffer cache, and can therefore use the non-buffered interface.

4   BAIO Implementation

The BAIO model has been implemented in the Linux kernel. The interface is provided in the form of a set of system calls. Additionally, a utility for configuring a disk device for BAIO has been implemented. A process using BAIO keeps track of additional information in its process control block, via a block descriptor table, which is similar to the file descriptor table in UNIX. Each table entry keeps track of information such as a pointer to the inode of the disk segment, a reference count indicating number of processes that have the disk segment open, and so forth. Additionally, a process also maintains a pointer to its service daemon. The service daemon maintains a queue of BAIO requests issued by its master process and are pending. Processes also maintain synchronization information for regulating access to the request queue which is shared between the master process and the service daemon.

When an application invokes the BAIO system call, the process adds the request to the request queue of its service daemon and issues the request to the buffer cache layer or the device driver, as was required by the application. It then returns control to the application immediately. The disk driver, on completion of the I/O, wakes up the service daemon. On waking up, the service daemon looks through its request queue to find out which of the requests have completed, and accordingly notifies the application. Notification can be done in two ways. The first is to set a status variable in the application address space. The second is to signal the application on completion of I/O. A signal-based scheme would substantially complicate the application code since it must be prepared to process completion of I/O requests in signal handlers. Moreover, signals may be lost if the application spends too much time in handling a request. The first scheme is simpler and more scalable than a signal based scheme, and therefore, has been chosen in our implementation.

The set of system calls which constitute the BAIO interface to the application level are:

1. `create_dseg`: takes a device name, segment name and a segment size as input, allocates a disk segment of the required size if possible
2. `open_dseg`: opens an existing disk segment specified by the name and device, and returns a descriptor to the segment
3. `baio`: this is the call used to perform I/O. Multiple read/write requests can be specified as part of a single BAIO call; includes a block descriptor as input
4. `baio_service`: this is the entry-point to the service daemon. The service daemon invokes this syscall and never returns.
5. `baio_mount`: this call takes a device name as argument, and if the device is configured for BAIO, mounts the device.

5   Performance Evaluation

The performance enhancement possible by allowing application control over various aspects of filesystem policies has been studied in great detail. For example, Cao *et al.* show that application-level control over file-caching can reduce execution time by around 45% [5]. In another study, Cao *et al.* measured that application-controlled file caching and prefetching reduced the running time by 3% to 49% for single-process workloads and by 5% to 76% for multiprocess workloads [4]. By exporting the entire filesystem policy to the user-level, we enable applications to obtain all these performance advantages.

The measurements we have taken mainly evaluate the impact of exposing the asynchrony of disks to applications. We compare the performance of our model with the UNIX raw disk implementation and another asynchronous I/O technique developed by SGI called Kernel Asynchronous I/O (KAIO). Experiments were performed with and without the effect of buffer cache. In the first set of experiments, we compared non-buffered BAIO, raw disk, and KAIO over raw disk. In the next, we compared buffered BAIO, ext2 and KAIO over ext2.

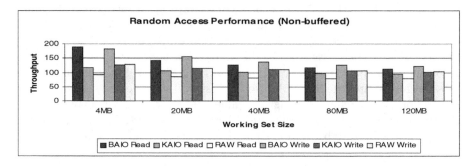

Fig. 2. Non-buffered random reads and writes.

Considering that an asynchronous interface will benefit those applications that can overlap disk I/O with useful processing, we have chosen a workload that resembles that in a web-server. There is a continuous stream of requests, and each request will involve a disk I/O and some processing on the data read/written. In the asynchronous interfaces, we pipeline the disk I/O and processing over several stages. Stage i will process the data corresponding to the (i-k)*th* request while issuing the I/O for the i*th* request. This way, by the time data is required for processing, it will already be available since I/O for that data was initiated quite in advance. In other words, the application does not spend any time waiting for I/O to complete. In our studies, as expected, we find that both BAIO and KAIO consistently outperform raw I/O and ext2. BAIO performs much better than KAIO as well. The reason for this is that while BAIO exposes the full asynchrony of the disk to applications, KAIO employs slave threads which invoke the blocking I/O routines. Hence the asynchrony of KAIO is limited to the number of slave threads in operation. Having a large number of threads is not a solution because, as the number of threads increases, system performance degrades due to the additional overhead of context switches and thread management.

The performance metric we have chosen in all the three graphs presented, is the average throughput of the application, i.e the number of requests serviced by the application in one second. Figures 2 and 3 show the behavior of random and sequential reads and writes without the effect of buffer cache (each read/write involved a 8K data transfer, from/to disk blocks randomly chosen within the working set size). In this case, BAIO with buffer cache disabled was compared with KAIO running over raw disk, and plain raw I/O. We observe that in the random access case, BAIO performs on an average 25% better than KAIO and 54% better than raw I/O. This is possible because BAIO allows the disk latency to be partially hidden by enabling applications to overlap disk I/O with processing, so that by the time the results of an I/O are required, they are most likely to be

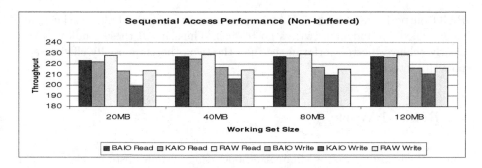

Fig. 3. Non-buffered sequential reads and writes.

already available, thereby eliminating the time the application spends in waiting for I/O. In sequential access, BAIO outperforms KAIO, but raw I/O performs marginally better than BAIO. The reason we believe this happens is because sequential raw I/O is inherently fast and not much is gained by asynchrony – instead the overhead of an external process intimating I/O completion and additional context switching brings down the performance of the asynchronous schemes. This has the obvious implication that exposing asynchrony is of greater benefit for a random access workload than one of sequential access.

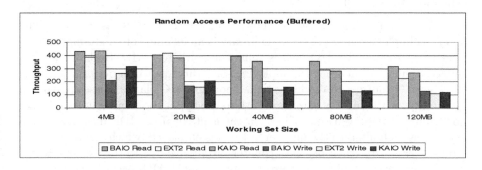

Fig. 4. Buffered random reads and writes.

In the next study, shown in Figure 4, we compared BAIO with all requests going through the buffer cache, KAIO over ext2 and plain ext2. We observe that BAIO consistently outperforms KAIO and raw I/O in random reads, but in random writes, ext2 and KAIO perform better. This is because of the delayed write strategy of ext2, which also applies to KAIO since the slave threads of KAIO internally use ext2 I/O calls. Since writes are fully buffered and only flushed to disk once at the completion of the experiment, many writes to the same block are absorbed by the buffer cache. Moreover, the sync is highly optimized because of the large number of blocks to write to disk, thereby facilitating optimal disk scheduling. In spite of all these, for very large working sets, exceeding 80 MB,

BAIO outperforms ext2 and KAIO, since at this size, the buffer cache tends to get filled up and more writes go to disk. Thus, in all these studies, BAIO clearly achieves better performance than the two other supposedly efficient and powerful interfaces.

6   Related Work

The idea of minimizing or eliminating operating system abstractions is as old as the concept of micro-kernels are. There have been numerous contributions that argue for a minimal operating system with policies exposed to the applications. Nucleus [7] and Exokernel [6] are two classic examples of such models. BAIO draws inspiration from these contributions and applies the principle in the restricted context of a filesystem. What is different in BAIO compared to prior works is that BAIO can fit into an existing operating system with very few modifications, as opposed to ideas like Exokernel which warranted a total "start from scratch" revamp. The performance advantages of exposing certain filesystem policies like caching and prefetching to the the application level have been studied by Cao et al. [5,4]. BAIO takes this to an extreme by allowing applications to dictate virtually every aspect of filesystem policy and operation, thereby providing applications with the highest possible performance advantage achievable through customization.

The idea of providing an asynchronous interface to disk I/O is not new. The FreeBSD implementation of the AIO library is an example of an attempt to provide asynchronous disk I/O, and is similar in spirit to KAIO. Both of these interfaces achieve asynchrony by employing slave threads, which perform the same blocking I/O routines of the filesystem. Though KAIO claims to use split-phase I/O to expose the full asynchrony of disk to the application, our studies indicate no significant improvement, at least in the context of the IDE disk. Hence the asynchrony is limited to the number of threads employed, which cannot increase indefinitely since it then becomes counter-productive due to the excessive overhead in context-switching and thread management. BAIO, because of its careful integration into the kernel and direct interaction with the disk device driver, outperforms both these models, as was shown in the performance study (KAIO claims to outperform AIO, so performing better than KAIO is equivalent to outperforming both).

7   Conclusion

In an attempt to overcome the potential performance limitations imposed by generic in-kernel filesystems, we have proposed a mechanism known as BAIO by which the application is given full control over the way its data is both organized and managed. Application-level semantic knowledge facilitates optimized policies and decisions that perform specifically well to the peculiarities of the application. In implementing BAIO, we take special care to ensure that the kernel does not determine any policy with regard to the management of application data on disk, other than enforcing protection boundaries across applications and maintaining the notion of ownership of disk segments. One of the main features of the BAIO model is that it integrates quite easily into an existing operating system, in contrast to extensible systems of the past which require that entirely new kernels be designed, developed, and deployed. A remaining challenge for BAIObased systems is to build and experiment with a broad range of specialized user-level filesystems, and to understand how concurrently running user-level filesystems interact with one another.

References

1. T. Anderson, B. Bershad, E. Lazowska, and H. Levy. Scheduler Activations: Effective Kernel Support for the User-Level Management of Parallelism. In *ACM TOCS*, 1992.
2. A. C. Arpaci-Dusseau and R. H. Arpaci-Dusseau. Information and Control in Gray-Box Systems. In *SOSP 18*, October 2001.
3. B. N. Bershad, S. Savage, P. Pardyak, E. G. Sirer, M. E. Fiuczynski, D. Becker, C. Chambers, and S. Eggers. Extensibility, Safety and Performance in the SPIN Operating System. In *SOSP 15*, December 1995.
4. P. Cao, E. W. Felten, A. R. Karlin, and K. Li. Implementation and Performance of Integrated Application-controlled File Caching, Prefetching, and Disk Scheduling. *ACM TOCS*, 14(4), November 1996.
5. P. Cao, E. W. Felten, and K. Li. Implementation and Performance of Application-controlled File Caching. In *OSDI 1*, November 1994.
6. D. R. Engler, M. F. Kaashoek, and J. O'Toole. Exokernel: An Operating System Architecture For Application-level Resource Management. In *SOSP 15*, 1995.
7. P. B. Hansen. The Nucleus of a Multiprogramming System. *Communications of the ACM*, 13(4), April 1970.
8. M. I. Seltzer and C. Small. Self-Monitoring and Self-Adapting Systems. In *HotOS '97*, Chatham, MA, May 1997.
9. M. Stonebraker. Operating System Support For Database Management. *Communications of the ACM*, 24(7), July 1981.

TWLinuX: Operating System Support for Optimistic Parallel Discrete Event Simulation

Subramania Sharma T.[1] and Matthew J. Thazhuthaveetil[2]

[1] Atheros India LLC.
Chennai, India
sharmat@atheros.com

[2] Department of Computer Science and Automation
Indian Institute of Science
Bangalore, India
mjt@csa.iisc.ernet.in

Abstract. Parallel or Distributed Discrete Event Simulation (PDES) refers to the concurrent execution of a single discrete event simulation application on a parallel or distributed computing system. Most available PDES implementations provide user level library support for writing distributed simulation applications. We discuss how OS support can be designed to facilitate optimistic PDES of large, complex simulation models. *TWLinuX* is our implementation of these concepts through modification of Linux. Through *TWLinuX*, a simple, low cost network of machines becomes a high performance discrete event simulation platform.

1 Introduction

Discrete Event Simulation(DES) is a widely used performance evaluation technique. In DES, the system under study is modeled by a set of state variables and a set of discrete events. Each event has an event handler which specifies how the state variables change on occurrence of that event. Simulation events are timestamped, and the simulation proceeds by processing the pending event with the smallest timestamp next. To do this, it executes the appropriate event handler. The processing of an event may generate one or more future events. The objective of PDES is to reduce simulation time by parallelizing the simulation [1]. This is done by decomposing the sequential simulation model into a set of smaller sub-models, called logical processes or simulation processes, which can be executed concurrently on the different nodes of a distributed system, or in parallel on the nodes of a parallel system. Logical processes running on different nodes communicate with each other by exchanging timestamped event messages.

This simple scheme is complicated by the possibility of *causality errors*; an event message may arrive at a logical process where messages with higher timestamp values have already been received and processed. PDES systems are classified as either optimistic or conservative depending on how they handle causality errors. In optimistic PDES, the causality error is allowed to happen, but the error is subsequently detected and handled using a recovery mechanism. When

B. Monien, V.K. Prasanna, S. Vajapeyam (Eds.): HiPC 2001, LNCS 2228, pp. 262–271, 2001.

an out-of-order or *straggler* message is received, the recovery mechanism rolls the state of the simulation process back to a point in its past, before the arrival of the straggler message. For this to be possible, the state of the simulation process must be saved every once in a while. Note that for a simulation process to truly be rolled back in time like this, it may be necessary to *undo* certain activities that it has done, such as sending messages or doing I/O.

Existing optimistic PDES implementations provide this functionality by placing restrictions on the programmer of the PDES application. An optimistic PDES implementation must provide support for saving and rolling back the state of simulation processes. This is usually done by saving the values of the state variables at regular intervals. For this purpose, many optimistic PDES implementations require the simulation modeler to explicitly identify the state variables used in the simulation program. To simplify state saving, some implementations prohibit the use of complex data structures involving pointers for state variables. We are not aware of any optimistic PDES implementation that provides rollback semantics for all forms of I/O.

In this paper, we describe our optimistic PDES implementation, *TWLinuX*, which removes the above mentioned restrictions by supporting optimistic PDES directly in the operating system kernel. Section 2 reviews existing PDES implementations. In Section 3, we motivate the idea of providing OS support for optimistic PDES. The key modifications we made to the Linux kernel are described in Section 4, followed by a discussion of *TWLinuX* performance in Section 5.

2   Related Work

Maisie, developed at UCLA, is a C-based simulation language that can be used for sequential or parallel execution of discrete event simulation models [2]. A Maisie program is a collection of entity definitions and C functions. An entity definition (or an entity type) describes a class of objects. An instance of an entity type may be created to model an object in the system. The language provides primitives to create an entity, define a message and wait for an event.

Warped is a simulation library developed at the University of Cincinnati [3]. The Warped kernel presents an interface to optimistic PDES based on the Time Warp synchronization protocol [4]. The Warped system is implemented in C++. Simulation entities are represented as objects that are derived from the Time Warp class. All state variable should be defined as object data members. The Warped state saving mechanism checkpoints these state variables, by copying the entire object at regular intervals. *GTW*, the Georgia Tech Time Warp, is another simulation package that provides support for PDES in the form of a library [5]. The GTW kernel is written in C, incorporating many optimized parallel simulation techniques.

Parasol, developed at Purdue University, supports PDES in the form of a user level simulation thread library [6]. A Parasol simulation consists of simulation threads that access simulation objects. Each simulation object is mapped to a specific computing node. Parasol provides thread migration, by which a simulation thread that accesses an object present at another node is moved to that node.

3   Why Provide Operating System Support for PDES?

Maisie, Warped, GTW and Parasol provide support for PDES through libraries. We believe that a strong case is to be made for providing OS support.

Support for large and complex state spaces: Large and complex simulation state spaces can be readily supported with simple OS support. This removes the restriction on the simulation application programmer to explicitly identify all state variables, or not to use dynamically allocated state variables.

Support for efficient state saving and restoration: State saving, also known as checkpointing, is required in optimistic PDES systems. When a straggler message is received, the simulation is rolled back by restoring a saved state copy. Techniques such as incremental state saving, lazy saving and lazy restoration can be used to reduce the saving/restoration overhead. In incremental saving, only that portion of simulation state that has been modified since the last state save is saved. Lazy saving and lazy restoration delay the saving and restoration operations until actually required. A PDES system that supports these features requires a mechanism which traps to the simulation kernel whenever simulation state is modified. Incremental saving is done in Parasol by associating a dirty flag with each simulation object, and setting the flag for an object when an event acts on it. A more efficient implementation could be built by providing OS support, using typical processor architecture support for virtual memory. Pages that were modified since the last save can be identified using the dirty bit in the page table, which is automatically set by the processor when a write is made to the page. The lazy saving/restoration for a page can then be done by making the page read-only; writes to it can then be handled by the protection fault handler.

Support for file I/O operations: As we have seen, when a simulation process is rolled back in time, certain file I/O operations that it has done may have to be undone. The semantics of read-only files are easy to maintain without added OS support. The file offset can be saved as part of state saving, and restored on a rollback. It is harder to support undo-able I/O on read-write or write-only files. This functionality can be efficiently incorporated into the operating system. To our knowledge, *TWLinuX* is the first PDES implementation which supports read-write files. Warped supports only write-only files.

4   TWLinuX Implementation

TWLinuX based on the Linux kernel [7] provides OS support for PDES. The *TWLinuX* simulation model is composed of several simulation objects. A simulation object is an entity in the model or an instance of an entity. An entity functionality is represented by a Logical Process (LP). For efficient communication, LPs that communicate frequently with each other can be grouped together to form a Logical Process Group (LPG). The LPGs are executed concurrently on the computing nodes involved in the distributed simulation. In *TWLinuX*, each LPG is implemented as one Linux process and each LP in an LPG is implemented as a separate Linux thread[1]. In this section, we provide an overview of the functionality that we have added to Linux in order to support PDES.

4.1 Scheduling

Linux uses a preemptive scheduling policy, associating a time quantum with each process. This policy does not use much of the available information about the state of simulation processes, but is a good choice as far as fairness to processes is concerned. In our context, it can be improved upon to reduce overall PDES program execution time. It has been suggested that a better policy for scheduling simulation processes would use simulation parameters, such as LVT (logical process Local Virtual Time), or the timestamp of the next event message [8]. A new scheduling class of processes is therefore added in the *TWLinuX* kernel to identify simulation processes in the system, and the scheduling policy for simulation processes makes use of LVT information. We currently use the logical process LVT as the scheduling parameter in computing simulation thread priority; a thread with a low LVT value has a high thread priority value.

4.2 State Saving

Basic Design. State saving is an integral part of optimistic PDES to support the rollback mechanism. It is implemented in the *TWLinuX* kernel using the support provided by the virtual memory manager and Intel x86 features. The state saving mechanism in *TWLinuX* is designed to support large and complex state spaces. Simulation threads can allocate dynamic memory in heap, as heap allocated data is also saved on state saves. State saving is complicated by this feature – it must incorporate a mechanism to follow pointers and save the corresponding data areas. This requires extensive book keeping. To overcome this problem, the *TWLinuX* state saving mechanism treats the state space as a stream of bytes present in the virtual address space, rather than a collection of data. In this way, the implementation is made independent of the structure and organization of data used to represent the state. *TWLinuX* associates a separate heap area with each thread to satisfy its dynamic memory allocation requests.

Incremental State Saving. Virtual memory areas of the stack and the heap of a *TWLinuX* simulation thread are saved during a state save. Due to program locality properties, accesses are likely to be limited to localities in the memory and hence the address space. Since a simulation thread probably modifies only a portion of its virtual memory area, the entire virtual address space need not be saved on each state save – only those portions that have been modified since the

---

[1] created using the *clone* system call

last state save need be saved. This reduces the state saving time and the storage space required for saved states. To achieve this, portions of the virtual memory space that have been were modified since the last state save must be identified.

The Linux virtual memory manager divides a virtual address space into pages of equal size. We therefore use pages as the basic unit of operation in state saving. Only modified pages are saved on a state save. The modified pages are easily identified using the *dirty* bit in page table entries, as the dirty bit is automatically set by the Intel x86 processor when a write is made to a page.

Implementation. The stack and the heap of the LP are virtual memory areas in the virtual address space of the simulation process. During state saving, pages that are allocated in the virtual memory areas of the stack and the heap are saved if their dirty bit is set. The dirty bit is then cleared. As an optimization for stack pages, only the useful portion of the stack page, *viz*, the area below the stack pointer, is saved. Saved states could be stored either in fast secondary storage, such as a swap disk, or in main memory. Use of the swap disk for this purpose increases the number of disk I/Os involved in PDES program execution. But, there is no guarantee that enough main memory will be available for storing saved states. *TWLinuX* dynamically decides whether to use the swap disk or main memory to save a given state based on the number of free memory pages available. The state of registers is also saved.

4.3   Rollback Implementation

When an LP receives an out-of-order or straggler event message, the most recent state saved with a timestamp less than the stragglers timestamp is identified. The state of the LP is restored using that saved state. *TWLinuX* uses a *delayed copying* restoration technique. On a rollback, the entire saved state is not copied immediately. Memory references of the LP are re-mapped to use the saved state. When the LP does a write that would modify the saved state copy, the relevant page of the saved state is restored. Thus, only those parts of the state that the LP changes are actually restored through copying from the saved state.

The *TWLinuX* delayed restoration technique for memory pages uses the copy-on-write feature of the Intel x86 processor and Linux virtual memory management. In Linux, the stack and heap are mapped as read/write, private virtual memory areas. *TWLinuX* makes memory pages that are used for state saving read-only. During restoration, page tables are modified to refer to this read-only saved copy. Thus, reads after rollback will get the correct values from the saved copy. When a write is made to any of these read-only pages, a protection exception will be raised and handled by the Linux page fault handler. The handler finds that the page is mapped to a writable private area and is read-only. This is done to support copy-on-write. The handler makes a copy of this page, makes the new page writable, and changes the page table to refer to this writable page. The *TWLinuX* LP thread then writes to the newly allocated page. Stack pages are copied only until the stack pointer as an optimization.

4.4   Input/Output Management

In optimistic PDES, I/O activity generated cannot be committed immediately on request. If the simulation process that initiated the I/O operation is subsequently rolled back to a time prior to the timestamp associated with the I/O operation, the I/O operation would have to be 'undone'. *TWLinuX* maintains a file list identifying the files that are currently in use by a simulation program. During state saving of an LP, the file list is traversed and the offset and the size of each file is included as part of the saved state. This is later used to restore file state on a roll back.

Read-only Files. The semantics of read only files are easy to maintain, as operations on them do not affect the state of the file. On a read, the file offset is incremented by the number of bytes read. For this type of file, file offsets must be saved during a state save. On a subsequent rollback, the offset can be reset to the saved value.

Write-only Files. A write into a write-only file changes the file state. Since they are difficult to undo, the data written and file offset are buffered within the kernel. The writes are committed only when the entire PDES simulation program has progressed to a time[2] greater than the timestamp associated with the concerned I/O operation. A programmer uses the printf, fwrite, fput or some write function to write to a file/terminal. All of these functions in turn make a *write* system call. In *TWLinuX*, the write I/O is caught at the system call level. When an LP thread makes a write system call in *TWLinuX*, a write request is created. The request contains the data to be written and the file offset. The write request is timestamped and a list of write requests is maintained for each write-only file. During rollback, writes made since the rollback time are canceled.

Read-Write Files. Maintaining rollback semantics for read-write files involves more overhead than for read-only or write-only files. Data written must be available immediately for reading even though the file write operation has not yet been committed. Linux file management uses a *page cache* supported by the virtual memory manager, to reduce the number of disk reads required. The caching mechanism reads the file into free memory pages and associates a file offset with the page. When a file is read using the read system call, the file page to be used is determined from the file offset. The file offset is changed to a offset within the page. The page cache is first searched for the required file page. If it is not present, it is read from the disk and added to the page cache. Subsequent reads to that file page will hit in the page cache. When a file is written using the write system call, the corresponding page is updated in page cache (if present).

---

[2] Optimistic PDES systems use Global Virtual Time (GVT) to estimate the overall simulation progress. *TWLinuX* includes GVT estimation activity. It uses GVT estimates in fossil collection, where saved states are deleted when it is certain that they are no longer required.

In *TWLinuX*, when an LP issues a write to a read-write file, it is buffered as for a write-only file. Since the LP should be able to read the data immediately. it is updated in the page cache also. A problem arises when the corresponding page is not present in the page cache. In this case, the corresponding file page is read from the disk if possible. (If the write is being made to the end of file and the write crosses the page boundary, a new empty page is used.) added to the page cache and updated. The process reads the updated data from the page cache. Also, any data that is overwritten by the write is buffered, so that writes can be canceled during rollback by restoring the old contents. Read-write files are rolled back in the same way as write-only files.

Standard I/O. Standard output and standard error are handled like write-only files; the data output is buffered in the *TWLinuX* write system call. Standard input is handled like a read-only file.

4.5   Miscellaneous

An API of 18 functions provides required functionality to the PDES application developer on *TWLinuX*. This includes LP and LPG creation functions, dynamic memory allocation and deallocation functions, in addition to message send and receive functions. The messaging library handles the problem of how to 'unsend' the messages that a simulation thread may have sent, as part of rolling back that simulation thread. Many optimistic PDES systems implement this functionality by introducing the notion of an *anti-message* which causes mutual annihilation when it meets the message that it was created to unsend. The *TWLinuX* library supports both aggressive and lazy anti-message cancellation strategies.

The changes that we made to Linux result in an increase in kernel size of about 7500 bytes. 26 kernel functions were added along with one system call through which the kernel functions are invoked by the user library functions. More details of the *TWLinuX* kernel functions and user API are available [10].

5   Performance Measurements

We used three PDES applications to evaluate *TWLinuX* performance: Colliding Pucks and a Queuing application from the University of Cincinnati [3], and Warpnet from Georgia Tech [5]. Our performance measurements were done on a network of Intel Pentium machines. The timing measurements were made using the Pentium hardware cycle counter. Other measurements were also made using carefully instrumented versions of the *TWLinuX* kernel. A report on our more comprehensive performance measurement studies on *TWLinuX* is available in a technical report [10]. Due to space restrictions, we discuss only three sets of measurements here: (i) speedup in PDES execution, (ii) performance of our state saving mechanism, and (iii) state saving and system call overheads,

5.1 Speedup

We used the Queuing application to compare *TWLinuX* performance with that of Warped running on identical hardware, as reported in Table 1. Note that *TWLinuX* outperforms Warped and also achieves speedups when more than one machine is used, i.e., execution time decreases as more machines are used.

Table 1. Performance comparison: Warped vs *TWLinuX*

Configuration: 8 LPs.	Execution Time (μs).	
	Warped	*TWLinuX*
1 LPG, 1 Machine.	4,419,446	9,204,838
2 LPGs, 2 Machines.	9,469,771	5,861,820
3 LPGs, 3 Machines.	16,766,279	4,546,951

We measured the speedup obtained by running the PDES applications on more than one machine. 16 LPs were used for Warpnet and Queuing, and 9 LPs for Pucks. The LPs were grouped into different numbers of LPGs and each LPG was run on a different machine. Figure 1 shows the execution times, normalized with respect to the execution time of 1 machine, 1 LPG experiment. Observe that *TWLinuX* achieves speedup for all the applications. Queuing shows super-linear speedup of about 2.5 when run on 2 machines. ,due to the fact that the machines used in this experiment were not identical one being a 200 MHz Pentium III and the other a 233 MHz Pentium III. This illustrate the point that may arise in running large scale PDES applications on a network of low cost machines using *TWLinuX*: a faster CPU would process events and advance LVT faster than the slower CPUs, which could lead to more straggler messages and rollbacks. There is clearly a need for intelligent load balancing across the nodes used in PDES activity based on their computing power.

5.2 State Saving and System Call Overheads

We used the Queuing application to measure state saving and system call overheads. Unlike the other two applications, Queueing does not suffer from rollbacks. We could therefore make measurements on a version of the *TWLinuX* kernel in which the state saving activity is entirely eliminated. Queuing was run with and without state saving, and in each case the execution time measured; results are shown in the third and fourth columns of Table 2.

Observe that the *TWLinuX* state saving techniques imposes an execution time overhead of only 2.28% and 3.29% for the two experimental runs reported in the table. In comparison, the highly efficient Parasol state saving mechanism is reported to involve time overheads of around 18% [9].

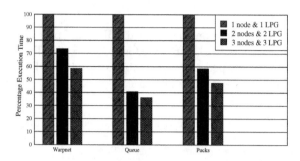

Fig. 1. Speedup obtained

On the flip side of the coin, we do pay a price for providing direct OS support
– the system call overhead for invoking the *TWLinuX* kernel functions. Using
kernel profiling, we found that a significant amount of time overhead is due to a
system call that updates the LVT (simulation thread Local Virtual Time), which
is maintained inside the kernel. This system call must be called once for every
event message processed. This major system call overhead can be eliminated
by keeping track of LVT in user address space. We therefore made additional
measurements on a modified version of the *TWLinuX* kernel in which this system
call is eliminated. The fifth column of Table 2. shows that the associated system
call overhead amounts to only 2.80% and 4.63% for the two experimental runs
reported in the table. Extrapolating forward for all system calls, we expect that
the *TWLinuX* system call overhead for the runs reported is 13.63% and 22.57%.

5.3   *TWLinuX* State Saving

TWLinuX uses an adaptive technique where backup storage for saved states is
selected either from disk swap area or main memory. We compared this adaptive
technique with that for the alternative of saving only in swap. The result in table
3 shows a substantial reduction in execution time, as adaptive saving technique
is able to utilize the free memory available for state saving most of the time.

Table 2. State saving and system call overhead in *TWLinuX*

Configuration: **Queuing**, 20 LPs, 2 LPGs, P200 (9 LPs), P333(11 LPs), Lazy cancellation.

Queuing: Parameters.	Number of pages used for state save	Execution Time (μs).			Number of system calls reduced (from)
		Default	Without state save	Without state save and sys call	
24 hours,1000 messages per hour	2,391	3,845,905	3,719,501	3,611,726	91,260 (429,300)
24 hours,5000 messages per hour	11,511	19,387,039	18,944,195	18,046,365	449,758 (2,141,039)

Table 3. State saving - Swap vs Adaptive technique

Configuration:	Execution Time (μs).	
	Swap	Adaptive
Warpnet, 16 LPs, 3 LPGs, 3 machines.	152,011,414	41,410,383
Queuing, 16 LPs, 3 LPGs, 3 machines	18,531,695	10,199,346
Pucks, 9 LPs. 2 LPGs, 2 machines	60,706,500	9,227,884

6   Conclusions

Existing PDES are implemented to support PDES through a user level library. Our PDES environment *TWLinuX*, which is implemented through modifications to Linux provides direct OS support. It supports large and complex simulation state spaces, efficient state saving/restoration techniques, and all modes of file operations with rollback semantics. Large scale DES requirements can be met on low cost networks of ordinary PCs by running this variant of Linux. Our experience with porting three existing PDES applications, Queueing, Warpnet and Pucks, to *TWLinuX* was very positive. Our current activity addresses scalability issues – providing acceptable performance to large PDES applications on large networks of machines.

References

1. R. Fujimoto. Parallel discrete event simulation. *Communications of the ACM.* 33(10):31-53. October 1990.
2. R. L. Bagrodia and W. Liao. Maisie: A Language for the Design of Efficient Discrete Event Simulations. *IEEE Transactions on Software Engineering.* 20(4):225-238. April 1994.
3. Warped - A TimeWarp Simulation Kernel. http://www.ece.uc.edu/~paw/warped
4. D. R. Jefferson. Virtual time. *ACM Transactions on Programming Languages and Systems.* 7(3):404-425. July 1985.
5. Georgia Tech Time Warp. http://www.cs.gatech.edu.computing/pads
6. R. Pasquini and V. Rego. Efficient process interaction with threads in PDES. In *Proceedings of the Winter Simulation Conference.* pages 451-458. 1998.
7. M. Beck. H. Bohme, M. Dziadzka, U. Kunitz, R. Magnus and D. Verworner. *Linux Kernels Internals.* Addison-Wesley. 2nd Edition. 1998.
8. C. Burdorf and J. Marti. Non-preemptive Time Warp Scheduling Algorithms. *ACM SIGOPS Operating Systems Review.* pages 7-18. Jan 1990.
9. R. Pasquini. *Algorithms for Improving the Performance of Optimistic Parallel Simulation.* PhD dissertation. Department of Computer Science, Purdue University. Aug 1999.
10. T. Subramania Sharma. High Performance Optimistic Parallel Discrete Event Simulation in a Network of Linux Machines. MSc(Engineering) dissertation. Department of Computer Science and Automation. Indian Institute of Science. Bangalore. July 2000.

Low-Cost Garbage Collection for Causal Message Logging*

JinHo Ahn, Sung-Gi Min, and ChongSun Hwang

Dept. of CS & Eng., Korea University
5-1 Anam-dong, Sungbuk-gu, Seoul 136-701, Republic of Korea
jhahn@disys.korea.ac.kr, sgmin@korea.ac.kr, hwang@disys.korea.ac.kr

Abstract. This paper presents two garbage collection schemes for causal message logging with independent checkpointing. The first scheme allows each process to autonomously remove useless log information in its storage by piggybacking only some additional information without requiring any extra message and forced checkpoint. The second scheme enables the process to remove a part of log information in its storage if more empty storage space is required after executing the first scheme. It reduces the number of processes to participate in the garbage collection by using the size of the log information of each process. Simulation results show that combining the two schemes significantly reduces the garbage collection overhead compared with the traditional schemes.

1 Introduction

Message logging protocols are classified into three categories: pessimistic, optimistic and causal [5]. Causal message logging approach has advantages of both pessimistic and optimistic approaches because it allows all processes to log received messages on stable storage asynchronously and no live processes to roll back in case of failures. Therefore, this approach is very attractive for providing fault-tolerance for parallel and distributed applications on the message passing systems due to these desirable properties. However, in this approach, each process should maintain a large amount of message logging information in its volatile storage and piggyback a part of the information on each sending message. Thus, it requires efficient garbage collection schemes to minimize the log management overhead and the amount of log information piggybacked on each sending message. Existing garbage collection schemes require a large number of additional messages and forced checkpoints such that the system always maintains a globally consistent state despite future failures [4].

In this paper, we present two garbage collection schemes, *Passive* and *Active*, to remove log information of processes. The *Passive* scheme enables each process to autonomously remove useless log information in its local storage by piggybacking only some additional information. It requires no extra message and

---

* This work was supported by University Research Program supported by Ministry of Information and Communication in South Korea under contract 2000-024-01.

B. Monien, V.K. Prasanna, S. Vajapeyam (Eds.): HiPC 2001, LNCS 2228, pp. 272–281, 2001.

forced checkpoint. However, it may still lead to overloading the storage buffers in some communication and checkpointing patterns. If more empty buffer space is required after executing the *Passive* scheme, each process performs the *Active* scheme to remove a part of the remaining log information in its storage. It selects a minimum number of processes to participate in the garbage collection by using an array to save the size of the log information by process. It can avoid the risk of overloading the storage buffers, but incurs more additional messages and forced checkpoints than the *Passive* scheme. In addition, the *Active* scheme reduces the number of additional messages and forced checkpoints needed by the garbage collection compared with the existing scheme [4].

Due to space limitation, there are no proofs about the correctness of and no formal descriptions of the proposed schemes in this paper. The interested reader can find the proofs and the descriptions in [1].

2   System Model

The distributed computation on the system, N, consists of $n(n>0)$ processes executed on nodes. Processes have no global memory and global clock. The system is asynchronous: Each process is executed at its own speed and communicates with each other only through messages with finite but arbitrary transmission delays. We assume that messages are reliably delivered in first-in first-out order, the communication network is immune to partitioning and there is a stable storage that every process can always access [5]. Additionally, we assume that nodes fail, in which they lose contents in their volatile memories and stop their executions, according to the fail stop model [5]. The execution of each process is piecewise deterministic [5]: At any point during the execution, a state interval of the process is determined by a non-deterministic event, which in this paper is delivering a received message to the appropriate application. The k-th state interval of process p, denoted by $si_p{}^k(k>0)$, is started by the delivery event of the k-th message m of p, denoted by $dev_p{}^k(m)$. Therefore, given p's initial state, $si_p{}^0$, and all the non-deterministic events that have occurred in p's execution, its corresponding state is uniquely determined. Events of processes in a failure-free execution are ordered using Lamport's *happened before* relation [5].

The determinant of a message consists of the identifiers of its sending process (sid) and receiving process(rid), send sequence number(ssn) and receive sequence number(rsn). In this paper, the determinant of message m and the set of determinants in process p's volatile storage are denoted by $det(m)$ and $d\text{-}set_p$. $sri(m)$ denotes the recovery information of a sent message m in its sender's volatile storage to retransmit the message to its receiver in case of the receiver's failure. It consists of rid, ssn and delivered data ($data$) of the message. $s\text{-}log_p$ is a set of recovery information of all messages process p has sent. $FTCML(f)$ denotes a sort of causal message logging protocols that allow the system to be recovered to a consistent state even in case of $f(1 \leq f \leq n)$ concurrent failures. $R\_MSG(Chk_p)$ is a set of all messages received before p takes its latest checkpoint Chk_p.

3 The Garbage Collection Schemes

The traditional causal message logging approach has two problems. First, each process needs to maintain in its volatile storage the determinants of its received messages and of the messages that the received messages depend on. Moreover, it piggybacks the determinants in its volatile storage on each sending message. Second, this approach must ensure the retransmission of *lost messages*, sent but not delivered messages, when processes crash and recover [5]. In this approach, the maintenance overhead of each process's *s-log* may have more significant influence on the performance degradation than that of each process's *d-set* during failure-free operation because the size of each message's *data* is generally much larger than that of its determinant. These two drawbacks may speed up overloading storage buffers of nodes and consuming high bandwidth on wired links. Thus, the causal message logging approach requires efficient garbage collection schemes to remove the log information while allowing the system to recover to a globally consistent state in case of future failures.

In this paper, we present two efficient garbage collection schemes to solve these problems.

3.1 The Passive Garbage Collection Scheme (*PGCS*)

The causal message logging approach with independent checkpointing has the property that each failed process has to be rolled back to the latest checkpoint and replay the received messages beyond the checkpoint by obtaining determinants for the process from the other processes. From the property, we can see that all the messages, received before process p takes its latest checkpoint, are useless for recovering p to be a consistent state in case of p's failure. Therefore, we can see that $det(m)$ and $sri(m)$ for all $p \in S$ and all $m \in R\_MSG(Chk_p)$ in $FTCML(f)$ are useless log information for recovering the entire system's globally consistent state. For example, in figure 1, process p saves $sri(m_1)$ in s-log_p and sends message m_1 to q. r also saves $sri(m_2)$ in s-log_r before sending message m_2 to q. After q receives m_1 and then m_2, it maintains $det(m_1)$ and $det(m_2)$ in d-set_q due to the feature of $FTCML(f)$. Then, it sends message m_3 with $(det(m_1), det(m_2))$ to r after saving $sri(m_3)$ in s-log_q and takes checkpoint Chk_q^{j+1}. When receiving m_3, r saves $(det(m_1), det(m_2))$ piggybacked on m_3 and $det(m_3)$ in d-set_r. In this case, m_1 and m_2 become useless messages due to Chk_q^{j+1}. Thus, the log information of the two messages, $(sri(m_1), det(m_1))$ and $(sri(m_2), det(m_2))$ are needed for future recovery no longer.

$PGCS$ is a scheme to eventually remove $det(m)$ and $sri(m)$ for all $p \in N$ and all $m \in R\_MSG(Chk_p)$ from the volatile storage without requiring any additional message and forced checkpoint. To achieve the goal, $PGCS$ uses two n-size vectors, $LSSN\_Bef\_Chkpt_p$ and $Stable\_LRSN_p$. First, $LSSN\_Bef\_Chkpt_p$ is a vector where $LSSN\_Bef\_Chkpt_p[q]$ is the ssn of the last message which was delivered to p from q before taking the latest local checkpoint of p. Every process p can autonomously remove useless

$sri(m)$s in $s\text{-}log_p$ based on $LSSN\_Bef\_Chkpt_p$. When p takes checkpoint Chk_p, for all $m \in R\_MSG(Chk_p)$, it updates $LSSN\_ Bef\_Chkpt_p[det(m).sid]$ to $det(m).ssn$ if $det(m).ssn$ is greater than $LSSN\_Bef\_Chkpt_p[det(m).sid]$. Whenever p send an application message to q, $LSSN\_Bef\_ Chkpt_p[q]$ is piggybacked on the message. Receiving the message from p, q can remove all useless $sri(m)$s from $s\text{-}log_q$ such that $sri(m).rid$ is p and $sri(m).ssn$ is less than or equal to $LSSN\_Bef\_Chkpt_p[q]$. Second, $Stable\_LRSN_p$ is a vector where $Stable\_LRSN_p[q]$ is the rsn of the latest among messages, delivered to q, whose determinants have been removed from $d\text{-}set_q$ due to the latest checkpoint of q or saved in the stable storage. p can locally remove useless $det(m)$s in $d\text{-}set_p$ based on $Stable\_LRSN_p$. When p takes its local checkpoint Chk_p, it removes $det(m)$ from $d\text{-}set_p$ for all $m \in R\_MSG(Chk_p)$ and updates $Stable\_LRSN_p[p]$ to the largest $det(m).rsn$. p piggybacks $Stable\_LRSN_p$ on each application message sent from p. When receiving the message from p, q updates $Stable\_LRSN_q$ to the latest by using $Stable\_LRSN_p$. Then, it can remove all useless $det(m)$s from $d\text{-}set_q$ such that for all $k \in N$ $(k \neq q)$, $det(m).rid$ is k and $det(m).rsn$ is less than or equal to $Stable\_LRSN_q[k]$.

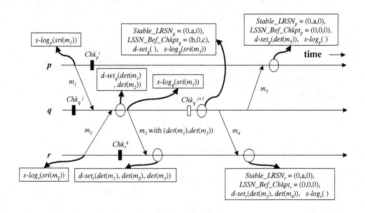

Fig. 1. An example for $PGCS$

For example, in figure 1, when q takes checkpoint Chk_q^{j+1}, it removes useless $det(m_1)$ and $det(m_2)$ from $d\text{-}set_q$, and then updates $Stable\_LRSN_q[q]$, $LSSN\_ Bef\_Chkpt_q[p]$ and $LSSN\_Bef\_Chkpt_q[r]$ to rsn of m_2 and ssn of m_1 and m_2 respectively. Suppose that rsn of m_2 and ssn of m_1 and m_2 are a, b and c. q saves $sri(m_4)$ in $s\text{-}log_q$ and sends m_4 with $LSSN\_Bef\_Chkpt_q[r]$ and $Stable\_LRSN_q$ to r. When receiving m_4 from q, r saves $det(m_4)$ in $d\text{-}set_r$ and updates $Stable\_LRSN_r$ to (0,a,0) by using $Stable\_LRSN_q$ piggybacked on m_4. Then, it removes the useless $det(m_1)$ and $det(m_2)$ from $d\text{-}set_r$ because rsn of m_2 is equal to $Stable\_LRSN_r[q](=a)$. It also removes $sri(m_2)$ from $s\text{-}log_r$ because ssn of m_2 is equal to $LSSN\_Bef\_Chkpt_q[r](=c)$ piggybacked on m_4. In the same manner, after q saves $sri(m_5)$ in $s\text{-}log_q$, it sends m_5 with

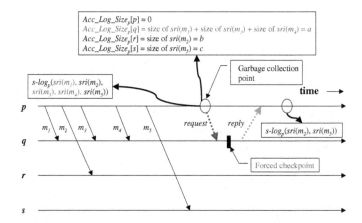

Fig. 2. An example for *AGCS* in case that $a \geq b \geq c$ and $a \geq$ the requested empty space

$LSSN\_Bef\_Chkpt_q[p]$ and $Stable\_LRSN_q$ to p. When receiving m_5, p saves $det(m_5)$ in $d\text{-}set_p$ and updates $Stable\_LRSN_p$ to (0,a,0) by using the piggy-backed $Stable\_LRSN_q$. Then, it removes $sri(m_1)$ from $s\text{-}log_p$ because ssn of m_1 is equal to $LSSN\_Bef\_Chkpt_q[p](=b)$ piggybacked on m_5.

As shown in the example, $PGCS$ allows each process to autonomously remove useless log information from its storage buffer by piggybacking only an n-size vector and a variable on each sending message without any additional message and forced checkpoint.

3.2 The Active Garbage Collection Scheme (*AGCS*)

In $PGCS$, each process p locally removes useless log information from its volatile storage based on $LSSN\_Bef\_Chkpt_p$ and $Stable\_LRSN_p$ and the information piggybacked on every received message. However, after a process has executed $PGCS$, its storage buffer may still be overloaded in some communication and checkpointing patterns. For example, in figure 2, p sends message m_1, m_2, m_3, m_4 and m_5 in this order after saving $sri(m_1)$, $sri(m_2)$, $sri(m_3)$, $sri(m_4)$ and $sri(m_5)$ in $s\text{-}log_p$ respectively. When receiving m_1, m_3 and m_4, q saves their determinants in $d\text{-}set_q$. In the same manner, r and s saves $det(m_2)$ in $d\text{-}set_r$ and $det(m_5)$ in $d\text{-}set_s$ respectively. In this checkpointing and communication pattern, p with $PGCS$ cannot autonomously decide whether log information of each sent message is useless for recovery of the receiver of the message by using the two vectors. Thus, p should maintain all the log information of the five messages in $s\text{-}log_p$.

To solve the problem, the causal message logging approach requires a scheme to allow each process to remove the log information in its volatile storage while ensuring system consistency in case of failures. This scheme should force the log information to become useless for future recovery to satisfy the goal.

In Manetho [4], a garbage collection scheme was proposed for this goal. In this scheme, to garbage collect every $sri(m)$ in $s\text{-}log_p$, p requests that $m.rid$ takes a checkpoint if $m.rid$ has indeed received m and taken no checkpoint since. Also, processes occasionally exchange the state interval indexes of their most recent checkpoints for garbage collecting the determinants in their volatile storages. However, this scheme may result in a large number of additional messages and forced checkpoints needed by the forced garbage collection. To illustrate how to remove the log information in the scheme, consider the example shown in figure 2. Suppose p intends to remove the log information in $s\text{-}log_p$ at the marked point. In this case, the scheme forces p to send checkpoint requests to q, r and s. When receiving the request, q, r and s take their checkpoints respectively. Then, the three processes send each a checkpoint reply to p. After receiving all the replies, p can remove $(sri(m_1), sri(m_2), sri(m_3), sri(m_4), sri(m_5))$ from $s\text{-}log_p$. From this example, we make one observation: if the requested empty space is less than or equal to the sum of sizes of $sri(m_1)$, $sri(m_3)$ and $sri(m_4)$, p has only to force q to take a checkpoint. This observation implies that the number of extra messages and forced checkpoints may be reduced if p knows sizes of the respective log information for q, r and s in its volatile storage. In this paper, we present $AGCS$ to obtain such information by maintaining an array, $Acc\_Log\_Size_p$, to save the size of the log information in the volatile storage by process. Thus, $AGCS$ can reduce the number of additional messages and forced checkpoints by using the vector compared with Manetho's scheme. $Acc\_Log\_Size_p$ is a vector where $Acc\_Log\_Size_p[q]$ is the sum of sizes of all $sri(m)$s in $s\text{-}log_p$ such that p sent message m to q and sizes of all $det(m)$s in $d\text{-}set_p$ such that m was delivered to q. Whenever p sends m to q, it increments $Acc\_Log\_Size_p[q]$ by the size of $sri(m)$. Whenever it receives m from a process, it also increments $Acc\_Log\_Size_p[n.rid]$ by the size of $det(n)$ for all $det(n)$s piggybacked on m, and then $Acc\_Log\_Size_p[p]$ by the size of $det(m)$. If p removes $sri(m)$s in $s\text{-}log_p$ or $det(m)$s in $d\text{-}set_p$ by executing $PGCS$, it decrements $Acc\_Log\_Size_p[q]$ by the sum of their sizes in case $m.rid$ is equal to q. When p needs more empty buffer space, it executes $AGCS$. It first chooses a set of processes, denoted by R_Set, which will participate in the forced garbage collection. It selects the biggest, $Acc\_Log\_Size_p[q]$, among the remaining elements of $Acc\_Log\_Size_p$ and then appends q to R_Set until the required buffer size is satisfied. Then, p sends a request message with the ssn of the last message, sent from p to q, to all $q \in R\_Set$ such that $sri(m).rid$ is q for $\exists sri(m) \in s\text{-}log_p$. It sends the other processes in R_Set each a request message with the highest rsn of among all $det(m)$s in $d\text{-}set_p$ such that $m.rid$ is q. When q receives the request message with the ssn from p, it checks whether the ssn is greater than $LSSN\_Bef\_Chkpt_q[p]$. If so, it should take a checkpoint and then send p a reply message with the rsn of the last message delivered to q. Otherwise, it has only to send p a reply message with the rsn. When p receives the reply message from q, it removes all $sri(m)$s in $s\text{-}log_p$ and all $det(m)$s in $d\text{-}set_p$ such that $det(m).rid$ is q. When q receives the request message with the rsn from p, it checks whether the rsn is greater than $Stable\_LRSN_q[q]$. If so, it should take a checkpoint or save all $det(n)$s in $d\text{-}set_q$, such that $n.rid$ is q, into stable storage according to each cost, and then send p a reply message with the rsn of the last message delivered to q. Otherwise, it has only to send p a reply message with the rsn. When p receives the reply message from q, it removes all $det(m)$s in $d\text{-}set_p$ such that $m.rid$ is q.

For example, in figure 2, when p attempts to execute $AGCS$ at the marked point after it has sent m_5 to s, it should create R_Set. In this figure, we can see that $Acc\_Log\_Size_p[q](= a)$ is the biggest among all the elements of $Acc\_Log\_Size_p$ due to $sri(m_1)$, $sri(m_3)$ and $sri(m_4)$ in $s\text{-}log_p$. Thus, it first selects and appends q to R_Set. Suppose that the requested empty space is less than or equal to a. In this case, it needs to select any process like r and s no longer. Therefore, p sends a checkpoint request message with $m_4.ssn$ to only q in R_Set. When q receives the request message, it should take a forced checkpoint like in this figure because the ssn included in the message is greater than $LSSN\_Bef\_Chkpt_q[p]$. Then, it sends p a reply message with the rsn of the last message delivered to q. When p receives a reply message from q, it can remove $sri(m_1)$, $sri(m_3)$ and $sri(m_4)$ from $s\text{-}log_p$. In this figure, $AGCS$ can choose a small number of processes to participate in the garbage collection based on $Acc\_Log\_Size_p$ compared with Manetho's scheme [4] - $AGCS$ may reduce the number of additional messages and forced checkpoints.

4   Performance Evaluation

In this section, we perform extensive simulations to compare a combined scheme (denoted by $CGCS$), having the advantages of the two proposed schemes $PGCS$ and $AGCS$, with Manetho's garbage collection Scheme ($MGCS$) [4] using PARSEC discrete-event simulation language [3]. Two peformance indices are used; the average number of additional messages (denoted by $NOAM$) and the average number of forced checkpoints (denoted by $NOFC$) required for garbage collection per process. The two indices indicate the overhead caused by garbage collection during failure-free operation. For the simulation, 20 processes have been executed for 72 hours per simulation run. Every process has a 128MB buffer space for storing its $s\text{-}log$ and $d\text{-}set$. The lower and upper bounds of the free buffer space are 28MB and 100MB respectively. The message size ranges from 1KB to 10MB. The interarrival times of message sending events and the scheduled checkpointing events follow the exponetial distribution with a mean T_{ms} and $T_{sc} = 5$ mins., respectively. Distributed applications for the simulation exhibit the following four communication patterns respectively [2]:

• **Serial Pattern**: All processes are organized in the serial manner and transfer messages one way. If every process except the first and the last ones, receives a message from its predecessor, it sends a message to its successor, and vice versa. The first process communicates with only its successor and the last one communicates with its predecessor only.

• **Circular Pattern**: A logical ring is structured for communication among processes in this pattern. Every process communicates with only two directly-connected neighbors.

- **Hierarchical Pattern**: A logical tree is structured for communication among processes in this pattern. Every process, except one root process, communicates with only one parent process and k child processes ($k \geq 0$). The root process communicates with its child processes only.
- **Irregular Pattern**: The communication among processes follows no special communication pattern.

Figure 3 shows $NOAM$ for the serial, circular, hierarchical and irregular communication patterns for the various T_{ms} values respectively. In the figures, we can see that $NOAM$s of the two schemes increase as T_{ms} decreases. The reason is that forced garbage collection is frequently performed because high inter-process communication rate causes the storage buffer of each process to be overloaded quickly. However, we can see in the figures that $NOAM$ of $CGCS$ is much lower than that of $MGCS$. $CGCS$ reduces about 41% $\sim$ 74% of $NOAM$ compared with $MGCS$. Figure 4 illustrates $NOFC$ for the various communication patterns for the various T_{ms} values respectively. In the figures, we can also see that $NOFC$s of the two schemes increase as T_{ms} decreases. The reason is that as the inter-process communication rate is higher, a process may takes a forced checkpoint when it performs forced garbage collection. In these figures, $NOFC$ of $CGCS$ is lower than that of $MGCS$. $CGCS$ reduces about 35% $\sim$ 76% of $NOFC$ compared with $MGCS$. The reductions of $NOAM$ and $NOFC$ of $CGCS$ result from the advantages of both $PGCS$ and $AGCS$. While each process maintains enough free space of its storage buffer, $PGCS$ allows the process to remove useless $sri(m)$s and $det(m)$s by piggybacking only an n-size vector and a variable on each sending message. In this case, $NOAM$ and $NOFC$ are both 0s. If the free buffer space is smaller than the lower bound even after $PGCS$ has been executed, $AGCS$ forces the process to remove a part of log information in the buffer until the free space becomes the upper bound. In this case, $NOAM$ and $NOFC$ of $AGCS$ are less than those of $MGCS$ because every process in $AGCS$ maintains an array to save the current size of the log information in its buffer by process and selects a minimum number of processes to participate in the garbage collection based on the array. Therefore, we can conclude from the simulation results that regardless of the specific communication patterns, $CGCS$ enables garbage collection overhead occurring during failure-free operation to be significantly reduced compared with $MGCS$.

5   Conclusion

In this paper, we presented two efficient schemes, $PGCS$ and $AGCS$, for efficiently garbage collecting log information of each process in causal message logging. In $PGCS$, each process autonomously removes the useless log information in its storage by piggybacking only an n-size vector and a variable on each sending message. Thus, $PGCS$ requires no additional message and forced checkpoint. However, even with $PGCS$, the storage buffer of a process may still be overloaded in some communication and checkpointing patterns. When more empty storage is required after executing $PGCS$, the process performs $AGCS$. In $AGCS$, each

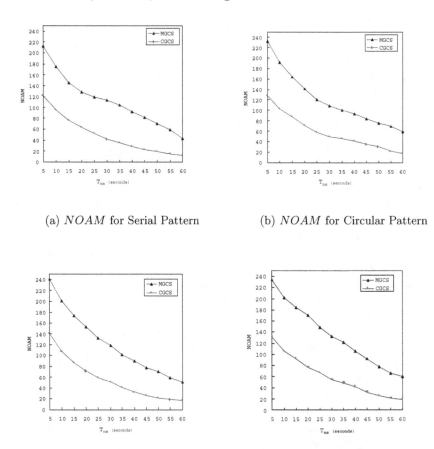

(a) *NOAM* for Serial Pattern   (b) *NOAM* for Circular Pattern

(c) *NOAM* for Hierarchical Pattern   (d) *NOAM* for Irregular Pattern

Fig. 3. *NOAM*s for the Four Communication Patterns

process maintains an array to save the size of the log information for every process in its storage by process. It chooses a minimum number of processes to participate in the forced garbage collection based on the array. Thus, it incurs more additional messages and forced checkpoints than *PGCS*. However, it can avoid the risk of overloading the storage buffers. Moreover, *AGCS* reduces the number of additional messages and forced checkpoints needed by the garbage collection compared with the existing scheme [4]. Experimental results indicate that combining *PGCS* and *AGCS* significantly reduces about 41% ∼ 74% of *NOAM* and 35% ∼ 76% of *NOFC* regardless of the communication patterns compared with *MGCS*.

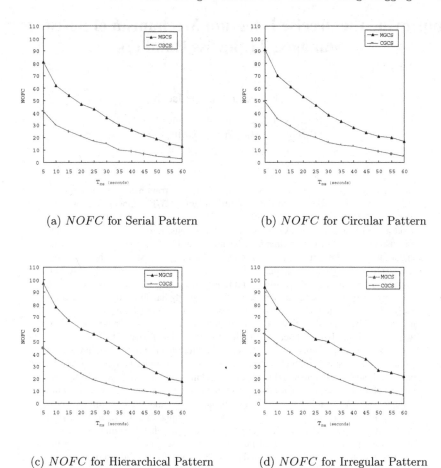

(a) *NOFC* for Serial Pattern   (b) *NOFC* for Circular Pattern

(c) *NOFC* for Hierarchical Pattern   (d) *NOFC* for Irregular Pattern

Fig. 4. *NOFC*s for the Four Communication Patterns

References

1. Ahn, J., Min, S., Hwang, C.: Reducing Garbage Collection Overhead in Causal Message Logging. Technical Report KU-CSE-01-36, Korea University (2000)
2. Andrews, G. R.: Paradigms for process interaction in distributed programs. ACM Computing Surveys, Vol. 23. (1991) 49–90
3. Bagrodia, R., Meyer, R., Takai, M., Chen, Y., Zeng, X., Martin, J., Song, H. Y.: Parsec: A Parallel Simulation Environments for Complex Systems. IEEE Computer, (1998) 77–85
4. Elnozahy, E. N., Zwaenepoel, W.: Manetho: Transparent rollback-recovery with low overhead, limited rollback and fast output commit. IEEE Transactions on Computers, Vol. 41. (1992) 526–531
5. Elnozahy, E. N., Alvisi, L., Wang, Y. M., Johnson, D. B.: A Survey of Rollback-Recovery Protocols in Message-Passing Systems. Technical Report CMU-CS-99-148, Carnegie-Mellon University (1999)

Improving the Precise Interrupt Mechanism of Software-Managed TLB Miss Handlers

Aamer Jaleel and Bruce Jacob

Electrical & Computer Engineering
University of Maryland at College Park
{ajaleel,blj}@eng.umd.edu

Abstract. The effects of the general-purpose precise interrupt mechanisms in use for the past few decades have received very little attention. When modern out-of-order processors handle interrupts precisely, they typically begin by flushing the pipeline to make the CPU available to execute handler instructions. In doing so, the CPU ends up flushing many instructions that have been brought in to the reorder buffer. In particular, many of these instructions have reached a very deep stage in the pipeline - representing significant work that is wasted. In addition, an overhead of several cycles can be expected in re-fetching and re-executing these instructions. This paper concentrates on improving the performance of precisely handling software managed translation lookaside buffer (TLB) interrupts, one of the most frequently occurring interrupts. This paper presents a novel method of in-lining the interrupt handler within the reorder buffer. Since the first level interrupt-handlers of TLBs are usually small, they could potentially fit in the reorder buffer along with the user-level code already there. In doing so, the instructions that would otherwise be flushed from the pipe need not be re-fetched and re-executed. Additionally, it allows for instructions independent of the exceptional instruction to continue to execute in parallel with the handler code. We simulate two different schemes of in-lining the interrupt on a processor with a 4-way out-of-order core similar to the Alpha 21264. We also analyzed the overhead of re-fetching and re-executing instructions when handling an interrupt by the traditional method. We find that our schemes significantly cut back on the number of instructions being re-fetched by 50-90%, and also provides a performance improvement of 5-25%.

1   INTRODUCTION

1.1   The Problem

With the continuing efforts to maximize instruction level parallelism, the out-of-order issue of instructions has drastically increased the utilization of the precise interrupt mechanism to handle exceptional events. Most of these exceptions are transparent to the user application and are only to perform "behind-the-scene" work on behalf of the programmer [27]. Such exceptions are handled via hardware or software means. In this paper, we will be concentrating on those handled via software. Some examples of such exceptions are unaligned memory access, instruction emulation, TLB miss handling.

Among these exceptions, Anderson, et al. [1] show TLB miss handlers to be among the most commonly executed OS primitives; Huck and Hays [10] show that TLB miss handling can account for more than 40% of total run time; and Rosenblum, et al. [18] show that TLB miss handling can account for more than 80% of the kernel's computation time. Recent stud-

B. Monien, V.K. Prasanna, S. Vajapeyam (Eds.): HiPC 2001, LNCS 2228, pp. 282–293, 2001.
Springer-Verlag Berlin Heidelberg 2001

ies show that TLB-related precise interrupts occur once every 100–1000 user instructions on all ranges of code, from SPEC to databases and engineering workloads [5, 18].

With the current trends in processor and operating systems design, the cost of handling exceptions precisely is also becoming extremely expensive; this is because of their implementation. Most high performance processors typically handle precise interrupts at commit time [15, 17, 21, 25]. When an exception is detected, a flag in the instructions' reorder buffer entry is set indicating the exceptional status. Delaying the handling of the exception ensures that the instruction didn't execute along a speculative path. While the instructions are being committed, the exception flag of the instruction is checked. If the instruction caused an exception and software support is needed, the hardware handles the interrupt in the following way:

The ROB is flushed; the exceptional PC[1] is saved; the PC is redirected to the appropriate handler

1. Handler code is executed, typically with privileges enabled

2. Once a return from interrupt instruction is executed, the exceptional PC is restored, and the program resumes execution

In this model, there are two primary sources of application-level performance loss: (1) while the exception is being handled, there is no user code in the pipe, and thus no user code executes—the application stalls for the duration of the handler; (2) after the handler returns control to the application, all of the flushed instructions are re-fetched and re-executed, duplicating work that has already been done. Since most contemporary processors have deep pipelines and wide issue widths, there may be many cycles between the point that the exception is detected and the moment that the exception is acted upon. Thus, as the time to detect an exception increases, so does the number of instructions that will be re-fetched and re-executed [17]. Clearly, the overhead of taking an interrupt in a modern processor core scales with the size of the reorder buffer, pipeline depth, issue-width, and each of these is on a growing trend.

1.2   A Novel Solution

If we look at the two sources of performance loss (user code stalls during handler; many user instructions are re-fetched and re-executed), we see that they are both due to the fact that the ROB is flushed at the time the PC is redirected to the interrupt handler. If we could avoid flushing the pipeline, we could eliminate both sources of performance loss. This has been pointed out before, but the suggested solutions have typically been to save the internal state of the entire pipeline and restore it upon completion of the handler. For example, this is done in the Cyber 200 for virtual-memory interrupts, and Moudgill & Vassiliadis briefly discuss its overhead and portability problems [15]. Such a mechanism would be extremely expensive in modern out-of-order cores, however; Walker & Cragon briefly discuss an *extended shadow registers* implementation that holds the state of every register, both architected and internal, including pipeline registers, etc. and note that no ILP machine currently attempts this [25]. Zilles, et al. discuss a multi-threaded scheme, where a new thread fetches the handler code [27].

We are interested instead in using existing out-of-order hardware to handle interrupts both precisely and inexpensively. Looking at existing implementations, we begin by questioning why the pipeline is flushed at all—at first glance, it might be to ensure proper execution with regard to privileges. However, Henry has discussed an elegant method to allow

---

1. Exceptional PC depends on the exception class. Certain interrupts, such as the TLB interrupts, require the exception causing instruction to re-execute thus set exceptional PC to be the PC of the exception causing instruction. Other interrupts, such as I/O interrupts, set the exception PC to be the PC of the next instruction after the exception causing instruction.

privileged and non-privileged instructions to co-exist in a pipeline [9]; with a single bit per ROB entry indicating the privilege level of the instruction, user instructions could execute in parallel with the handler instructions.

If privilege level is not a problem, what requires the pipe flush? Only *space*: user instructions in the ROB cannot commit, as they are held up by the exceptional instruction at the head. Therefore, if the handler requires more ROB entries than are free, the machine would deadlock were the processor core to simply redirect the PC without flushing the pipe. However, in those cases where the entire handler could fit in the ROB in addition to the user instructions already there, the processor core could avoid flushing the ROB and at the same time also such deadlock problems.

Our solution to the interrupt problem, then, is simple: if at the time of redirecting the PC to the interrupt handler there are enough unused slots in the ROB, we in-line the interrupt handler code without flushing the pipeline. If there are not sufficient empty ROB slots, we handle the interrupt as normal. If the architecture uses reservation stations in addition to a ROB [7, 26] (an implementation choice that reduces the number of result-bus drops), we also have to ensure enough reservation stations for the handler, otherwise handle interrupts as normal.

Though the mechanism is applicable to all types of interrupts (with relatively short handlers), we focus on only one interrupt in this paper—that used by a software-managed TLB to invoke the first-level TLB-miss handler. We do this for several reasons:

1. As mentioned previously, TLB-miss handlers are invoked *very* frequently (once per 100-1000 user instructions)

2. The first-level TLB-miss handlers tend to be short (on the order of ten instructions) [16, 12]

3. These handlers also tend to have deterministic length (i.e., they tend to be straight-line code—no branches)

This will give us the flexibility of software-managed TLBs without the performance impact of taking a precise interrupt on every TLB miss. Note that hardware-managed TLBs have been non-blocking for some time: e.g., a TLB-miss in the Pentium-III pipeline does not stall the pipeline—only the exceptional instruction and its dependents stall [24]. Our proposed scheme emulates the same behavior when there is sufficient space in the ROB. The scheme thus enables software-managed TLBs to reach the same performance as non-blocking hardware-managed TLBs without sacrificing flexibility [11].

1.3   Results

We evaluated two implementations of the in-lined mechanism, *append*: inserting the handler after existing code, and *prepend*: inserting the handler before existing code, on a processor model of an out-of-order core with specs similar to the Alpha 21264 (4-way out-of-order, 150 physical registers, up to 80 instructions in flight, etc.). No modifications are required of the instruction-set; this could be implemented on existing systems transparently—i.e., without having to rewrite any of the operating system.

An in depth analysis of the instructions flushed on an interrupt shows that 20-30% of those flushed have finished execution, 50-60% are waiting for execution units, and the remaining are waiting to be decoded and register rename. This shows significant waste in both execution time and energy consumption.

The schemes cut the TLB-miss overhead by 10–40% [28], the number of instructions flushed by 50-90%. When applications generate TLB misses frequently, this reduction in overhead amounts to a performance improvement of 5-25% in execution time in the *prepend* scheme and 5-10% in the *append* scheme. Our scheme only considers the data-TLB misses; we will be considering instruction-TLB misses next, as mentioned in our future work section.

2   BACKGROUND

2.1   Reorder Buffers and Precise Interrupts

Most contemporary pipelines allow instructions to execute out of program order, thereby taking advantage of idle hardware and finishing earlier than they otherwise would have—thus increasing overall performance. To provide precise interrupts in such an environment typically requires a reorder buffer (ROB) or a ROB-like structure in which instructions are brought in at the tail, and retired at the head [20, 21]. The reorder buffer queues up partially-completed instructions so that they may be retired in-order, thus providing the illusion that all instructions are executed in sequential order—this simplifies the process of handling interrupts precisely.

There have been several influential papers on precise interrupts and out-of-order execution. In particular, Tomasulo [22] gives a hardware architecture for resolving inter-instruction dependencies that occur through the register file, thereby allowing out-of-order issue to the functional units; Smith & Pleszkun [20] describe several mechanisms for handling precise interrupts in pipelines with in-order issue but out-of-order completion, the reorder buffer being one of these mechanisms; Sohi & Vajapeyam [21] combine the previous two concepts into the register update unit (RUU), a mechanism that supports both out-of-order instruction issue and precise interrupts (as well as handling branch misspeculations).

2.2   The Persistence of Software-Managed TLBs

It has been known for quite some time that hardware-managed TLBs outperform software-managed TLBs [11, 16]. Nonetheless, most modern high-performance architectures use software-managed TLBs (eg. MIPS, Alpha, SPARC, PA-RISC), not hardware-managed TLBs (eg. IA-32, PowerPC), largely because of the increased flexibility inherent in the software-managed design [12], and because redesigning system software for a new architecture is non-trivial. Simply redesigning an existing architecture to use a completely different TLB is not a realistic option. A better option is to determine how to make the existing design more efficient.

2.3   Related Work

In our earlier work, we presented the *prepend* method of handling interrupts [28]. When the processor detects a TLB miss, it checks to see if enough space exists within the reorder buffer and enough resources exist, if so, it sets the processor to INLINE mode, resets the head and tail pointer, and starts fetching handler code into the empty space of the reorder buffer. Section 3 of this paper presents this scheme again for ease.

Torng & Day discuss an imprecise-interrupt mechanism appropriate for handling interrupts that are transparent to application program semantics [23]. The system considers the contents of the instruction window (i.e., the reorder buffer) part of the machine state, and so this information is saved when handling an interrupt. Upon exiting the handler, the instruction window contents are restored, and the pipeline picks up from where it left off. Though the scheme could be used for handling TLB-miss interrupts, it is more likely to be used for higher-overhead interrupts. Frequent events, like TLB misses, typically invoke low-overhead interrupts that use registers reserved for the OS, so as to avoid the need to save or restore any state whatsoever. Saving and restoring the entire ROB would likely change TLB-refill from a several-dozen-cycle operation to a several-hundred-cycle operation.

Qiu & Dubois recently presented a mechanism for handling memory traps that occur late in the instruction lifetime [17]. They propose a tagged store buffer and prefetch mechanism to hide some of the latency that occurs when memory traps are caused by events and structures distant from the CPU (for example, when the TLB access is performed near to the memory system, rather than early in the instruction-execution pipeline). Their mechanism is orthogonal to ours and could be used to increase the performance of our scheme, for example in multiprocessor systems.

Zilles, Emer, & Sohi recently presented a multithreaded mechanism to handle precise interrupts [27]. They propose the creation of a second thread that fetches the exceptional handler. After the second thread is done finishing the handler, the main thread continues fetching user level instructions. Their mechanism is similar to ours, with the added necessity that the processor be able to create and handle multiple threads.

Walker & Cragon [25] and Moudgill & Vassiliadis [15] present surveys of the area; both discuss alternatives for implementation of precise interrupts. Walker describes a taxonomy of possibilities, and Moudgill looks at a number of imprecise mechanisms.

3   IN-LINE INTERRUPT HANDLING

We present two methods of in-lining the interrupt handler within the reorder buffer. Both of our schemes exploit the property of a reorder buffer: instructions are brought in at the tail, and retired from the head [20]. If there is enough room between the head and the tail for the interrupt handler to fit, we essentially inline the interrupt by either inserting the handler before the existing user-instructions or after the existing user-instructions. Inserting the handler instructions after the user-instructions, the *append* scheme, is similar to the way that a branch instruction is handled: the PC is redirected when a branch is predicted taken, similarly in this scheme, the PC is redirected when a TLB miss is encountered. Inserting the handler instructions before the user-instructions, the *prepend* scheme [28], uses the properties of the head and tail pointers and inserts the handler instructions before the user-instructions. The two schemes differ in their implementations, the first scheme being easier to build into existing hardware. To represent our schemes in the following diagrams, we are assuming a 16-entry reorder buffer, a four-instruction interrupt handler, and the ability to fetch, enqueue, and retire two instructions at a time. To simplify the discussion, we assume all instruction state is held in the ROB entry, as opposed to being spread out across ROB and reservation-station entries. A detailed description of the two in-lining schemes follows:

1. *Append in-line mode*: Figure 1 illustrates the *append* scheme of inlining the interrupt handler. In the first state [state (a)], the exceptional instruction has reached the head of the reorder buffer and is the next instruction to commit. Because it has caused an exception at some point during its execution, it is flagged as exceptional (indicated by asterisks). The hardware responds by checking to see if the handler would fit into the available space—in this case, there are eight empty slots in the ROB. Assuming the handler is four instructions long, it would fit in the available space. The hardware turns off user-instruction fetch, sets the processor mode to INLINE, and begins fetching the first two handler instructions. These have been enqueued into the ROB at the tail pointer as usual, shown in state (b). In state (c) the last of the handler instructions have been enqueued, the hardware then resumes fetching of user code as shown in state (d). Eventually when the last handler instruction has finished execution and has updated the TLB, the processor can reset the flag on the excepted instruction and retry the operation, shown in state (e).

Note that, though the handler instructions have been fetched and enqueued after the exceptional instruction at the head of the ROB, the handler is nonetheless allowed to affect the state of that exceptional instruction (which logically precedes the handler, according to its relative placement within the ROB). Though this may seem to imply out-of-order instruction commit, it is current practice in the design of modern high-performance processors. For example, the Alpha's TLB-write instructions modify the TLB state once they have finished execution and not at instruction-commit time. In many cases, this does not represent an inconsistency, as the state modified by such handler instructions is typically trans-

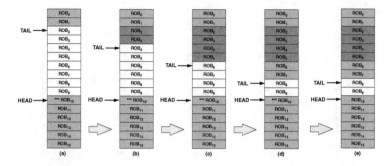

Fig. 1. **Append Scheme:** This figure illustrates the in-lining of a 4-instruction handler, assuming that the hardware fetches and enqueues two instructions at a time. The hardware stops fetching user-level instructions (light grey) and starts fetching handler instructions (dark grey) once the exceptional instruction, identified by asterisks, reaches the head of the queue. When the processor finishes fetching the handler instructions, it can resume fetching the user instructions When the handler instruction updates the TLB, the processor can reset the flag of the excepted instruction and it can reaccess the TLB.

parent to the application—for example, the TLB contents are merely a hint for better address translation performance.

2. **Prepend in-line mode** [28]: Figure 2 illustrates the *prepend* scheme of inlining the interrupt handler. In the first state, [state (a)], the exceptional instruction has reached the head of the reorder buffer. The hardware checks to see if it has enough space, and if it does, it saves the head and tail pointer into temporary registers and moves the head and tail pointer to four instructions before the current head, shown in (b). At this point the processor is put in INLINE mode, the PC is redirected to the first instruction of the handler, and the first two instructions are fetched into the pipe. They are enqueued into the tail of the reorder buffer as usual, shown in (c). The hardware finishes fetching the handler code [state (d)], and restores the tail pointer to its original position, and continues fetching user instructions from where it originally stopped. Eventually, when the last handler instruction fills the TLB, the flag of the excepted instruction can be removed and the exceptional instruction may re-access the TLB [state (e)]. This implementation effectively does out-of-order committing of handler instructions, but again, since the state modified by such instructions is transparent to the application, there is no harm in doing so.

The two schemes presented differ slightly in the additional hardware needed to incorporate them into existing high performance processors. Both the schemes require additional hardware to determine if there are enough reorder buffer entries available to fit the handler code. Since the *prepend* scheme exploits the properties of the head and tail pointers, additional registers are required to save the old values of the head and tail pointers. As we shall see later, incorporating these additional registers will allow for the *prepend* scheme to outperform the *append* scheme by 20-30%. There are a few implementation issues concerning the in-lining of interrupt handlers. They include the following:

1. *The hardware knows the handler length.* To determine if the handler will fit in the reorder buffer, the hardware must know the length of the handler. If there aren't enough slots in the reorder buffer, the interrupt must be handled by the traditional method [28]. If speculative in-lining is used, as mentioned in our future works section, this attribute is not required, but the detection and recovery from a deadlock must be incorporated.

2. *There should be a privilege bit per ROB entry.* Since both user and kernel instructions coexist withing the reorder buffer when inlining; to prevent security holes, a privilege

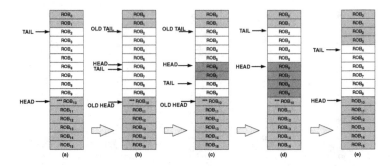

Fig. 2. **Prepend scheme:** This figure illustrates the in-lining of a 4-instruction handler, assuming that the hardware fetches and enqueues two instructions at a time. The hardware stops fetching user-level instructions (light grey), saves the current head and tail pointers, resets the head and tail pointers and starts fetching handler instructions (dark grey) once the exceptional instruction, identified by asterisks, reaches the head of the queue. When the entire handler is fetched, the old tail pointer is restored and the normal fetching of user instruction resumes.

bit must be attached to each instruction, rather than having a single mode bit that applies to all instructions in the pipe [9].

3. *Hardware needs to signal the exceptional instruction when the handler is finished.* When the handler has finished updating the TLB, it should undo any TLBMISS exceptions found in the pipeline, and restore those instructions affected to a previous state so they can re-access the TLB & cache. The signal can be the update of the TLB state while in INLINE mode [28].

4. *After loading the handler, the "return from interrupt" instruction must be killed, and fetching resumes at **nextPC**, which is unrelated to **exceptionalPC**.* When returning from a interrupt handler, the processor must NOP the "return from interrupt" instruction, and resume fetching at some completely unrelated location in the instruction stream at some distance from the exceptional instruction [28].

5. *In-lined handler instructions shouldn't affect the state of user registers.* Since handler instructions are brought in after the excepted instruction but commit before the excepted instruction, we have to make sure that when they commit, they don't wrongly update and release user registers. To fix this, when mapping the first handler instruction, the handler instruction should receive a copy of the current committed register file state rather than the register state of the previous instruction. Additionally, when a user instruction is being mapped after the handler is completely fetched, it should copy the register state from a previous user instruction, whose location can be stored in a temporary register. The logic here amounts to a MUX [28].

6. *The hardware might need to know the handler's register requirements.* If at the time the TLB miss is discovered, the processor will need to make sure it isn't stalled in one of the critical paths of the pipeline, eg. register renaming. A deadlock situation might occur if there aren't enough free physical registers available to map existing instructions prior to and including those in the register renaming phase. To prevent this, the processor can do one of two things. (a) handle the interrupt the traditional method, or (b) flush all instructions in the fetch and decode stage and set nextPC to the earliest instruction in the decode/map pipeline stage. As mentioned, since most architectures reserve a handful of registers for handlers to avoid the need to save and restore user state, the handler will not stall at the mapping stage. In architectures that do **not** provide such registers, the hardware will need to ensure adequate physical register availability

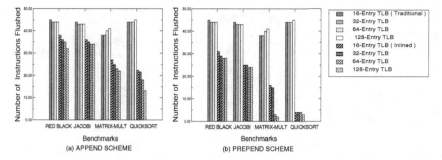

Fig. 3. This figure compares the average number of instructions flushed when handling a TLB miss by the traditional method to those of the two inlined schemes. The append scheme isn't able to reduce the number of instructions being flushed as it takes up space within the reorder buffer, while the prepend scheme is able to reduce the number of instructions being flushed significantly.

before vectoring to the handler code [28]. For our simulations, we only simulated scheme (a).

7. *Branch mispredictions in user code should not flush handler instructions.* If, while in INLINE mode, a user-level branch instruction is found to have been mispredicted, the resulting pipeline flush should not effect the handler instructions already in the pipeline. This means that the hardware should overwrite **nextPC** (described above) with the correct branch target, it should invalidate the appropriate instructions in the ROB, and it should be able to handle holes in the ROB contents. The *append* scheme will have to account for this, while the *prepend* scheme doesn't have to worry about this as all the handler instructions are physically before the interrupted instruction.

Overall, the hardware design is relatively simple, requiring beyond this a status bit that identifies when the processor is handling interrupts in this manner. Otherwise, the design of the processor is unmodified.

4   THE PERFORMANCE OF IN-LINING INTERRUPTS

4.1   Simulation Model

We model an out-of-order processor similar to the Alpha 21264. It has 64K/64K 2-way L1 instruction and data caches, fully associative 16/32/64/128 entry separate instruction and data TLBs with an 8KB page size. It can issue up to four instructions per cycle and can hold 80 instructions in flight at any time. It has a 72-entry register file (32 each for integer and floating point instructions, and 8 for privileged handlers), 4 integer functional units, and 2 floating point units. The model also provides 82 free renaming-registers, 32 reserved for integer instructions and 32 for floating point instructions. The model also has a 21 instruction TLB miss handler. The model doesn't have any renaming registers reserved for privileged handlers as they are a class of integer instructions. Therefore, the hardware must know the handler's register needs as well as length in instructions. We chose this for two reasons: (1) the design mirrors that of the 21264; and (2) the performance results would be more conservative than otherwise.

Like the Alpha 21264 and MIPS R10000 [7, 26], our model uses a reorder buffer as well as reservation stations attached to the different functional units—in particular, the floating-point and integer instructions are sent to different execution queues. Therefore, both ROB space and execution-queue space must be sufficient for the handler to be in-lined, and

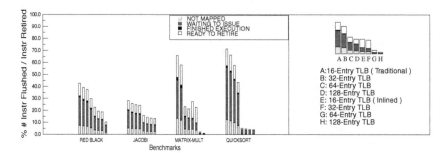

Fig. 4. This figure shows the stages in which the instructions were before they were flushed. The graphs show that 15-30% of the instructions have already finished execution, 50-60% are waiting to be dispatched to execution units, and the remaining are awaiting decoding and register renaming. In-lining using the prepend scheme cuts back significantly on the number of instructions being flushed.

instruction-issue to the execution queues stalls for user-level instructions during the handler execution. The page table and TLB-miss handler are modeled after the MIPS architecture [14, 12] for simplicity.

4.2   Benchmarks

While the SPEC 2000 suite might seem a good source for benchmarks, as it is thought to exhibit a good memory behavior, the suite demonstrates TLB miss rates that are three orders of magnitude lower than those of realistic high-performance applications. In his WWC-2000 Keynote address [2], John McCalpin presents, among other things, a comparison between SPEC 2000 and a set of more real-world high-performance programs. The reason why SPEC applications don't portray "real world applications" is because they tend to access memory in sequential fashion [2].

McCalpin's observations are important because our previous work suggests that the more often the TLB requires management, the more benefits one sees from handling the interrupt by the in-line method [28]. Therefore, we use a handful of benchmarks that display typically non-sequential access to memory and have correspondingly higher TLB miss rates than SPEC 2000. The benchmark applications include quicksort, red-black, Jacobi, and matrix-multiply.

4.3   Results

We first take a look at the number of instructions that are flushed on average when a TLB miss is detected. Figure 3 shows that when handling TLB misses traditionally, the reorder buffer, at the time that a TLB miss is detected, is only 50% full. The figure shows that with our schemes of in-lining the interrupt, *prepend* scheme performs the best in terms of reducing the number of instructions being flushed. Our studies show that roughly 60% of the time TLB interrupts in our benchmarks were able to benefit from in-lining using the *append* scheme, while 80-90% of the time interrupts were able to benefits using the *prepend* scheme. This can be explained in the fact that the *append* scheme retains the miss handler code within the reorder buffer after having finished refilling the TLB, thus occupying reorder buffer space, while in the *prepend* scheme the handler instructions exit the reorder buffer, thus restoring the amount of free space within the reorder buffer. We find that in both schemes, 10-20% of the time the handler cannot be in-lined due to insufficient physical registers available to map the user instructions already present in the pipeline. As men-

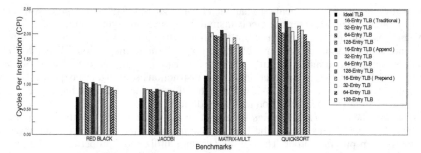

Fig. 5. The figure shows the execution time (cycles-per-user-instructions) of a perfect TLB, 16/32/64/128-entry traditionally handled TLBs, append in-lined TLBs, and prepend in-lined TLBs.

tioned earlier, the in-lined interrupt method can still be made to work if we allow for partial flushing of pipeline stages. Most modern high performance processors, like the Alpha 21264, currently allow for such techniques.

We further investigate the properties of all the instructions that were flushed due to TLB misses. Figure 4 shows the stages at which the instructions were flushed when the TLB miss handler was executed. The first four bars show the statistics for the traditional method of handling interrupts using 16/32/64/128 entry TLBs, while the last four bars show the same with the *prepend* scheme. The x-axis shows the different benchmarks, and the y-axis represents the ratio of the total number of instructions flushed (because of TLB misses alone) to the number of user instructions retired. The graph first of all shows that in the traditional scheme, the number of instructions (including speculative) flushed are about 20-70% of the instructions that are retired, therefore, representing significant work wasted. Of the instructions being flushed: 20-30% of the instructions being flushed have already finished execution; 50-60% are waiting to be dispatched to execution units, and the others are awaiting instruction decoding and register renaming. This overhead causes more time and energy to be consumed in re-fetching and re-executing the instructions. The *prepend* in-lined scheme reduces the number of instructions being flushed by 50-90% and the *append* in-lined scheme by 25-50%. Thus the in-lined methods significantly reduce the tremendous overhead in handling TLB interrupts using traditional means.

Additionally, we compare the performance of perfect TLBs, traditional software-managed TLBs, and in-lined TLBs. Figure 5. shows the *append* scheme reducing the execution time by 5-10% and the *prepend* scheme reducing the execution time by 5–25% for the same-size TLB. The significant performance difference again is due to the interrupt handler retaining space within the reorder buffer in the *append* scheme. Both the schemes provide performance improvements both in terms of the number of instructions flushed and execution time. The favoring of one scheme over the other depends on which scheme is easier to integrate into existing hardware. We showed earlier that the *append* scheme was easier to build into existing hardware, but also suggested the additional minimal logic for the *prepend* scheme.

5   CONCLUSIONS & FUTURE WORK

The general purpose precise interrupt mechanisms in use for the past few decades have received very little attention. With the current trends in processor and operating systems design, the overhead of re-fetching and re-executing instructions is severe for applications that incur frequent interrupts. One example is the increased use of the interrupt mechanism to perform memory management—to handle TLB misses in today's microprocessors. This is putting pressure on the interrupt mechanism to become more lightweight.

We propose the use of in-line interrupt handling, where the reorder buffer is not flushed on an interrupt unless there isn't enough space for the handler instructions. This allows the user application to continue executing while an interrupt is being serviced. For a software-managed TLB miss, this means that only those instructions stall that are dependent on the instruction that misses the TLB. All other instructions continue executing, and are only held up at commit (by the instruction that missed the TLB).

We present the *append* and *prepend* schemes that allow the in-lining of the interrupt handler. The *append* scheme inserts the handler instructions after the user-instructions, thus retiring them in program order. The *prepend* scheme utilizes the properties of the reorder buffer and inserts the handler instructions before the user-instructions, thus retiring them out of program order without any side affects. We concluded that though both schemes are very similar and provide a degree of performance improvement, because the *prepend* scheme restores reorder buffer space, it provides better performance and also significantly reduces the amount of instructions being flushed.

Our simulations show that in-lined interrupt handling cuts the overhead by 10–40% [28], with a performance improvement of 5-25% for the *prepend* scheme and a 5-10% improvement for the *append* scheme. Additionally, our simulations revealed the overhead of handling TLB misses traditionally: 20-30% of the instructions being flushed will be re-executed again, and 80% of the instructions being flushed will need to be re-decoded. The overhead in terms of performance and power consumption is significantly high. The in-lined schemes reduce the number of instructions being flushed by 25-50% in the *append* scheme and 50-90% in the *prepend* scheme.

We are currently exploring the performance of in-lined handling of interrupts from the power consumption perspective. Flushing the pipeline, re-fetching and re-executing the instructions will definitely consume more power than not doing so. We are also looking into instruction-TLB interrupts, and the non-speculative in-lining of handler code. It is possible to begin fetching the handler into the ROB without first checking to see if there is enough room or resources. This requires a check for deadlock, and the system responds by handling a traditional interrupt when deadlock is detected—flush the pipe and resume at the handler. This allows support for variable-length TLB-miss handlers, such as the Alpha's.

References

[1]   T. E. Anderson, H. M. Levy, B. N. Bershad, and E. D. Lazowska. "The interaction of architecture and operating system design." In *Proc. Fourth Int'l Conf. on Architectural Support for Programming Languages and Operating Systems (ASPLOS'91)*, April 1991, pp. 108–120.

[2]   J. McCalpin. *An Industry Perspective on Performance Characterization: Applications vs Benchmarks.* Keynote address at Third Annual IEEE Workshop on Workload Characterization, Austin TX, September 16, 2000.

[3]   B. Case. "AMD unveils first superscalar 29K core." *Microprocessor Report*, vol. 8, no. 14, October 1994.

[4]   B. Case. "x86 has plenty of performance headroom." *Microprocessor Report*, vol. 8, no. 11, August 1994.

[5]   Z. Cvetanovic and R. E. Kessler. "Performance analysis of the Alpha 21264-based Compaq ES40 system." In *Proc. 27th Annual International Symposium on Computer Architecture (ISCA'00)*, Vancouver BC, June 2000, pp. 192–202.

[6]   L. Gwennap. "Intel's P6 uses decoupled superscalar design." *Microprocessor Report*, vol. 9, no. 2, February 1995.

[7]   L. Gwennap. "Digital 21264 sets new standard." *Microprocessor Report*, vol. 10, no. 14, October 1996.

[8] D. Henry, B. Kuszmaul, G. Loh, and R. Sami. "Circuits for wide-window superscalar processors." In *Proc. 27th Annual International Symposium on Computer Architecture (ISCA'00)*, Vancouver BC, June 2000, pp. 236–247.

[9] D. S. Henry. "Adding fast interrupts to superscalar processors." Tech. Rep. Memo-366, MIT Computation Structures Group, December 1994.

[10] J. Huck and J. Hays. "Architectural support for translation table management in large address space machines." In *Proc. 20th Annual International Symposium on Computer Architecture (ISCA'93)*, May 1993, pp. 39–50.

[11] B. L. Jacob and T. N. Mudge. "A look at several memory-management units, TLB-refill mechanisms, and page table organizations." In *Proc. Eighth Int'l Conf. on Architectural Support for Programming Languages and Operating Systems (ASPLOS'98)*, San Jose CA, October 1998, pp. 295–306.

[12] B. L. Jacob and T. N. Mudge. "Virtual memory in contemporary microprocessors." *IEEE Micro*, vol. 18, no. 4, pp. 60–75, July/August 1998.

[13] B. L. Jacob and T. N. Mudge. "Virtual memory: Issues of implementation." *IEEE Computer*, vol. 31, no. 6, pp. 33–43, June 1998.

[14] G. Kane and J. Heinrich. *MIPS RISC Architecture*. Prentice-Hall, Englewood Cliffs NJ, 1992.

[15] M. Moudgill and S. Vassiliadis. "Precise interrupts." *IEEE Micro*, vol. 16, no. 1, pp. 58–67, February 1996.

[16] D. Nagle, R. Uhlig, T. Stanley, S. Sechrest, T. Mudge, and R. Brown. "Design tradeoffs for software-managed TLBs." In *Proc. 20th Annual International Symposium on Computer Architecture (ISCA'93)*, May 1993.

[17] X. Qiu and M. Dubois. "Tolerating late memory traps in ILP processors." In *Proc. 26th Annual International Symposium on Computer Architecture (ISCA'99)*, Atlanta GA, May 1999, pp. 76–87.

[18] M. Rosenblum, E. Bugnion, S. A. Herrod, E. Witchel, and A. Gupta. "The impact of architectural trends on operating system performance." In *Proc. 15th ACM Symposium on Operating Systems Principles (SOSP'95)*, December 1995.

[19] M. Slater. "AMD's K5 designed to outrun Pentium." *Microprocessor Report*, vol. 8, no. 14, October 1994.

[20] J. E. Smith and A. R. Pleszkun. "Implementation of precise interrupts in pipelined processors." In *Proc. 12th Annual International Symposium on Computer Architecture (ISCA'85)*, Boston MA, June 1985, pp. 36–44.

[21] G. S. Sohi and S. Vajapeyam. "Instruction issue logic for high-performance, interruptable pipelined processors." In *Proc. 14th Annual International Symposium on Computer Architecture (ISCA'87)*, June 1987.

[22] R. M. Tomasulo. "An efficient algorithm for exploiting multiple arithmetic units." *IBM Journal of Research and Development*, vol. 11, no. 1, pp. 25–33, 1967.

[23] H. C. Torng and M. Day. "Interrupt handling for out-of-order execution processors." *IEEE Transactions on Computers*, vol. 42, no. 1, pp. 122–127, January 1993.

[24] M. Upton. *Personal communication*. 1997.

[25] W. Walker and H. G. Cragon. "Interrupt processing in concurrent processors." *IEEE Computer*, vol. 28, no. 6, June 1995.

[26] K. C. Yeager. "The MIPS R10000 superscalar microprocessor." *IEEE Micro*, vol. 16, no. 2, pp. 28–40, April 1996.

[27] C.B. Zilles, J.S. Emer, and G.S. Sohi, "Concurrent Event-Handling Through Multithreading", *IEEE Transactions on Computers*, 48:9, September, 1999, pp 903-916.

[28] Jaleel, Aamer and Jacob, Bruce. "In-line Interrupt Handling for Software Managed TLBs". *Proc. 2001 IEEE International Conference on Computer Design (ICCD 2001)*, Austin TX, September 2001.

Hidden Costs in Avoiding False Sharing in Software DSMs*

K.V. Manjunath[1] and R. Govindarajan[2]

[1] Hewlett Packard ISO, Bangalore 560 052, India
manjukv@india.hp.com
[2] Department of Computer Science and Automation and
Supercomputer Education and Research Centre,
Indian Institute of Science. Bangalore 560 012, India
govind@csa.iisc.ernet.in

Abstract. Page based software DSMs experience high degrees of false sharing especially in irregular applications with fine grain sharing granularity. The overheads due to false sharing is considered to be the dominant factor limiting the performance of software DSMs. Several approaches have been proposed in the literature to reduce/eliminate false sharing. In this paper, we evaluate two of these approaches, *viz.,* the Multiple Writer approach and Emulated Fine Grain Sharing (EmFiGS) approach. Our evaluation strategy is two pronged: firstly, we use a novel implementation independent analysis which uses overhead counts to compare the different approaches. The performance of EmFiGS approach is significantly worse, by a factor of 1.5 to as much as 90 times, compared to the Multiple Writer Approach. In many cases, EmFiGS approach performs worse that even a single writer lazy release protocol which experiences very high overheads due to false sharing. Our analysis shows that the benefits gained by eliminating false sharing are far outweighed by the performance penalty incurred due to the reduced exploitation of spatial locality in the EmFiGS approach. The implementation independent nature of our analysis implies that any implementation of the EmFiGS approach is likely to perform significantly worse than the Multiple Writer approach. Secondly, we use experimental evaluation to validate and complement our analysis. Experimental results match well with our analysis. Also the execution times of the application follow the same trend as in our analysis, reinforcing our conclusions.

1 Introduction

Software Distributed Shared Systems [1,2], which rely on the virtual memory mechanism provided by the Operating System for detecting accesses to shared data, support sharing granularity of page size. The large page size results in excessive *false sharing*, especially in fine grain irregular applications [3]. Different methods have been proposed in literature

---

* This work was done when the first author was a research student at the Department of Computer Science and Automation, Indian Institute of Science, Bangalore. This work was supported by research grants from IBM, Poughkeepsie, NY and IBM Solutions Research Centre, New Delhi, under IBM's Shared University Program.

B. Monien, V.K. Prasanna, S. Vajapeyam (Eds.): HiPC 2001, LNCS 2228, pp. 294–303, 2001.

to reduce the effects of false sharing in page-based software DSMs. Two basic approaches followed in these methods are (i) allowing concurrent writers to a page and (ii) providing fine grain granularity through emulation without additional architectural support. We refer to these approaches as Multiple Writer approach and Emulated Fine Grain Sharing (EmFiGS) approach respectively. We call the latter *emulation* because it provides fine grain sharing over a coarse grain sharing system and as a consequence incurs higher cost even for a fine grain coherence miss.

Lazy Multiple Writer Protocol (LMW) [4] implemented in most state-of-the-art software DSMs [5,6] is a typical example of the Multiple Writer approach. It allows concurrent writers to a page by providing mechanisms to merge the modifications at synchronization points. LMW is considered heavy-weight because of the Twin/Diff creation overheads incurred to maintain and update the modifications. Writer Owns Protocol [7] improves upon LMW by performing run time re-mapping of sub pages such that all sub pages in a page are written by the same process. The sharing granularity is still maintained at page level, but by re-mapping parts of page appropriately, false sharing is eliminated. Millipage/Multiview [8] follows the EmFiGS approach. It decreases the sharing granularity to less than a page size by mapping multiple smaller size virtual pages to a physical page. Millipage is implemented with single-writer protocol for efficient operation. Whereas Multiple Writer approach *tolerates* false sharing, EmFiGS approach *eliminates* false sharing.

While it is true that these approaches are successful in reducing false sharing, these work do not sufficiently address the following questions: (i) what are the additional costs incurred, if any, in reducing false sharing? and (ii) is the reduction in false sharing overheads significantly higher than the additional costs, thereby leading to overall performance improvement of the application? Also, while there have been performance evaluation of individual implementations of these approaches, there has been been no complete comparative analysis of these approaches which is required to understand the inter-play of overheads. This motivates our work on performance analysis of methods to overcome false sharing in software DSMs.

Our performance evaluation strategy is two pronged. Firstly, we present a novel implementation independent analysis to obtain the counts of overheads incurred under different protocols. The overhead counts provide a basis for comparison of the different methods. Our analysis indicates that the overheads incurred by the EmFiGS approach are significantly higher by a factor of 4 or more as compared to Multiple Writer approach. This is due to the decrease in exploitation of spatial locality in the EmFiGS approach. As our analysis is independent of any specific implementation and captures the overheads that are intrinsic to the protocol, we conclude that any implementation of the EmFiGS approach is likely to incur more true sharing overheads than the savings in false sharing.

Secondly, we augment this comparison with execution time results of the benchmarks under an actual implementation of the methods on the same platform. The experimental evaluation validates and complements our manual analysis. From our experimental evaluation, we see that the decreased exploitation of spatial locality manifesting as increased page faults and messages in EmFiGS approach, resulting in a degradation of 1.5 to as much as 90 times compared to the Multiple Writer approach.

Section 2 discusses the protocols studied and the overheads considered in our analysis. Section 3 describes the manual analysis in detail and also presents the results of the analysis. Section 4 deals with the experimental results. Concluding remarks are provided in Section 5.

2   Protocols and Overheads

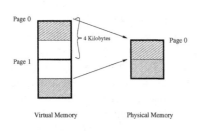

Fig. 1. Illustration of Millipage

This section discusses the protocols studied and the overheads considered in our analysis. We have considered those protocols in our analysis that sufficiently represent the two approaches to overcome false sharing software DSMs, *viz.,* the multiple writer approach and the EmFiGS approach. First of all, we consider the Lazy Single Writer(LSW) [9] protocol as the base case, because LSW can incur high overheads in the presence of false sharing. LSW implements the Lazy Release Consistency (LRC) [4] memory consistency model and allows only one process to write to a shared page at any time. Detailed descriptions of different memory consistency models can be found in [10]. We consider the Lazy Multiple Writer Protocol (LMW) in our study to represent the multiple writer approach. LMW also implements LRC model, but allows multiple concurrent processes to write to non-overlapping regions of a shared page. LMW uses *twinning and diffing* [5] to maintain modifications done to a shared page by different processes, and merges these modifications at a subsequent synchronization point. LMW is considered to be a heavy-weight protocol [11] because of the twinning and diffing overheads.

Array Bloating is our alternative method for implementing the EmFiGS approach, which is similar in behaviour to Millipage/Multiview [8]. In this method, we *pad and align* the elements of shared arrays so that a virtual page holds fewer elements (possibly only one) than it can accommodate, and non-overlapping regions of the array accessed by two different processes do not lie on the same virtual page. Further, multiple virtual pages are mapped to a single physical page as shown in Figure 1. Virtual to physical page mapping is done in a manner similar to Millipage technique. We use the term *Bloat Factor* to refer to the number times the data structure is bloated. To be precise, *bloat factor* refers to the number of virtual pages which are mapped to the same physical page (refer to Figure 1). The *bloat factor* also indicates the amount of data allocated per page. Higher the *bloat factor*, lesser the data allocated per page and hence lesser the false sharing. Access to appropriate elements of the array is accomplished by modifying the *sh_malloc* statements and instrumenting the array indices in the application program. Array bloating is less generalized than Millipage because it cannot be applied to irregular data structures with pointer references. In our analysis, we have considered array bloating implemented with LSW (ABLSW) and Sequential consistency protocol (ABSC) as methods following EmFiGS approach.

2.1   Overheads

The overheads considered in our analysis are listed below. Each of these overheads would be incurred a number of times (referred to as **overheads counts**) and hence contribute certain time (referred to as **overheads time**) to the execution time. Our manual analysis deals with overheads counts, while the experimental evaluation reports overheads time.

Page fault kernel overhead : Page based software DSMs rely on the page fault mechanism to detect accesses to a shared page. Access to a page without proper read/write permissions raises a page fault exception. The context is switched to the kernel mode exception handler. The exception handler in turn raises the SIGSEGV signal and the SIGSEGV signal handler provided by the DSM software gets control,*i.e.,* the context is switched back to the used mode. This overhead due to OS kernel involvement in a page fault is referred to as page fault kernel overhead.

Mprotect : This refers to the `mprotect` system calls that are invoked.

Buffer copy : This is the overhead incurred in copying data from message data structures to the page and vice versa. This is over and above the various levels of copying that might be required by the underlying messaging layer which is implementation dependent.

Twin creation, Diff creation, and Diff apply : These overheads are specific to LMW and are incurred in maintaining modifications to a shared page.

Message Overhead : A process sends messages to other processes on various events like page fault, lock acquire, or barrier synchronization. A message arriving at a process causes some consistency protocol actions to be performed and a reply is sent to the requesting processor. Further, since the messages arrive asynchronously at the receiver, some overheads are incurred at the messaging layer due to polling/interrupt handling. We refer to the time elapsed between the sending of the request and receiving the reply as message time, which includes the round trip message latency, the message processing time at receiver and other messaging layer overheads.

Synchronization/Barrier Overhead : This includes the calls to lock acquire/release and barrier synchronization and the associated overheads involved in performing the consistency protocol actions.

The above overheads represent all overheads that occur in page-based DSM systems. That the list is exhaustive can be inferred from our experimental results (described in Section 4) where the fraction of execution time of the benchmarks contributed by these overheads accounts for rest of the execution time other than application time.

3   Manual Analysis

3.1   The Method

We illustrate the method of our analysis with a simple example. We are interested in those statements of the application program that can possibly cause any of the overheads listed in Section 2.1. These are the statements containing shared memory references and synchronization points. In software DSM programs, synchronization is achieved by an explicit function call to the underlying software layer. We inspect the benchmark source code to separate these two kinds of statements, *viz.,* shared memory accesses and synchronization points, and also the control structures of the program using an approach similar to program slicing. Figure 2(a) shows such a sliced program. Assume we have two processes and let the array *shared_a*[1024] span a single virtual page. By analyzing the source code, we then arrive at a sequence of accesses to actual shared pages in different processes, as shown in Figure 2(b).

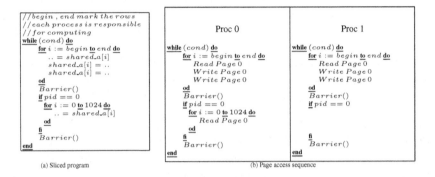

(a) Sliced program   (b) Page access sequence

Fig. 2. Sliced program and page access sequence

Figure 3 shows the overheads calculation for LSW for a single iteration of the access. We consider different temporal interleavings of the sequences of shared accesses to arrive at the maximum and minimum overheads incurred (shown in Figure 3(a) and 3(b) respectively). Figure 3(c) summarizes the minimum and maximum overheads. Similar steps are followed in manual analysis of the benchmarks for LMW, ABLSW and ABSC protocols. In this illustration, both Proc 0 and Proc 1 have a readable copy of the page initially. For a detailed explanation of the analysis, refer [12].

3.2   Results

We have analysed three applications from SPLASH-2 suite,*viz.*, BARNES (1024 bodies, 400 iterations), WATER-SPATIAL (64 molecules, 40 iterations), and RADIX (16384 keys and 256 radix). The results of the analysis on the overheads counts in different protocols for the three applications are presented in Table 1. We have considered the applications running on 4 processes and the overheads shown are incurred on process 0. The overheads incurred on other processes exhibit a similar behavior. Each row in the table indicate the number of times a particular overhead is incurred in the different protocols.

The consistent trend we see in all the applications is that the overheads in ABLSW and ABSC are significantly higher than in LMW or in best case LSW. This is somewhat surprising given that false sharing is completely eliminated in ABLSW and ABSC (only one body, molecule, or key per shared page), by reducing the sharing granularity to the level of basic units of memory access in the applications. The number of page fault kernel overheads and the message overheads incurred in ABLSW are higher by at least a factor of 2 in WATER-SPATIAL and by a factor of 10 in other applications compared to LMW.

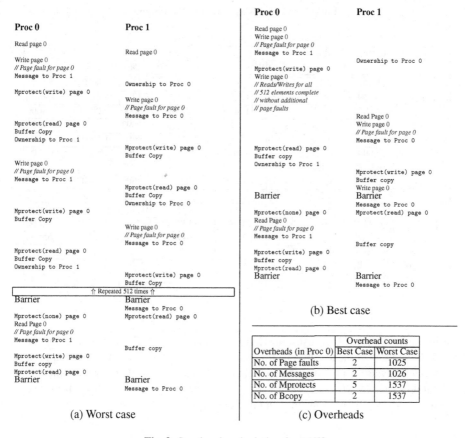

Fig. 3. Overheads calculation for LSW

Why should ABLSW or the EmFiGS approach incur such high page fault overheads given that it eliminates false sharing completely? To answer this question, first let us note that the applications we have analyzed do contain true sharing behavior wherein a single process (such as Proc 0) reads the entire shared array after force calculation on all bodies/molecules were complete (similar to the read in Proc 0 between the two barriers in the example shown in Figure 2). In the Multiple Writer approach, a single page fault causes the entire page to be validated, *i.e.*, a single page fault brings in the entire page (several bodies/molecules/keys), whereas in the EmFiGS approach, due to the emulation of fine grain sharing, each page fault brings in a lesser amount of data (only one body, molecule, or key) on a true sharing miss. That is, the amount of *spatial locality* exploited reduces with reduction in sharing granularity. This reduction in exploitation of spatial locality is the principal reason for ABLSW and ABSC having more page faults and consequently more associated overheads like messages and `mprotects` than LMW or best case LSW. It can be seen that if the costs of the overheads of twin/diff create and diff apply (which are not incurred in LSW, ABLSW, or ABSC) are not significant, LMW is likely to perform better than other methods.

Table 1. Overheads in Benchmark Programs

Overheads	Applns.	LMW	LSW		ABLSW		ABSC	
			Best case	Worst case	Best case	Worst case	Best case	Worst case
Page Faults	BARNES	3362	4797	34809	54407	54407	58507	58507
	WATER	19200	18400	140000	34800	73200	66000	104400
	RADIX	1049	1372	4194652	28000	28000	28000	28000
Messages	BARNES	3684	3567	35055	31693	31693	58589	58589
	WATER	18000	22000	143600	41200	79600	194000	347600
	RADIX	1050	1053	3146032	28708	28708	60000	60000
Barriers	BARNES	82	82	82	82	82	82	82
	WATER	3602	3602	3602	3602	3602	3602	3602
	RADIX	22	22	22	22	22	22	22
Mprotect	BARNES	6683	9512	54120	108732	108732	116932	116932
	WATER	38800	35600	208400	83200	147200	153594	236800
	RADIX	2754	2680	4195960	61496	61496	65536	65536
Twin/Diff	BARNES	7339	–	–	–	–	–	–
	WATER	32400	–	–	–	–	–	–
	RADIX	2734	–	–	–	–	–	–

In conclusion, we observe that the EmFiGS approach (ABLSW and ABSC) incur significantly higher overheads (by a factor of as much as 90) than Multiple Writer approach. Since our analysis is implementation independent, we conclude that *any* implementation of EmFiGS such as ABLSW, ABSC, or even Millipage/Multiview will incur more true sharing overheads than the savings in false sharing.

4   Experimental Evaluation

This section presents our experimental evaluation of the benchmarks. Our experimental platform is a 12 node IBM SP2 connected by a high performance switch. The page size is 4 K Bytes. We used CVM [6], an open source software DSM system for our experiments. CVM supports LMW, LSW and Sequential Consistency protocols. We implemented the array bloating technique in CVM. Although in principle array bloating can be achieved using compiler instrumentation, we manually instrumented the benchmarks to support array bloating. Also, we have suitably modified CVM to obtain detailed break-ups of the overhead times.

4.1   Validation of Manual Analysis

Table 2 shows the overheads incurred by BARNES in the 4 processor case under different protocols. Results for other applications and for 2, 8, and 12 processor cases follow a similar trend[12] and are not reported here due to space limitation. The column marked "Expt." shows the measured values of the overheads incurred by node 0 when the applications were run on CVM.

Table 2. Overheads in BARNES — Comparison of Analysis vs. Experimental

Overheads	LMW		LSW			ABLSW		ABSC	
	Analysis	Expt.	Best case	Worst case	Expt.	Analysis	Expt.	Analysis	Expt.
Page Faults	3362	3425	4797	34809	15393	54407	55833	58507	59104
Messages	3684	3963	3567	35055	14654	31693	31202	58589	89126
Barriers	82	82	82	82	82	82	82	82	82
Mprotect	6683	12004	9512	54120	37219	108732	148077	116932	154817
Twin/Diff	7339	8002	–	–	–	–	–	–	–

We see that the experimental values of all the overheads except mprotect match well and fall within 10% of those obtained from our manual analysis. A part of this deviation is due to the simplifying assumption — molecules/bodies do not move from one process to another in our analysis. The overheads in LSW measured in the experiments fall within the minimum and maximum values calculated in our analysis, and are closer to the best case overhead values. Comparing the overheads measured from our experimental evaluation, we can observe that LMW incurs significantly less overhead than ABLSW (by a factor of 2.5 to 15) or ABSC (by a factor of 4 to 20). LSW, despite experiencing considerable false sharing, incurs less overheads, less by a factor of 1.5 to 3 than ABLSW in all applications. Next, we address the question, "how do these overhead counts translate into execution times?" in the following section.

4.2 Execution Time Results

Figure 4 shows detailed split-ups of the execution time of the applications in the 4 processor case. The performance results for 2 and 8 processor cases are reported in [12]. The total execution time of an application is broken down into the computation time (*i.e.*, time spent in the computation steps of the application — shown as "Computation" in graph legend, measured independently from a single processor run with $\frac{1}{4}th$ the problem size) and the overheads time. The overheads time is in turn broken down into the various overheads listed in Section 2.1. All timings are normalized with respect to the total execution time of the applications running under the LMW protocol in respective cases.

All the applications follow the same trend as in the manual analysis, *i.e.*, ABLSW and ABSC perform significantly worse than LMW by a factor of 1.5 to as much as 90 times. Further, contrary to the popular belief about the "heavy-weightedness" of multiple writer protocols, we observe that the Twin/Diff overheads (not visible in the graphs) contribute to less than 1% of the total overheads. Decreased false sharing overheads leveraged by the negligible overheads in supporting multiple writers causes LMW to perform better than ABLSW and ABSC. As the overheads in LSW fall closer to the best case (as seen from Table 2), we see that LSW also performs better than ABLSW and ABSC by 1.2 to 20 times. In ABLSW and ABSC, as discussed in Section 3.2, emulated fine grain sharing increases the number of true sharing faults, which necessitates more page request messages. This is the reason for higher overheads in ABLSW and ABSC.

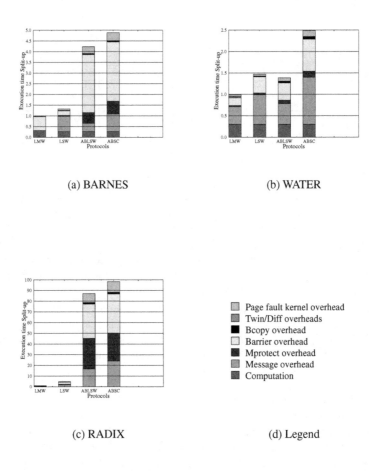

(a) BARNES

(b) WATER

(c) RADIX

(d) Legend

Fig. 4. Normalized Execution time split-up

5   Conclusions

In this paper, we present the performance evaluation of two approaches to overcome false sharing, *viz.*, the Multiple Writer approach and EmFiGS approach, in page-based software DSM systems. Our evaluation is two pronged : (i) a novel implementation independent approach which reports overheads counts in different methods and (ii) an experimental evaluation of the methods implemented on top of CVM, a page-based software DSM running on IBM-SP2. Our results indicate that even though false sharing is eliminated completely in EmFiGS approach, its overheads are found to be significantly higher by a factor of 1.5 to 90 times the overheads in multiple writer approach. We attribute the high overheads in EmFiGS approach to the decreased exploitation of spatial locality. By emulating finer granularity of sharing, the EmFiGS approach incurs more true sharing overheads than the savings in false sharing. In contrast, the overheads incurred by multiple writer approach to support multiple writers is insignificant, which is contrary to the popular belief. Further, EmFiGS approach performs worse than LSW (Lazy release Single Writer) protocol, which experiences the false sharing overheads in full.

References

1. Li, K., Hudak, P.: Memory coherence in shared virtual memory systems. In: Proc. of the 5th Annual ACM Symp. on Principles of Distributed Computing (PODC'86). (1986) 229–239
2. Bennett, J.K., Carter, J.B., Zwaenepoel, W.: Munin: Shared memory for distributed memory multiprocessors. Technical Report COMP TR89-91, Dept. of Computer Science, Rice University (1989)
3. Torellas, J., Lam, M., Henessey, J.: False sharing and spatial locality in multiprocessor caches. IEEE Transactions on Computers **6** (1994) 651–663
4. Keleher, P., Cox, A.L., Zwaenepoel, W.: Lazy release consistency for software distributed shared memory. In: Proc. of the 19th Annual Int'l Symp. on Computer Architecture (ISCA'92). (1992) 13–21
5. Amza, C., Cox, A.L., Dwarkadas, S., Keleher, P., Lu, H., Rajamony, R., Yu, W., Zwaenepoel, W.: TreadMarks: Shared memory computing on networks of workstations. IEEE Computer **29** (1996) 18–28
6. Keleher, P.: The CVM manual. Technical report, University of Maryland (1996)
7. Freeh, V.W., Andrews, G.R.: Dynamically controlling false sharing in distributed shared memory. In: Proc. of the Fifth IEEE Int'l Symp. on High Performance Distributed Computing (HPDC-5). (1996) 403–411
8. Itzkovitz, A., Schuster, A.: MultiView and Millipage: Fine-grain sharing in page-based DSMs. In: Proc. of the 3rd Symp. on Operating Systems Design and Implementation (OSDI'99). (1999) 215–228
9. Keleher, P.: The relative importance of concurrent writers and weak consistency models. In: Proc. of the 16th Int'l Conf. on Distributed Computing Systems (ICDCS-16). (1996) 91–98
10. Adve, S.V., Gharachorloo, K.: Shared memory consistency models: A tutorial. IEEE Computer **29** (1996) 66–76
11. Amza, C., Cox, A.L., Dwarkadas, S., Zwaenepoel, W.: Software DSM protocols that adapt between single writer and multiple writer. In: Proc. of the 3rd IEEE Symp. on High-Performance Computer Architecture (HPCA-3). (1997) 261–271
12. Manjunath, K.V., Govindarajan, R.: Performance analysis of methods that overcome false sharing effects in software DSMs. Technical Report TR-HPC-08 /2001, Indian Institute of Science, Bangalore (2001) http://hpc.serc.iisc.ernet.in/Publications/kvm2001.ps

Heterogeneous Computing: Goals, Methods, and Open Problems*

Tracy D. Braun[1], Howard Jay Siegel[2], and Anthony A. Maciejewski[2]

[1] NOEMIX, Inc.
1425 Russ Blvd. Ste. T-110
San Diego, CA 92101-4717 USA
tdbraun@noemix.com
http://members.home.net/brauntd
[2] Electrical and Computer Engineering Department
Colorado State University
Fort Collins, CO 80523-1373 USA
{hj, aam}@colostate.edu
http://www.engr.colostate.edu/{~hj, ~aam}

Abstract. This paper discusses the material to be presented by H. J. Siegel in his keynote talk. Distributed high-performance heterogeneous computing (HC) environments are composed of machines with varied computational capabilities interconnected by high-speed links. These environments are well suited to meet the computational demands of large, diverse groups of applications. One key factor in achieving the best performance possible from HC environments is the ability to assign effectively the applications to machines and schedule their execution. Several factors must be considered during this assignment. A conceptual model for the automatic decomposition of an application into tasks and assignment of tasks to machines is presented. An example of a static matching and scheduling approach for an HC environment is summarized. Some examples of current HC technology and open research problems are discussed.

1 Introduction

Existing high-performance computers sometimes achieve only a fraction of their peak performance capabilities on some tasks [14]. This is because different tasks can have very different computational requirements that result in the need for different machine capabilities. A single machine architecture may not satisfy all the computational requirements of different tasks equally well. Thus, the use of a heterogeneous computing environment is more appropriate.

---

* This research was supported in part by the DARPA/ITO Quorum Program under the NPS subcontract numbers N62271-97-M-0900, N62271-98-M-0217, and 62271-98-M-0448, and under the GSA subcontract number GS09K99BH0250. Some of the equipment used was donated by Intel and Microsoft.

B. Monien, V.K. Prasanna, S. Vajapeyam (Eds.): HiPC 2001, LNCS 2228, pp. 307–318, 2001.

This paper summarizes and extends some of the material in [6,26], and corresponds to the keynote presentation H. J. Siegel will give at the conference. He will give an overview of some research in the area of heterogeneous computing (HC), where a suite of different kinds of machines are interconnected by high-speed links. Such a system provides a variety of architectural capabilities, orchestrated to perform tasks with diverse execution requirements by exploiting the heterogeneity of the system [10,26,29]. An HC system may consist of a set of high-performance machines. A cluster composed of different types (or models or ages) of machines also constitutes an HC system. Alternatively, a cluster could be treated as a single machine in an HC suite. An HC system could also be part of a larger grid [12].

An application is assumed to be composed of one or more independent (i.e., non-communicating) tasks. It is also assumed that some tasks may be further decomposed into two or more communicating subtasks. The subtasks have data dependencies among them, but are able to be assigned to different machines for execution.

Consider Fig. 1, which shows a hypothetical example of an application program with various components that are best suited for execution on different machine architectures [14]. The example application in Fig. 1 consists of one task, decomposed into four consecutive subtasks. The application executes for 100 time units on a baseline workstation, where each subtask is best suited to the machine architecture and takes the amount of time indicated underneath it in the figure.

By performing the entire application on a cluster of workstations, the execution time of the large cluster-oriented subtask may decrease from 35 to 0.3 time units. The overall execution time improvement for the entire application is only by about a factor of two because the other subtasks may not be well suited for a cluster architecture.

Alternatively, the use of four different machine architectures, each matched to the computational requirements of the subtask to which it was assigned, can result in an execution 50 times as fast as the baseline workstation. This is because each subtask is executing on the high performance architecture for which it is best suited. The execution time shown below the bars for the HC suite in Fig. 1 include inter-machine communication overhead for each subtask to pass data to the next subtask. This inter-machine communication overhead is not needed in the single machine implementations of the task.

The construction of an HC suite with all four of these types of machines is, of course, more costly than just a single workstation or a single cluster of workstations. Thus, the steady state workload of applications must be sufficient to justify the system cost.

A key factor in achieving the best performance possible from HC environments is the ability to match (assign) the tasks and subtasks to the machines and schedule (order) the tasks and subtasks on each machine in an effective and efficient manner. The matching and scheduling of tasks and subtasks is defined as a mapping.

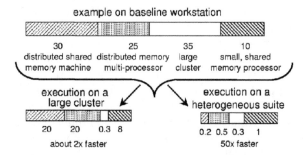

example on baseline workstation

Fig. 1. Hypothetical example of the advantage of using a heterogeneous suite of machines, where the heterogeneous suite execution time includes inter-machine communication overhead (based on [14]). Not drawn to scale

Two different types of mapping are static and dynamic. <u>Static</u> mapping is performed when the tasks are mapped in an off-line planning phase, e.g., planning the schedule for tomorrow. <u>Dynamic</u> mapping is performed when the tasks are mapped in an on-line, real-time fashion, e.g., when tasks arrive at random intervals and are mapped as they arrive. In either case, the mapping problem has been shown, in general, to be NP-complete [7,11,18]. Thus, the development of heuristic techniques to find near-optimal mappings is an active area of research, e.g. [3,5,6,10,17,25,27,31].

A conceptual model for automatic HC is introduced in Sect. 2. As an example of current research in static matching and scheduling, Sect. 3 presents a greedy-based approach. Section 4 gives a brief sampling of some HC environments and applications. Open problems in the field of HC are discussed in Sect. 5.

This research is supported by the DARPA/ITO Quorum Program project called <u>MSHN</u> (Management System for Heterogeneous Networks) [16]. MSHN is a collaborative research effort among Colorado State University, Purdue University, the University of Southern California, NOEMIX, and the Naval Postgraduate School. One objective of MSHN is to design and evaluate mapping heuristics for different types of HC environments.

2   A Conceptual Model for HC

One of the long-term goals of HC research is to develop software environments that will automatically map and execute applications expressed in a machine-independent, high-level language. Developing such environments will facilitate the use of HC suites by (1) increasing software portability, because programmers need not be concerned with the constitution of the HC suite, and (2) increasing the possibility of deriving better mappings than users themselves derive with *ad hoc* methods. Thus, it will improve the performance of and encourage the use of HC in general.

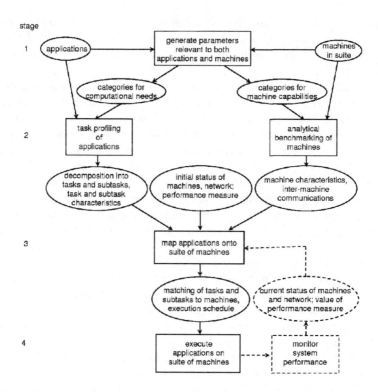

Fig. 2. Model for the support needed for automating the use of HC systems (based on [29]). Ovals indicate information and rectangles represent actions. The dashed lines represent the components needed to perform dynamic mapping

A conceptual model for such an environment using an HC suite of dedicated machines is described in Fig. 2. The conceptual model consists of four stages. It builds on the model presented in [29] and is referred to as a "conceptual" model because no complete automatic implementation currently exists.

In stage 1, using information about the expected types of application tasks and about the machines in the HC suite, a set of parameters is generated that is relevant to both the computational requirements of the applications and the machine capabilities of the HC system. For example, if none of the expected applications include floating point operations, there is no need to characterize the floating point performance of each machine in the suite. For each parameter relevant to both the expected applications and the expected suite of machines, categories for computational characteristics and categories for machine architecture features are derived.

Stage 2 consists of two components, task profiling and analytical benchmarking. Task profiling decomposes the application into tasks (and possibly subtasks), each of which is computationally homogeneous. Different tasks (and subtasks) may have different computational needs. The computational requirements for each task (and subtask) are then quantified by profiling the code and data. Analytical benchmarking is used to quantify how effectively each of the available machines in the suite performs on each of the types of computations being considered.

One of the functions of stage 3 is to be able to use the information from stage 2 to derive the estimated execution time of each task and subtask on each machine in the HC suite (as well as other quality of service (QoS) attributes, such as security level [23]), and the associated inter-machine communication overhead. Then, these results, along with the machine and inter-machine network initial loading and "status" (e.g., machine/network casualties) are used to generate a mapping of tasks and subtasks to machines based on certain performance measures [23] (e.g., minimizing the overall task execution time).

Stage 4 is the execution of the given application. In systems where the mapping is done statically in stage 3, estimated information about all of the tasks and subtasks to execute is known in advance. In dynamic mapping systems, general information about the applications to execute might be known a *priori* (e.g., from benchmarking in stage 2), but specific information (e.g., the exact set of applications or when an application will be submitted for execution) may not be known in advance. Thus, in dynamic mapping systems the task completion times and loading/status of the machines/network are monitored (shown by the dashed lines in Fig. 2). This information may be used to reinvoke the mapping of stage 3 to improve the machine assignment and execution schedule, or to alter the mapping based on changing user needs.

Automatic HC is a relatively new field. Frameworks for task profiling, analytical benchmarking, and mapping have been proposed, however, further research is needed to make this conceptual model a reality [26,29].

3 Static Mapping Heuristics

3.1 Introduction

As mentioned in Sect. 1, the heuristics to map tasks to machines can execute statically (off-line) or dynamically (on-line). There are several trade-offs between these two approaches. Static heuristics typically can use more time to determine a mapping because it is being done off-line, e.g., for production environments; but static heuristics must then use estimated values for parameters such as when a machine will be available. In contrast, dynamic heuristics operate on-line, therefore they must make scheduling decisions in real-time, but have feedback for actual parameter values instead of estimates for many system parameters.

As an example of current HC research on mapping statically, a greedy approach from [6] is summarized. Examples of other static mapping heuristics are given in [5,10,18,26,31]. Examples of dynamic mapping heuristics in HC environments can be found in [3,25,26].

Static mapping is utilized for many different purposes. Static mapping is used in large production environments to plan work to perform over a future interval, e.g., mapping tasks to execute the next day. Static mapping is also used for "what-if" predictive studies. For example, a system administrator might want to estimate the benefits of adding a new machine to the HC suite to justify purchasing it. Static mapping can also be used to do a post-mortem analysis of the performance of on-line dynamic mappers (where the static mapper would assume advance knowledge of all the applications to be executed).

The static mapping heuristics in this section are evaluated using simulated execution times for an HC environment. Because these are static heuristics, it is assumed that an estimate of the expected execution time for each task on each machine is known prior to execution and contained within an *ETC* (expected time to compute) matrix [1,2]. The assumption that these estimated expected execution times are known is commonly made when studying mapping heuristics for HC systems (e.g., [9,15,20,30]). Approaches for doing this estimation based on task profiling and analytical benchmarking of real systems are discussed in [22,26,29].

3.2   Problem Description

Assume the tasks being mapped to the HC environment have the following additional characteristics, increasing the complexity of the basic mapping problem: priorities, deadlines, multiple versions, and user-assigned preferences. Some tasks are also decomposed into subtasks with inter-subtask data dependencies. Each of these characteristics is described below.

In this study, each task t_i will have one of four possible weighted priorities, $p_i \in \{1, 4, 16, 64\}$ (each value is equally likely to be assigned to a task in the simulation studies conducted). Values of $p_i = 64$ represent the most important or highest priority tasks.

This research assumes an oversubscribed system, i.e., there are not enough resources available to satisfy all the requirements of all the tasks. To model this, the tasks in the simulation studies conducted are assigned an arrival time and a deadline. Let the arrival time for task t_i be denoted a_i, and let the deadline for task t_i be denoted D_i.

Deadlines are assigned to each task as follows. First, for task t_i, the median execution time, med_i, of the task on all machines in the HC suite is found. Next, each task t_i is randomly assigned a deadline factor, δ_i, where $\delta_i \in \{1, 2, 3, 4\}$ (each value has an equal likelihood of being assigned). Finally, the deadline for task t_i, D_i, is assigned as $D_i = (\delta_i \times med_i) + a_i$. All the subtasks within a task have that task's arrival time and deadline.

Because the system is oversubscribed, there may be tasks that cannot complete before their deadline. A deadline characteristic function, d_i, indicating whether or not each task t_i is able to finish executing before its deadline, D_i, based on the mapping used, is

$$d_i = \begin{cases} 1 & : \quad \text{if } t_i \text{ finishes at or before } D_i \\ 0 & : \quad \text{otherwise.} \end{cases} \tag{1}$$

Several types of applications can be executed in more than one format or version. For example, weather information may be computed using varying sensor grid densities. Each task version may have different resource requirements and execution characteristics. It is assumed that each task has three versions available to execute, but at most one version of any given task is executed. It is assumed here that version i is always preferred over version $i+1$ of a task, but also requires more execution time (e.g., uses more resources).

The estimated expected execution time for task t_i, on machine m_j, using version v_k would be $ETC(i,j,k)$. Thus, based on the previous assumption, $ETC(i,j,0) > ETC(i,j,1) > ETC(i,j,2)$. To generate the simulated execution times in the ETC matrix, the coefficient-of-variation-based (CVB) method from [1] was used. The HC environment in the simulation study had eight machines.

The intuition behind allowing execution of lower preference versions of a task is that, typically, the lower preference versions will have reduced requirements, e.g., reduced execution times. Let r_{ik} be the normalized user-defined preference for version v_k of task t_i [23]. For the simulation studies:

$$\begin{aligned}
r_{i0} &= 1 &&\text{(most preferred)}\\
r_{i1} &= r_{i0} \times U[0,\ 0.99] &&\text{(medially preferred)}\\
r_{i2} &= r_{i1} \times U[0,\ 0.99] &&\text{(least preferred)}.
\end{aligned} \tag{2}$$

where $U[w,x]$ is a uniform random (floating point) number sampled from $[w,\ x]$. The lowest version assigned to any subtask in a task is the version (and preference) enforced on all of the other subtasks within the task.

In instances where either a non-decomposed task or a communicating subtask (within a task) is being mapped, the term m-task (mappable task) is used. The number of m-tasks in the simulation study was $T = 2000$. Approximately 1000 of the m-tasks are non-decomposed, and approximately 1000 of the m-tasks are subtasks. The size and inter-dependencies of the subtasks within a task were generated randomly. The number of subtasks within a task was $U[2,5]$. Thus, there were approximately 1000 non-decomposed tasks, and 285 tasks with subtasks. For evaluation purposes, there were $T_{eval} \approx 1285$ tasks.

3.3   Simulation Study

Greedy techniques perform well in many situations, and have been well-studied (e.g., [8,18]). One greedy technique, Min-min, has been shown to perform well in many situations (e.g., [5,6,18,31]), and is applied to this mapping problem.

To rate the quality of the mappings produced by the heuristics, a post-mapping evaluation function, E, is used. Assume that if any version of task t_i completes, it is version v_k (k may differ for different tasks). Then, let E be defined as

$$E = \sum_{i=0}^{T_{eval}-1} (d_i \times p_i \times r_{ik}). \tag{3}$$

Here, the mapping problem is considered a maximization problem (i.e., higher values of E represent better mappings). Analysis of Eqn. 3 reveals the upper bound (UB) is $UB = \sum_{i=0}^{T_{eval}-1} p_i$.

To describe Min-min, it is useful to define f_i, the task fitness for m-task t_i,

$$f_i = -(d_i \times p_i \times r_{ik}), \tag{4}$$

where t_i is executed using version v_k. The task fitness f_i represents (the negative of) the contribution of each task to the post-mapping evaluation function, E. Because Min-min is a minimization heuristic, it can be used to maximize E by simply applying it with $-E$ as the minimization criterion.

To compute d_i in the above equations, the task's completion time is found. Let $mav(i,j)$ be the earliest time (after m-task t_i's arrival time) at which machine m_j (1) is not reserved, (2) is available for a long enough interval to execute t_i, and (3) can receive all input data if t_i is a subtask. Then, the completion time of m-task t_i, version v_k, on machine m_j, denoted $ct(i,j,k)$, is $ct(i,j,k) = mav(i,j) + ETC(i,j,k)$.

The Min-min heuristic from [18] that was adapted and applied to this mapping problem is outlined now. Let U be the set of all unmapped m-tasks. Let UP be the set of all unmapped m-tasks such that if the m-task is a subtask, all of that subtask's predecessor subtasks have been mapped. The first phase of Min-min (the first "min") finds the best machine (i.e., minimum f_i) for each m-task in UP, and then stores these m-task/machine pairs in the set $Made$. The m-tasks within $Made$ are referred to as candidate tasks.

After phase one has completed, phase two of Min-min selects the candidate task from $Made$ with the minimum f_i over all candidate tasks, and maps this m-task to its corresponding machine. This task is then removed from U. Phase one and two are repeated until all m-tasks are mapped (or removed from consideration because they cannot meet their deadline).

A comparison of how Min-min performs for two different arrival rates against two other mapping heuristics is shown in Fig. 3. The results are averaged over 50 different ETC matrices. The range bars show the 95% confidence interval for each average [19].

The other techniques shown in Fig. 3 are a simple FIFO-based greedy technique (MCF), and a genetic algorithm (GA) [6]. Recall that a higher value of the post-mapping evaluation function E represents a better mapping. Also note that because an oversubscribed system is assumed, the UB is unachievable. As shown in Fig. 3, Min-min does very well, achieving about 95% of the performance of the GA but with a significantly shorter running time.

For each arrival rate, the same number of tasks are available to map, thus UB is the same. However, for the moderate arrival rate, tasks arrive over a larger interval of time (and have deadlines over a larger interval of time). Hence, there is less contention for machines, and the heuristics perform better. Other variations of Min-min, other scenarios (including different weighted priorities), other heuristics (including the genetic algorithm), and their experimental results are defined and compared in [6].

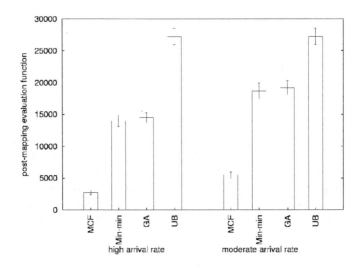

Fig. 3. Comparison of the Min-min technique against two other techniques for 2000 m-tasks with high and moderate arrival rates (based on [6]). Results are averaged over 50 ETC matrices

4   Environments and Applications

Examples of HC environments that have actually been deployed are: (1) the Purdue University Network Computing Hub, a wide area network computing system which can be used to run a selection of software tools via a World Wide Web browser [21]; (2) the Globus meta-computing infrastructure toolkit, a set of low-level mechanisms that can be built upon to develop higher level HC services [12]; and (3) SmartNet, a mapping framework that can be employed for managing jobs and resources in an HC environment [13].

Example applications that have demonstrated the usefulness of HC include: (1) a three-dimensional simulation of mixing and turbulent convection at the Minnesota Supercomputer Center [24]; (2) the shipboard anti-air warfare program (HiPer-D) used at the Naval Surface Warfare Center for threat detection, engagement, and missile guidance (e.g., [17]); and (3) a simulation of colliding galaxies performed by solving large n-body and large gas dynamics problems at the National Center for Supercomputing Applications [28].

5   Open Research Problems

HC is a relatively new research area for the computer field. Interest in HC systems continues to grow in the research, industrial, and military communities. However, there are many open problems that remain to be solved before HC can be made available to application programmers in a transparent way. Some of these problems are outlined below.

The realization of the automatic HC environment envisioned in Fig. 2 requires further research in many areas. Machine-independent languages with user-specified directives are needed to (1) allow compilation of a given application into efficient code for any machine in the HC suite, (2) aid in decomposing applications into tasks and subtasks, and (3) facilitate determination of task and subtask computational requirements. Methods must be refined for measuring the loading and status of the machines in the HC suite and the network, and for estimating task and subtask completion times. Also, the uncertainty present in the estimated parameter values, such as task completion times, should be taken into consideration in determining the mappings.

Another area of research is that of modeling the application tasks to execute on the HC suites [1,4]. Individual tasks that execute once (e.g., a simulation) have different requirements than continuously-running tasks (e.g., monitoring and analyzing sensor information). Methods for allowing feedback communications among subtasks tasks are necessary.

Several HC environments have inherent QoS requirements that must be met. For example, these requirements might involve priority semantics, bandwidth requirements, guaranteed processor time for certain users, or real-time response capabilities. Research is being conducted on how to incorporate all of these components into HC environments.

Hierarchical scheduling techniques are under development to allow for HC suites to be scaled up to very large sizes. Also, in some environments, distributed mapping heuristics (as opposed to centralized ones) are necessary. Incorporating multi-tasking is an area of ongoing HC research. Most operating systems support multi-tasking at the individual processor level, but how to incorporate this into the mapping process, allocate processor time among different tasks, and still leverage this capability with other QoS requirements is being investigated. Furthermore, the security issues inherent with multi-user, distributed, peer-to-peer environments where users are executing tasks on other machines must be addressed.

In summary, with the use of existing HC systems, significant benefits have been demonstrated. However, the amount of effort currently required to implement an application on an HC system can be substantial. Future research on the above open problems will improve this situation, make HC accessible to more users, and allow HC to realize its inherent potential.

Acknowledgments. The authors thank S. Ali, R. F. Freund, and L. L. Vandenberg for their comments. A version of this manuscript appeared as an invited keynote paper in the *International Conference on Parallel and Distributed Processing Techniques and Applications (PDPTA 2001)*.

References

1. Ali, S., Siegel, H. J., Maheswaran, M., Hensgen, D., Ali, S.: Representing task and machine heterogeneities for heterogeneous computing systems. Tamkang Journal of Science and Engineering. **3** (2000) 195–207
2. Armstrong, R.: Investigation of Effect of Different Run-Time Distributions on SmartNet Performance. Master's Thesis, Department of Computer Science, Naval Postgraduate School, Monterey, CA (1997)
3. Banicescu, I., Velusamy, V.: Performance of scheduling scientific applications with adaptive weighted factoring. In: 10th IEEE Heterogeneous Computing Workshop (HCW 2001), in the CD-ROM Proceedings of the 15th International Parallel and Distributed Processing Symposium (IPDPS 2001). (2001) HCW_06
4. Braun, T. D., Siegel, H. J., Beck, N., Bölöni, L., Maheswaran, M., Reuther, A. I., Robertson, J. P., Theys, M. D., Yao, B.: A taxonomy for describing matching and scheduling heuristics for mixed-machine heterogeneous computing systems. In: IEEE Workshop on Advances in Parallel and Distributed Systems, in the Proceedings of the 17th IEEE Symposium on Reliable Distributed Systems. (1998) 330–335
5. Braun, T. D., Siegel, H. J., Beck, N., Bölöni, L. L., Maheswaran, M., Reuther, A. I., Robertson, J. P., Theys, M. D., Yao, B., Hensgen, D., Freund, R. F.: A comparison of eleven static heuristics for mapping a class of independent tasks onto heterogeneous distributed computing systems. Journal of Parallel and Distributed Computing. **61** (2001) 180–837
6. Braun, T. D.: Heterogeneous Distributed Computing: Off-line Mapping Heuristics for Independent Tasks and for Tasks with Dependencies, Priorities, Deadlines, and Multiple Versions. Ph.D. Thesis, School of Electrical and Computer Engineering, Purdue University, West Lafayette, IN (2001)
7. Coffman, Jr., E. G. (ed.): Computer and Job-Shop Scheduling Theory. John Wiley & Sons, New York, NY (1976)
8. Cormen, T. H., Leiserson, C. E., Rivest, R. L.: Introduction to Algorithms. MIT Press, Cambridge, MA (1992)
9. Dietz, H. G., Cohen, W. E., Grant, B. K.: Would you run it here... or there? (AHS: Automatic heterogeneous supercomputing). In: International Conference on Parallel Processing. **II** (1993) 217-221
10. Eshaghian, M. M. (ed.): Heterogeneous Computing. Artech House, Norwood, MA (1996)
11. Fernandez-Baca, D.: Allocating modules to processors in a distributed system. In: IEEE Transactions Software Engineering. **SE-15** (1989) 1427–1436
12. Foster, I., Kesselman, C. (eds.): The Grid: Blueprint for a New Computing Infrastructure. Morgan Kaufmann, San Francisco, CA (1999)
13. Freund, R. F., Gherrity, M., Ambrosius, S., Campbell, M., Halderman, M., Hensgen, D., Keith, E., Kidd, T., Kussow, M., Lima, J. D., Mirabile, F., Moore, L., Rust, B., Siegel, H. J.: Scheduling resources in multi-user, heterogeneous computing environments with SmartNet. In: 7th Heterogeneous Computing Workshop (HCW '98). (1998) 184–199
14. Freund, R. F., Siegel, H. J.: Heterogeneous processing. In: IEEE Computer. **26** (1993) 13–17
15. Ghafoor, A., Yang, J.: Distributed heterogeneous supercomputing management system. In: IEEE Computer. **26** (1993) 78–86

16. Hensgen, D. A., Kidd, T., Schnaidt, M. C., St. John, D., Siegel, H. J., Braun, T. D., Maheswaran, M., Ali, S., Kim, J.-K., Irvine, C., Levin, T., Wright, R., Freund, R. F., Godfrey, M., Duman, A., Carff, P., Kidd, S., Prasanna, V., Bhat, P., Alhusaini, A.: An overview of MSHN: A management system for heterogeneous networks. In: 8th Heterogeneous Computing Workshop (HCW '99). (1999) 184–198

17. Huh, E.-N., Welch, L. R., Shirazi, B. A., Cavanaugh, C. D.: Heterogeneous resource management for dynamic real-time systems. In: 9th IEEE Heterogeneous Computing Workshop (HCW 2000). (2000) 287–294

18. Ibarra, O. H., Kim, C. E.: Heuristic algorithms for scheduling independent tasks on nonidentical processors. Journal of the ACM. **24** (1977) 280–289

19. Jain, R.: The Art of Computer Systems Performance Analysis Techniques for Experimental Design, Measurement, Simulation, and Modeling. John Wiley & Sons, New York, NY (1991)

20. Kafil, M., Ahmad, I.: Optimal task assignment in heterogeneous distributed computing systems. In: IEEE Concurrency. **6** (1998) 42–51

21. Kapadia, N. H., Fortes, J. A. B.: PUNCH: An architecture for web-enabled wide-area network-computing. Cluster Computing: The Journal of Networks, Software Tools and Applications. **2** (1999) 153–164

22. Khokhar, A., Prasanna, V. K., Shaaban, M., Wang, C. L.: Heterogeneous computing: Challenges and opportunities. In: IEEE Computer. **26** (1993) 18–27

23. Kim, J.-K., Kidd, T., Siegel, H. J., Irvine, C., Levin, T., Hensgen, D. A., St. John, D., Prasanna, V. K., Freund, R. F., Porter, N. W.: Collective value of QoS: A performance measure framework for distributed heterogeneous networks. In: 10th IEEE Heterogeneous Computing Workshop (HCW 2001), in the CD-ROM Proceedings of the 15th International Parallel and Distributed Processing Symposium (IPDPS 2001). (2001) HCW_08

24. Klietz, A. E., Malevsky, A. V., Chin-Purcell, K.: A case study in metacomputing: Distributed simulations of mixing in turbulent convection. In: 2nd Workshop on Heterogeneous Processing (WHP '93). (1993) 101–106

25. Maheswaran, M. Ali, S., Siegel, H. J., Hensgen, D., Freund, R. F.: Dynamic mapping of a class of independent tasks onto heterogeneous computing systems. Journal of Parallel and Distributed Computing. **59** (1999) 107–121

26. Maheswaran, M., Braun, T. D., Siegel, H. J.: Heterogeneous distributed computing. In: Webster, J. G. (ed.): Encyclopedia of Electrical and Electronics Engineering. John Wiley & Sons, New York, NY. **8** (1999) 679–690

27. Michalewicz, Z., Fogel, D. B.: How to Solve It: Modern Heuristics. Springer-Verlag, New York, NY (2000)

28. Norman, M. L., Beckman, P., Bryan, G., Dubinski, J., Gannon, D., Hernquist, L., Keahey, K., Ostriker, J. P., Shalf, J., Welling, J., Yang, S.: Galaxies Collide on the I-way: An example of heterogeneous wide-area collaborative supercomputing. The International Journal of Supercomputer Applications and High Performance Computing. **10** (1996) 132–144

29. Siegel, H. J., Dietz, H. G., Antonio, J. K.: Software support for heterogeneous computing. In: Tucker, Jr., A. B. (ed.): The Computer Science and Engineering Handbook. CRC Press, Boca Raton, FL (1997) 1886–1909

30. Singh, H., Youssef, A.: Mapping and scheduling heterogeneous task graphs using genetic algorithms. In: 5th Heterogeneous Computing Workshop (HCW '96). (1996) 86–97

31. Wu, M.-Y., Shu, W., Zhang, H.: Segmented min-min: A static mapping algorithm for meta-tasks on heterogeneous computing systems. In: 9th IEEE Heterogeneous Computing Workshop (HCW 2000). (2000) 375–385

Maximum Achievable Capacity Gain through Traffic Load Balancing in Cellular Radio Networks: A Practical Perspective*

Swades De and Sajal K. Das

Center for Research in Wireless Mobility and Networking (CReWMaN)
Department of Computer Science and Engineering
University of Texas at Arlington
Arlington, TX 76019-0015, USA
{skde,das}@cse.uta.edu

Abstract. An analytical framework is developed showing the maximum possible capacity gain through load balancing in cellular radio networks. Performance of the best-known scheme, namely, load balancing with selective borrowing (LBSB), is also analyzed for comparison purpose. Through an example of 3-tier cellular architecture it is demonstrated that the maximum achievable capacity gain of an ideal load balancing scheme could be as much as 70%. In an identical scenario, the LBSB scheme achieves a gain only up to 30%.

1 Introduction

In a cellular wireless network channel bandwidth is a scarce resource. Even if the network design takes into account the average traffic pattern in an area, time dependent variation of traffic creates imbalance in the load in different parts of the network. This time dependent congestion is sometimes defined as *hot spot* [2]. While approaches like cell splitting or sectorization can be adopted to support more traffic, they do not alleviate the hot spot problem. Since it is quite unlikely that the entire cellular network in a particular region becomes overloaded at the same time, if the hotness of part of the network could be evenly distributed throughout the less-hot areas, the system may be able to meet the grade of service (GoS) requirements even without additional bandwidth.

Several schemes have been proposed to combat the unevenness of traffic pattern in cellular radio networks (see e.g., [2], [4]). In a dynamic channel assignment (DCA) scheme [1], channels are allocated to cells from a central pool on the basis of traffic demand. However, this scheme is constrained by the computational overhead and channel locking requirements. In a hybrid channel assignment (HCA) scheme [5], besides fixed allocation of majority of channels to the cells, a certain number of channels are kept in the central pool to be used

* This work is partially supported by grants from Nortel Networks and Texas Telecommunication Engineering Consortium (TxTEC).

B. Monien, V.K. Prasanna, S. Vajapeyam (Eds.): HiPC 2001, LNCS 2228, pp. 321–330, 2001.
© Springer-Verlag Berlin Heidelberg 2001

by different cells on demand as per traffic requirement. This scheme reduces the computational overhead, however, it is unable to smooth out the traffic imbalance effectively. Performance of load sharing techniques, such as in [3],[6], are limited by the cell area overlap and co-channel interference. Performance enhancement of channel borrowing without locking (CBWL) [4] is also limited due to its reduced power application of the borrowed channels. Although load balancing with selective borrowing (LBSB) [2] improves over [6] and [4] by structured borrowing, it is still constrained with the co-channel interference and is effective only for uniform traffic and even mobility pattern. We observe that none of the existing schemes can achieve perfect load balancing because of co-channel interference problem associated with the concept of frequency re-use in a cellular system.

In this paper, we evaluate analytically the maximum possible capacity gain that can be achieved through load balancing in a cellular radio system. We hypothize that there exists an ideal load balancing scheme in which the co-channel interference problem is completely avoided. Through an analogy of hexagonal cellular structure with a set of connected water reservoirs, we study the dynamics of the ideal load balancing system and obtain the load-balanced traffic pattern in the cells. In order to compare the performance of this ideal load balancing system, we consider the load balancing with selective borrowing (LBSB) scheme [2], which is known to provide the maximum system capacity gain. An example 3-tier cellular structure with a given traffic distribution is used to demonstrate that the maximum achievable capacity gain of the ideal scheme could be as much as 70%. In an identical scenario, the LBSB scheme is observed to achieve a gain only up to 30%.

2   Load Balancing Dynamics

Load balancing in a cellular communication system aims to distribute the traffic among different cells and attempts to maintain nearly the same grade of service (GoS) throughout the network. A conceptual plot of load balancing is shown in Fig. 1, where the traffic pattern is assumed to be uniformly decreasing from a congested cell to its less-congested surrounding neighbors. Refer to Fig. 2(a) for an example cellular architecture where the congested center cell, $\mathcal{A}$, is surrounded by two tiers of relatively less-congested cells. Note in Fig. 1 that the ideal load curve rotates around a pivot (O), whereas that in an interference-limited scheme moves upward as the traffic load balancing is attempted.

We observe that sharing traffic among the neighboring cells can be a close analog to a number of water reservoirs connected through outlets. In our study of traffic load balancing, we consider a 3-tier hexagonal cell structure as shown in Fig. 2(a). Although in reality, the traffic distribution in cells can be quite uneven, for analytical tractability we consider the traffic in all cells within a tier are equal and uniformly distributed within a cell. In the 3-tier cell model the central cell ($\mathcal{A}$) is highly congested, the next tier ($\mathcal{B}$) of 6 cells are less congested, and the outer tier ($\mathcal{C}$) of 12 cells are even less congested. The analogy is depicted

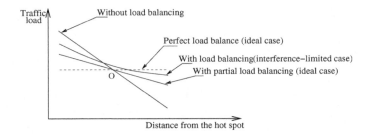

Fig. 1. Load balancing concept.

in Fig. 2(b), where the most congested center cell is represented as reservoir $\mathcal{A}$; neighboring less-congested tier $\mathcal{B}$ cells are represented as reservoir $\mathcal{B}$; and the least congested tier $\mathcal{C}$ cells are represented as reservoir $\mathcal{C}$. All cells are assumed to be of equal size, with diameter D. Since there are 6 (or 12) tier $\mathcal{B}$ (or $\mathcal{C}$) cells, the diameter of reservoir $\mathcal{B}$ (or $\mathcal{C}$) is $\sqrt{6}$ (or $\sqrt{12}$) times that of reservoir $\mathcal{A}$. Representing traffic in each tier as water levels, the "dashed" lines show the initial water level without any load balancing, while the "dot-dashed" lines show the water level with load balancing. The respective initial water (traffic) levels in reservoirs (tiers) $\mathcal{A}$, $\mathcal{B}$, and $\mathcal{C}$ are denoted as T_{a0}, T_{b0}, and T_{c0}, and the respective final reachable water (traffic) level after load balancing are denoted as T_{af}, T_{bf}, and T_{cf}.

Let the amount of load balancing in a cell be denoted by a fraction p ($0 \leq p \leq 1$), which determines the height and bore of the outlets of the reservoirs. When $p = 0$, the outlet in each reservoir is at the highest water level and bore of the outlet is 0, indicating no water flow (i.e., no load balancing). When $p = 1$, the outlet in each reservoir is at height 0 and bore is equal to the height of entire water column in that reservoir, so that any amount of necessary flow (i.e., perfect load balancing) is possible. For $0 < p < 1$, there may be only limited amount of flow possible, thus limiting total balance of water levels. As we shall see in subsequent discussions, for a given traffic distribution in the three tiers, beyond a "threshold" value of p, perfect water level (i.e., traffic load) balancing is possible.

With the analogy above, we study the dynamics of traffic load balancing. The rate of flow of traffic from a hot cell primarily depends on two factors: the fraction p, and the traffic intensities of the neighboring cells. As mentioned above, the rate of load balancing is controlled by p, if it is below a certain threshold. On the other hand, at sufficiently high p, the rate is controlled by the difference in traffic intensities among the cells in different tiers. Let T_a, T_b, and T_c be, respectively, the instantaneous traffic intensities in cell $\mathcal{A}$, tier $\mathcal{B}$ cells, and tier $\mathcal{C}$ cells. The governing equations characterizing the load balancing dynamics (rate of traffic flow) are as follows:

$$\frac{dT_a}{dt} = -k \cdot \min\left\{(T_a - T_{af}), (T_a - T_b)\right\} \tag{1}$$

$$\frac{dT_b}{dt} = \frac{k}{6} \cdot \min\left\{(T_a - T_{af}), (T_a - T_b)\right\} - k \cdot \min\left\{(T_b - T_{bf}), (T_b - T_c)\right\} \tag{2}$$

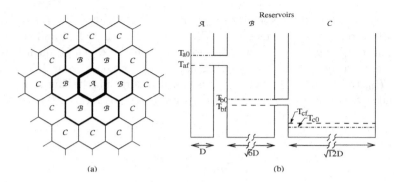

(a)   (b)

Fig. 2. (a) The cell model considered in our load balancing analysis (b) Water reservoir analog of 3-tier cell model. In this example, only limited amount of load balancing is possible.

$$\frac{dT_c}{dt} = \frac{k}{2} \cdot \min\left\{(T_b - T_{bf}), (T_b - T_c)\right\} \tag{3}$$

where k is a constant of proportionality. Since load balancing in the cells is controlled by the same controller (mobile switching center or MSC), k is the same system-wide. Also, since the rate of flow of traffic from a congested cell to a less-congested one is always positive and cannot be more than the difference in traffic levels in the two adjacent cells, k lies between 0 and 1.

Depending on the value of p, there may be four different cases leading to the steady-state traffic pattern in different cells.

Case 1: p is sufficiently low such that $T_a - T_{af} < T_a - T_b$ and $T_b - T_{bf} < T_b - T_c$. Solutions for T_a, T_b, and T_c in this case are

$$T_a(t) = T_{af} + \alpha e^{-kt} \tag{4}$$
$$T_b(t) = T_{bf} + \beta_1 e^{-kt} + \beta_2 t e^{-kt} \tag{5}$$
$$T_c(t) = T_{c0} + \gamma_1 + \gamma_2 e^{-kt} + \gamma_3 t e^{-kt} \tag{6}$$

where α, β_1, β_2, γ_1, γ_2, and γ_3 are integration constants, determined by k. Since $0 < k < 1$, the steady-state values of traffic intensities are obtained as $T_a(\infty) = T_{af}$, $T_b(\infty) = T_{bf}$, and $T_c(\infty) = T_{c0} + \gamma_1$.

Case 2: p is intermediate such that $T_a - T_{af} < T_a - T_b$ and $T_b - T_{bf} > T_b - T_c$. This situation may arise when $T_{a0} - T_{b0} > T_{b0} - T_{c0}$, the cell $\mathcal{A}$ is heavily loaded, and tier $\mathcal{B}$ and tier $\mathcal{C}$ cells are almost equally and lightly loaded. Substitution in Eqs. (1) – (3) give the following solutions

$$T_a(t) = T_{af} + \alpha' e^{-kt} \tag{7}$$
$$T_b(t) = \gamma_1' - \gamma_2' e^{-kt} - 2\gamma_3' e^{-\frac{3k}{2}t} \tag{8}$$
$$T_c(t) = \gamma_1' + \gamma_2' e^{-kt} + \gamma_3' e^{-\frac{3k}{2}t} \tag{9}$$

where α', γ_1', γ_2', and γ_3' are integration constants, determined by k. The steady-state solutions are given by $T_a(\infty) = T_{af}$ and $T_b(\infty) = T_c(\infty) = \gamma_1'$. Observe that the steady-state traffic in tier $\mathcal{B}$ and tier $\mathcal{C}$ cells are equal.

Case 3: p is intermediate such that $T_a - T_{af} > T_a - T_b$ and $T_b - T_{bf} < T_b - T_c$. This situation may arise when $T_{a0} - T_{b0} < T_{b0} - T_{c0}$, the cell $\mathcal{A}$ and tier $\mathcal{B}$ cells are almost equally loaded, and initial load in each tier $\mathcal{C}$ cell is much lower. In practice, however, either Case 2 or Case 3, but not both, can occur. If $T_{a0} - T_{b0} = T_{b0} - T_{c0}$, then neither of these two cases will occur. Substituting in Eqs. (1) – (3) and noting that the attainable steady-state value of $T_b(t)$ is T_{bf}, we obtain

$$T_a(t) = \gamma_1'' + \gamma_2'' e^{-\frac{3}{2}kt} + \gamma_3'' e^{-\frac{2}{3}kt} \tag{10}$$

$$T_b(t) = \gamma_1'' - \frac{\gamma_2''}{2} e^{-\frac{3}{2}kt} + \frac{\gamma_3''}{3} e^{-\frac{2}{3}kt} \tag{11}$$

$$T_c(t) = T_{c0} + \alpha'' + \frac{\gamma_2''}{3k} e^{-\frac{3}{2}kt} - \frac{\gamma_3''}{2k} e^{-\frac{2}{3}kt} \tag{12}$$

where α'', γ_1'', γ_2'', and γ_3'' are integration constants. From Eqs. (10) – (12) steady-state traffic intensities are obtained as $T_a(\infty) = T_b(\infty) = T_{bf}$ and $T_c(\infty) = T_{c0} + \alpha''$. Observe that in the steady-state, the cell $\mathcal{A}$ and tier $\mathcal{B}$ cells are load-balanced.

Case 4: p is sufficiently high such that $T_a - T_{af} > T_a - T_b$ and $T_b - T_{bf} > T_b - T_c$. Corresponding solutions of the Eqs. (1) – (3) yield

$$T_a(t) = \alpha_1''' + \alpha_2''' e^{m_1 t} + \alpha_3''' e^{m_2 t} \tag{13}$$

$$T_b(t) = \alpha_1''' + \beta_2''' e^{m_1 t} + \beta_3''' e^{m_2 t} \tag{14}$$

$$T_c(t) = \alpha_1''' + \gamma_2''' e^{m_1 t} + \gamma_3''' e^{m_2 t} \tag{15}$$

where α_1''', α_2''', α_3''', β_2''', β_3''', γ_2''', γ_3''', m_1, and m_2 are integration constants. The steady-state traffic intensities are obtained as $T_a(\infty) = T_b(\infty) = T_c(\infty) = \alpha_1'''$.

Steady-state values of traffic intensities in all the above four cases, which are utilized as boundary conditions to obtain the undetermined integration constants, are derived in the next section. integration constants.

3 Steady-State Solutions

Recall that the respective initial traffic intensities in cell $\mathcal{A}$, and in each of tier $\mathcal{B}$ and tier $\mathcal{C}$ cells are T_{a0}, T_{b0}, and T_{c0}. Let ΔT_a denote the amount of traffic transferred from cell $\mathcal{A}$ to tier $\mathcal{B}$ cells, and ΔT_b denote the transfer from each tier $\mathcal{B}$ cell to tier $\mathcal{C}$ cells. Then, the respective final traffic levels in cell $\mathcal{A}$, tier $\mathcal{B}$ cells, and tier $\mathcal{C}$ cells are given by

$$T_{af} = T_{a0} - \Delta T_a \tag{16}$$

$$T_{bf} = T_{b0} + \frac{\Delta T_a}{6} - \Delta T_b \tag{17}$$

$$T_{cf} = T_{c0} + \frac{\Delta T_b}{2}. \tag{18}$$

Let us now revisit the four cases in Section 2.

Case 1: $T_a - T_{af} < T_a - T_b$ and $T_b - T_{bf} < T_b - T_c$. In this case, the maximum transfer of traffic from any tier can be achieved. Maximum traffic transferred from cell $\mathcal{A}$ is $\Delta T_a = pT_{a0}(1 - \text{ß}_{bf})$, where ß_{bf} is the steady-state call blocking probability in a tier $\mathcal{B}$ cell. Similarly, maximum traffic transferred from each of the tier $\mathcal{B}$ cells is $\Delta T_b = \frac{p}{2}[T_{b0}(1 - \text{ß}_{cf})]$, where ß_{cf} is the steady-state call blocking probability in a tier $\mathcal{C}$ cell. Each of tier $\mathcal{B}$ cells receives $\Delta T_a/6$ amount of traffic from cell $\mathcal{A}$, while each of tier $\mathcal{C}$ cells receives $\Delta T_b/2$ amount of traffic. From Eq. (18) we obtain $\text{ß}_{cf} = \text{f}(T_{cf})$, where $\text{f}(\cdot)$ is governed by Erlang-B formula [7]. By numerical computation, we solve for ß_{cf} and hence T_{cf}. Using Eqs. (16) and (17), T_{af} and T_{bf} are similarly obtained.

Case 2: $T_a - T_{af} < T_a - T_b$ and $T_b - T_{bf} > T_b - T_c$. The expression for ΔT_a remains the same as in Case 1. However, $\Delta T_b = \frac{2}{3}[T_{b0} - T_{c0} + \frac{p}{6}T_{a0}(1 - \text{ß}_{bf})]$, from which ß_{bf} and hence T_{bf} can be obtained. Subsequent substitution in Eqs. (16) and (18) gives T_{af} and T_{cf}.

Case 3: $T_a - T_{af} > T_a - T_b$ and $T_b - T_{bf} < T_b - T_c$. The expression for ΔT_a is modified as $\Delta T_a = \frac{6}{7}[T_{a0} - T_{b0} + \frac{p}{2}T_{b0}(1 - \text{ß}_{cf})]$, from which T_{af} and T_{bf} can be obtained. The expression for ΔT_b remains the same as in Case 1, which leads to ß_{cf} and hence T_{cf}.

Case 4: $T_a - T_{af} > T_a - T_b$ and $T_b - T_{bf} > T_b - T_c$. The steady-state traffic level in all the tiers is given by $T_{a0} - \Delta T_a = T_{b0} + \frac{\Delta T_a}{6} - \Delta T_b = T_{c0} + \frac{\Delta T_b}{2} = T_f$, where $T_f = \frac{T_{a0} + 6T_{b0} + 12T_{c0}}{19}$.

4   Performance Analysis of LBSB Scheme

Having discussed the absolute upper limit of capacity gain due to traffic load balancing, we now turn to analyze the maximum traffic capacity gain achievable with the currently existing load balancing schemes, among which the load balancing with selective borrowing (LBSB) scheme [2] is claimed to offer the best performance. We follow the same line of approach as in [2], however, with certain modifications to aid our performance comparison. In particular, we take into account the hotness (congestion) levels of surrounding cells in our analysis in order to study the benefit of a load balancing scheme. Note that instead of attempting to balance the traffic loads in different cells, the LBSB scheme releases the required amount of traffic from the hot cell just to meet the predefined GoS limit, irrespective of the coldness of surrounding cells.

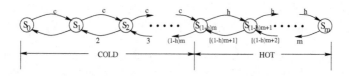

Fig. 3. Modified $M/M/m/m$ queue representation of the channel states in a hot cell.

The channel states in a cellular system with LBSB scheme is modeled as a modified $M/M/m/m$ queue, shown in Fig. 3. State S_i, $0 \le i \le m$, means i

channels in a cell are occupied, m is the number of wireless channels per cell, and h is a predefined hotness threshold parameter. When the available number of channels in a cell goes below $\lfloor hm \rfloor$, the cell looks for borrowing channels (for details, refer to [2]). Let the call arrival rate in a cell, as seen by the base station in the 'cold' region is λ_c, and that in the 'hot' region is λ_h. $\lambda_c = \lambda_1 + \lambda_2 + \lambda'$ and $\lambda_h = \lambda_1 + \lambda_2 - \lambda''$, where $\lambda_1 =$ net hand-off rate, $\lambda_2 =$ new call arrival rate, $\lambda' =$ channel lending rate, and $\lambda'' =$ channel borrowing rate. The service rates (μ) in both the 'cold' and 'hot' regions are the same. Solution of state equations in the Markov chain yields the probability P_i of being in the i-th state as

$$
P_i = \begin{cases} \frac{\rho_c^i}{i!} P_0 & \text{for } 0 \leq i \leq (1-h)m \\[2mm] \rho_c^{(1-h)m} \frac{\rho_h^{(i-(1-h)m)}}{i!} P_0 & \text{for } (1-h)m < i \leq m \end{cases} \tag{19}
$$

where $\rho_c = \lambda_c/\mu$ and $\rho_h = \lambda_h/\mu$ are utilization factors, and P_0 is given by

$$
P_0 = \left[1 + \sum_{i=1}^{(1-h)m} \frac{\rho_c^i}{i!} + \rho_c^{(1-h)m} \sum_{i=(1-h)m+1}^{m} \frac{\rho_h^{(i-(1-h)m)}}{i!} \right]^{-1} \tag{20}
$$

The call blocking probability, also called GoS, in a cell when all m channels are occupied is obtained as

$$
P_m = \left(\frac{1}{m!} \right) \rho_c^{(1-h)m} \rho_h^{hm} \cdot P_0 \tag{21}
$$

5   Numerical Results

In this section we present the comparative performance results of the ideal load balancing and the LBSB scheme. The following parameter values are considered in our comparison results. Number of wireless channels per cell $m = 50$. Referring to Fig. 2, original traffic intensity in a tier $\mathcal{B}$ cell, $T_{b0} = 40.25$ (call blocking probability or GoS=2%) and that in a tier $\mathcal{C}$ cell, $T_{c0} = 37.90$ (GoS=1%). Traffic intensity in cell $\mathcal{A}$ increases from $T_{a0} = 40.25$ *Erlang* to 70.0 *Erlang*. Note that our performance evaluation does not account for physical layer issues. In other words, the obtained results are based on perfect physical channel conditions.

5.1   Performance Results of the Ideal Load Balancing System

Based on the analyses in Sections 2 and 3, we study the performance of the ideal load balancing scheme where the co-channel interference is assumed nonexistent. Fig. 4 shows the variation of load balancing for different hotness levels in cell $\mathcal{A}$. Observe the saturation in gain in traffic load balancing beyond a certain "threshold" value of p. For example, with $T_{a0} = 60$ *Erlang*, saturation gain occurs at $p = 0.36$.

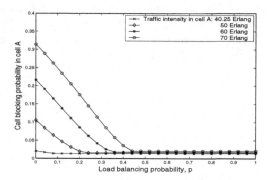

Fig. 4. Call blocking probability in cell A versus ideal load balancing probability (p). $T_{b0} = 40.25$ *Erlang*, $T_{c0} = 37.90$ *Erlang*, $m = 50$.

In Fig. 5, a comparison of call blocking performance of the ideal load balancing system with respect to the one without any load balancing is shown. The improvement in call blocking performance in cell $\mathcal{A}$ may be noted here. For example, with $p = 0.24$, for $T_{a0} = 50$ *Erlang* improvement in call blocking performance is 85%, while for $T_{a0} = 70$ *Erlang* the improvement is 54%.

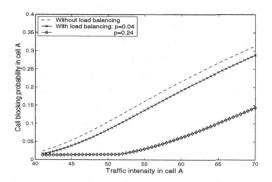

Fig. 5. Degree of hotness in cell $\mathcal{A}$ versus call blocking probability in the ideal load balancing system. $T_{b0} = 40.25$ *Erlang*, $T_{c0} = 37.90$ *Erlang*, $m = 50$.

5.2   Performance Results of the LBSB Scheme

For the best performance of the LBSB scheme, uniform traffic mobility to (from) a cell from (to) its neighbors is assumed which aids structured channel borrowing. For the cell $\mathcal{A}$, let the utilization in cold state be ρ_{ca} and that in the hot state be ρ_{ha}. For each of the tier $\mathcal{B}$ and tier $\mathcal{C}$ cells, the respective parameters are ρ_{cb}, ρ_{hb}, and ρ_{cc}, ρ_{hc}. This utilization (ρ) is a measure of degree of hotness (or coldness) of a cell, which is equivalent to the traffic intensity, T_{x0}, as denoted in the analysis of the ideal load balancing system, where x stands for a, b, or c. The parameters taken for performance analysis in the LBSB scheme are re-written as follows. $\rho_{cb} = 40.25$ *Erlang* and $\rho_{cc} = 37.90$ *Erlang*, whereas ρ_{ca} increases from 40.25 *Erlang*. With these parameters, the achievable capacity gain is computed numerically keeping the network-wide GoS within 2%.

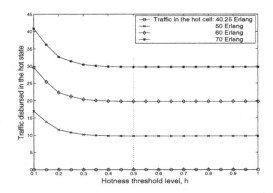

Fig. 6. Effect of hotness threshold on the mitigation of 'heat' in a cell to reach 2% GoS.

The effect of threshold parameter, h, on the amount of traffic to be released at different hotness levels is shown in Fig. 6. The minimum (steady-state) value is attained nearly at $h = 0.5$, which indicates the optimum threshold for the maximum capacity gain with the LBSB scheme. Higher value of h is not selected to avoid loading the neighboring cells unnecessarily. Intuitively, at very low h, the channel demand for traffic disbursement to mitigate the hotness of a cell increases suddenly. This leads to network-wide poor GoS performance. On the other hand, at relatively high h the process is more gradual which aids the overall network performance.

5.3   Comparison

Table 1 shows the capacity gain in cell $\mathcal{A}$ for different load balancing probability (p) in the ideal system. The original call blocking probabilities in tier $\mathcal{B}$ and tier $\mathcal{C}$ cells are 2% and 1%, respectively. The corresponding capacity gain with the LBSB scheme is 12.05 *Erlang*, 30% over the nominal value of 40.25 *Erlang*. In the ideal system, the maximum achievable gain is noted (at $p = 0.42$) as 28.20 *Erlang*, a 70% capacity increase.

A qualitative explanation for the equivalent traffic increase due to channel locking in the LBSB scheme to avoid co-channel interference is given as follows. To meet a GoS threshold, if cell $\mathcal{A}$ releases e_a *Erlang* traffic to neighboring tier $\mathcal{B}$ cells, then an additional $2e_a$ amount of equivalent traffic is generated in tier $\mathcal{C}$ cells due to channel locking requirement. Furthermore, if each tier $\mathcal{B}$ cell requires to release e_b *Erlang* traffic to the tier $\mathcal{C}$ cells, additional equivalent traffic generated in each of tier $\mathcal{B}$ and tier $\mathcal{C}$ cells is e_b. These amounts, e_a and e_b, grow

Table 1. Performance comparison of LBSB scheme with the ideal load balancing system

Ideal system		LBSB scheme
Load balancing probability, p	Capacity gain (*Erlang*)	Capacity gain (*Erlang*) for $h = 0.5$
0.04	1.80	
0.08	3.60	
0.12	5.40	12.05
0.16	7.60	
⋮	⋮	
0.42	28.20	

as the degree of hotness in cell $\mathcal{A}$ increases, thus limiting the capacity gain of the cells to meet a certain GoS limit (say, 2%).

6   Conclusion

In this paper, we evaluated the absolute maximum achievable system capacity gain via load balancing in cellular radio networks. To this end, we analyzed the system dynamics to obtain the steady-state performance. We compared the performance of the idealized load balancing system with the existing LBSB scheme and showed that the co-channel interference sufficiently limits the capacity gain in the conventional load balancing schemes. It might therefore be worthwhile to investigate a possible load balancing scheme which can potentially minimize the co-channel interference, thus allowing better utilization of the scarce wireless channel resource.

References

1. Cox, D.C., Reudink, D.O.: Increasing channel occupancy in large scale mobile radio systems: Dynamic channel reassignment. IEEE Trans. on Vehicular Tech. **22** (1973) 218–223.
2. Das, S.K., Sen, S.K., Jayaram, R.: A dynamic load balancing strategy for channel assignment using selective borrowing in cellular mobile environment. Wireless Networks. **3** (1997) 333–347.
3. Eklundh, B.: Channel utilization and blocking probability in a cellular mobile telephone system with directed retry. IEEE Trans. on Commun. COM-34, **4** (1986) 329–337.
4. Jiang, H., Rappaport, S.S.: CBWL: A new channel assignment and sharing method for cellular communication systems. IEEE Trans. on Vehicular Tech. **43** (1994) 313–322.
5. Kahwa, T.J., Georganas, N.D.: A hybrid channel assignment scheme in large-scale, cellular-structured mobile communication systems. IEEE Trans. on Commun. COM-26, **4** (1978) 432–438.
6. Karlsson, J., Eklundh, B.: A cellular mobile telephone system with load sharing - an enhancement of directed retry. IEEE Trans. on Commun. COM-37, **5** (1989) 530–535.
7. Rappaport, T.: Wireless Communications: Principles and Practice. Prentice Hall. (1996).

Performance Evaluation of Mobile Agents for E-commerce Applications

Rahul Jha and Sridhar Iyer

Kanwal Rekhi School of Information Technology, Indian Institute of Technology Bombay, Powai, Mumbai - 400 076.

{rahul, sri}@it.iitb.ac.in
http://www.it.iitb.ac.in

Abstract. Mobile agents are emerging as a promising paradigm for the design and implementation of distributed applications. While mobile agents have generated considerable excitement in the research community, they have not translated into a significant number of real-world applications. One of the main reasons for this is the lack of work that quantitatively evaluates (i) the effectiveness of one mobile agent framework over another, and (ii) the effectiveness of mobile agents versus traditional approaches. This paper contributes towards such an evaluation. We identify the underlying mobility patterns of e-commerce applications and discuss possible client-server (CS) and mobile agent (MA) based implementation strategies for each of these patterns. We have compared the performance of three mobile agent frameworks, viz., Aglets, Concordia and Voyager, for their suitability for an e-commerce application development. We have performed experiments to quantitatively evaluate the effectiveness of CS versus MA strategies, for the identified mobility patterns. We used Java sockets for the CS implementation and the ObjectSpace VoyagerTM (Voyager) framework for the MA implementation. In this paper, we present our observations and discuss their implications.

1   Introduction

The emergence of e-commerce applications has resulted in new net-centric business models. This has created a need for new ways of structuring applications to provide cost-effective and scalable models.

Mobile Agent (MA) systems have for some time been seen as a promising paradigm for the design and implementation of distributed applications. A mobile agent is a program that can *autonomously migrate* between various nodes of a network and perform computations on behalf of a user. Some of the benefits provided by MAs for structuring distributed applications include reduction in network load, overcoming network latencies, and support for disconnected operations [10]. The use of MAs has been explored for a variety of applications such as information retrieval, workflow management systems and e-commerce applications [7, 11, 12, 13].

While there are some experimental as well as commercial deployments of MAs for e-commerce applications (e.g. MAgNET [3], Gossip [www.tryllian.com]), they have still not translated into a significant number of real-world applications. We believe

B. Monien, V.K. Prasanna, S. Vajapeyam (Eds.): HiPC 2001, LNCS 2228, pp. 331–340, 2001.
© Springer-Verlag Berlin Heidelberg 2001

that one of the main reasons for this is the lack of work that quantitatively evaluates (i) the effectiveness of one mobile agent framework over another, and (ii) the effectiveness of mobile agents versus traditional approaches. This paper contributes towards such an evaluation.

We identify the underlying mobility patterns of e-commerce applications and discuss possible client-server (CS) and MA based implementation strategies for each of these patterns. We identify various application parameters that influence performance, such as, size of CS messages, size of the MA, number of remote information sources, etc., and have performed experiments to study their effect on application performance. We have used the *user turnaround time* (time elapsed between a user initiating a request and receiving the results), as the metric for performance comparison. We have performed experiments for MA using Aglets, Concordia and Voyager frameworks and Java socket for CS implementation. In this paper, we present our observations and conclusions regarding the choice of an agent framework and implementation strategy, for an e-commerce application.

While there exist some qualitative and quantitative studies of MAs in other application domains [2, 5, 7, 12, 13], to the best of our knowledge, there is no literature on the performance evaluation of MAs in the domain of e-commerce. Also, there is no prior work on identifying mobility patterns for e-commerce applications, or on evaluating mobile agent frameworks for these mobility patterns.

Section 2 provides a classification of identified mobility patterns, and section 3 discusses various implementation strategies for these mobility patterns. Section 4 describes the experimental setup. Section 5 presents a qualitative and quantitative evaluation of the three MA frameworks chosen for our study. Section 6 presents quantitative evaluation experiments of mobile agent and client-server implementation strategies. Section 7 concludes with a discussion on the implications of using mobile agents for e-commerce applications.

2 Mobility Patterns for Mobile Agents

Mobility in MAs can be characterized by the set of destinations that an MA visits, and the order in which it visits them. Hence we identify the following parameters to characterize the mobility of an MA:

1. **Itinerary :** The set of sites that an MA needs to visit. This could either be *static* (fixed at the time of MA initialization), or *dynamic* (determined by the MA logic).
2. **Order :** The order in which an MA visits the sites in its itinerary. This may also be either *static* or *dynamic*.

Based on these parameters, we distinguish MA applications in e-commerce as possessing one of the following mobility patterns:

Static Itinerary (SI)

The itinerary of the MA is fixed at the time of initialization and does not change during execution of the application. We further distinguish such applications into:

- **Static Itinerary Static Order (SISO)**
  The order in which an MA visits the sites in its itinerary is static and fixed at the time of initialization. An example application is that of an auction MA, which may be required to visit a set of auction sites in a pre-specified order.
- **Static Itinerary Dynamic Order (SIDO)**
  The order in which an MA visits the sites in its itinerary is decided dynamically by the MA. An example application is that of a shopping MA which may visit a set of online shops, in arbitrary order, to find the minimum price for a product.

Dynamic Itinerary (DI)

The itinerary of the MA is determined dynamically by the MA itself. Obviously, at least the first site in the itinerary needs to be fixed at the time of initialization. An example application is that of a shopping MA that is required to find a particular product. A shop that does not have the product may recommend an alternative shop, which is included in the MA's itinerary dynamically. It may be noted that dynamic itinerary always implies dynamic order.

3 Implementation Strategies for Applications

We identify four implementation strategies that may be adopted by a typical e-commerce application:

1. **Sequential CS :** This is based on the traditional client-server paradigm. The client makes a request to the first server and after processing the reply, makes a request to the second server and so on, till the list of servers to be visited is exhausted. This strategy is illustrated in figure 1(a).

2. **Sequential MA :** In this case, a single MA moves from its source of origin (client) to the first site (server) in its itinerary. It performs the processing at the server and then moves to the next site and so on, till it has visited all the sites in its itinerary. This strategy is illustrated in figure 1(b).

3. **Parallel CS :** This is also based on the client-server paradigm. However, instead of sequential requests, the client initiates several *parallel* threads of execution, each of which concurrently makes a request to one of the servers and processes the reply. This strategy is illustrated in figure 1(c).

4. **Parallel MA :** In this case, the client initiates multiple MAs, each of which visits a subset of the servers in the itinerary. The MAs then return to the client and collate their results to complete the task. This strategy is illustrated in figure 1(d).

It is also possible to use combinations of the above strategies. In our experiments, we restrict ourselves to these four strategies only.

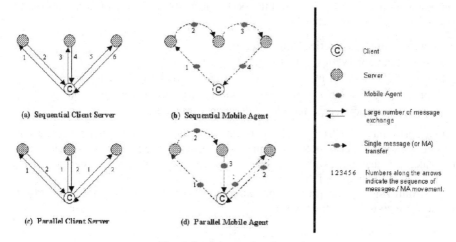

(a) Sequential Client Server   (b) Sequential Mobile Agent

(c) Parallel Client Server   (d) Parallel Mobile Agent

© Client

Server

Mobile Agent

Large number of message exchange

Single message (or MA) transfer

1 2 3 4 5 6 Numbers along the arrows indicate the sequence of messages / MA movement.

Fig. 1. Implementation Strategies

The feasibility of these implementation strategies is dependent on the mobility patterns of an e-commerce application identified in section 2.We have quantitatively evaluated all the strategies identified in this section for an e-commerce application scenario. The next section describes the chosen e-commerce application, the experimental setup and the performance metric for our performance evaluation experiments.

4 Experimentations

For our performance evaluation experiments, we have chosen a typical e-commerce application, viz., that of *product discovery*. In this application, a single client searches for information about a particular product from the catalogs of several on-line stores.

Our experiment focuses on scenarios where the client requires a highly customized, non-standard search, which the on-line store does not support. This would require the client to fetch a relevant subset of the catalog and implement a search at its end. A product search of this kind could be a "Java powered" wristwatch with "ETA quartz movement". "Java powered" being a non-standard feature, would result in all watches, which match rather standard "ETA quartz movement" featured being fetched to the client's end. The amount of data fetched by the client further increases when multiple shops are searched.

We have implemented such an application using all four strategies mentioned in Section 3 and have evaluated the performance of these implementation strategies.

4.1 Experimental Setup

The experiments were carried out on P-III, 450 MHz workstations connected through a 10 Mbps LAN with typical academic environment loads. We have implemented the MAs using three MA frameworks: Voyager, Aglets and Concordia. The CS implementation consisted of Java client and server applications, using sockets for communication. We have compared the performance of these implementations on the basis of the following parameters:

- number of stores (varies from 1 to 26);
- size of catalog (varies from 100 KB to 1 MB);
- size of client-server messages (varies proportionately with catalog size);
- processing time for servicing each request (varies from 20 ms to 1000 ms);
- network latency (typical academic load on 10Mbps LAN);

Our performance metric is the *user turnaround time*, which we define as the time elapsed between a user initiating a request and receiving the results. This includes the time taken for agent creation, time taken to visit/collect catalogs and the processing time to extract the required information.

The following section presents our qualitative and quantitative evaluation results for the three chosen MA framework.

5   Choice of Mobile Agent Framework

We have performed qualitative and quantitative performance evaluation across three Java based mobile agent frameworks viz., IBM's Aglets [1, 9], Mitsubishi's Concordia [14] and Object Space's Voyager [6]. These MA frameworks provide support for basic agent behaviors such as agent creation, agent dispatch, agent disposal, agent arrival, agent cloning and agent communication [5, 9]. All of the three MA frameworks studied implement weak agent mobility and uses object serialization for agent transfer from one node to another.

5.1   Qualitative Evaluation

Table 1 presents a comparison of Aglets, Concordia and Voyager across some of the mobility features of an MA framework, which influence application's performance.

Table 1. Qualitative comparison of agent frameworks

Features	Aglets	Concordia	Voyager
Multicast support	No	No	Yes
Agent persistence support	No	Yes	Yes
Remote agent creation support	No	No	Yes
Proxy update on MA migration	Yes	No	Yes
Messaging modes between MA	Oneway, synchronous, future	Events	Oneway, synchronous, future
Java messages to MA	Transparent	No	No
Garbage collection	No	No	Yes

5.2 Quantitative Evaluation

We have performed performance evaluation across these three mobile agent frameworks to examine agent messaging and agent migration costs for the product discovery application.

- **Message transfer cost:** A message of fixed size was sent across two nodes on the same LAN and the user turnaround time for this was measured. This was done while varying the number of message exchanges, from 10 to 500.

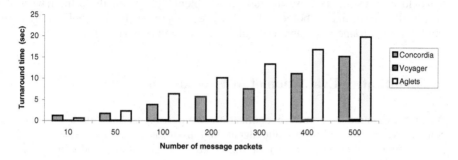

Fig. 2. Message exchange cost for Aglets, Concordia and Voyager

- **Agent shipment cost:** The processing delay at the server was kept constant and the user turnaround time of each of the MA framework was measured for different scenarios of product discovery from 1 to 26 shops.

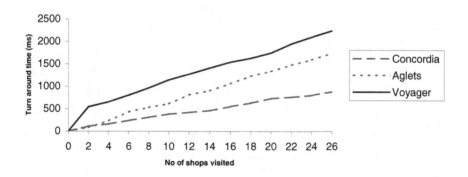

Fig. 3. Agent shipment cost for Aglets, Concordia and Voyager

5.3 Observations

The results of our MA framework evaluation experiments are shown in Fig. 2 and Fig.3. Some key observations are:

- Voyager had the least message transfer cost, followed by Concordia (Fig. 2).
- Concordia had the least agent shipment cost followed by Aglets (Fig. 3).
- Aglets performed better than Concordia when the number the number of messages was less than 50; beyond that Concordia performed better (Fig. 2).
- Aglets performed better when the number of shops to be visited were less than 4; beyond that Concordia showed better performance (Fig. 3).

Our experiments provide directions towards choice of an agent framework. The performance results depend upon the internal implementations of agent frameworks. It would be a useful exercise to study the implementations of these frameworks to identify the performance bottlenecks. However, at present the internals of Voyager and Concordia frameworks are not available for verification and study.

6 Quantitative Evaluation of MA versus CS

We use the same e-commerce application setup (as described in Section 4) for quantitative evaluation of mobile agent and traditional CS approaches. The CS implementation consists of a server that sends a catalog on request and a multi-threaded client that requests a catalog from one or more servers. The client and the server have been implemented using Java sockets. Voyager framework was used for MA implementation of these evaluation experiments. We have performed experiments to determine:

- **Effect of catalog size on turnaround time:** The processing delay at the server was kept constant and catalog sizes of 100KB, 200KB, 500KB and 1MB were used. This was done for different scenarios of product discovery from 1 to 26 shops.

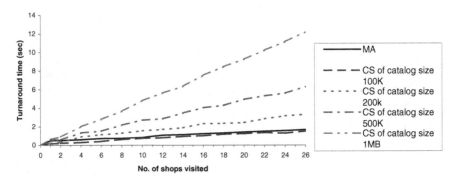

Fig. 4. Effect of catalog size on turnaround time for sequential MA & sequential CS

- **Effect of server processing delay on turnaround time:** The catalog size was kept constant at 1MB and the server delay was varied from 20ms to 500ms to 1000ms. The user turnaround time was measured for different scenarios of product discovery from 1 to 26 shops.

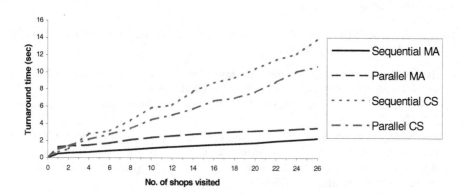

Fig. 5. Turnaround time for a processing delay of 20 ms (catalog size of 1 MB)

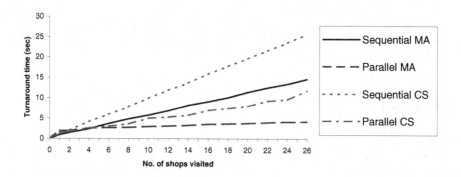

Fig. 6. Turnaround time for a processing delay of 500 ms (catalog size of 1 MB)

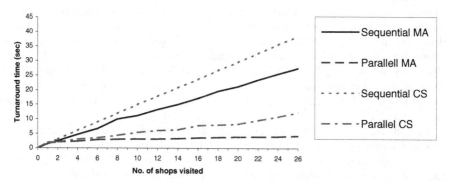

Fig. 7. Turnaround time for a processing delay of 1000 ms (catalog size of 1 MB)

6.1 Observations

The results of our MA versus CS evaluation experiments are shown in the graphs of Fig 4 to 7. Some key observations are:

- The performance of MA remains the same for different catalog sizes while the performance of CS degrades with increase in catalog size (Fig. 4).
- CS implementations perform better than MA implementations for catalog sizes less than 100 KB (Fig. 4).
- MA performs better than CS when the catalog size is greater than 200KB and number of shops to visit is greater than or equal to 3 (Fig. 4).
- MA performs better than all other strategies, for small processing delays (20 ms) and large (1MB) catalog size (Fig. 5).
- Parallel implementations perform better than sequential implementations when the number of shops to visit is greater than or equal to 6 and the processing delay is greater than or equal to 500ms (Fig. 6).
- Parallel MA performs better than parallel CS for higher processing delays (1000ms) and large (1MB) catalog size (Fig. 7).
- *Performance crossover points* i.e. parameter values for which MA starts performing better than CS implementation can be found for a given set of e-commerce application parameters (Fig. 4, 5, 6, 7).

7 Conclusions

We have identified the underlying mobility patterns of e-commerce applications and discussed possible implementation strategies for these patterns using the client-server and mobile agent paradigms. We have evaluated the agent shipment and messaging cost for three chosen MA frameworks and have performed experiments to evaluate the different implementation strategies for MA and CS implementations.

Our experiments suggest that CS implementations are suitable for applications where a "small" amount of information (less than 100 KB) is retrieved from a "few" remote servers (less than 4), having "low" processing delays (less than 20ms). However, most real-world e-commerce applications require large amount of information to be retrieved and significant processing at the server. MA's scale effectively as the size of data to be processed and the number of servers the data is obtained from increase. Scalability being one of the needs of net-centric computing, we find that MAs are an appropriate technology for implementing e-commerce applications.

Parallel implementations are effective when processing delay (greater than 1000ms) contributes significantly to the turnaround time. Our experiments also identify *performance crossover points* for different implementation strategies; this could be used to switch between implementations for performance critical applications. We feel that a *hybrid model* employing all the four identified implementation strategies would result in high performance gain for large-scale distributed applications.

Our experience suggests that mobility patterns play an important role in deciding the implementation strategy to be used for performance critical applications. The selection of an implementation strategy from those feasible for a given application could be based on several criteria such as ease of implementation, performance, availability of technology, etc. We are in the process of carrying out further experiments towards identifying the "ideal" implementation strategy given an application's characteristics.

References

1. Y.Aridov and D.Lang, "Agent design patterns: element of agent application design". In proceedings of Autonomous Agents 1998, pp108-115. ACM, 1998.
2. Antonio Carzaniga, Gian Pietro Picco and Giaovanni Vigna, "Designing Distributed Applications with Mobile Code Paradigms", Proceedings of the 19th International Conference on Software Engineering (ICSE '97) , pp. 22–32, ACM Press, 1997.
3. P. Dasgupta, N. Narasimhan, L.E. Moser and P.M. Melliar-Smith, "MAgNET: Mobile Agents for Networked Electronic Trading", IEEE Transactions on Knowledge and Data Engineering, Special Issue on Web Applications, July-August 1999.
4. Alfonso Fuggetta, Gian Pietro Picco and Giovanni Vigna, "Understanding Code Mobility", IEEE Transactions on Software Engineering, vol. 24(5), 1998.
5. Carlo Ghezzi and Giovanni Vigna, "Mobile code paradigms and technologies: A case study", Proceedings of the First International Workshop on Mobile Agents, Berlin, Germany, vol. 1219 of Lecture Notes on Computer Science, Springer, April 1997.
6. G. Glass, "ObjectSpace Voyager Core Package Technical Overview", Mobility: process, computers and agents, Addison-Wesley, Feb. 1999.
7. Dag Johansen, "Mobile Agent Applicability", Proceedings of the Mobile Agents 1998, Berlin, Springer-Verlag, Lecture notes in computer science; vol. 1477, ISBN 3-540-64959-X, (1998), pp. 9-11 September, 1998.
8. David Kotz and Robert S. Gray. "Mobile code: The future of the Internet", In Proceedings of the Workshop on Mobile Agents in the Context of Competition and Cooperation (MAC3) at Autonomous Agents '99, pages 6-12, Seattle, Washington, USA, May 1999.
9. D.Lange and M. Oshima, "Mobile Agents with Java: The Aglet API" In Special issue on Distributed World Wide Web Processing: Applications and Techniques of Web Agents, Baltzer Science Publishers, 1998.
10. Danny B. Lange and Mitsuru Oshima, "Seven Good Reasons for Mobile Agents", Communications of ACM, vol. 42, no. 3, March 1999.
11. Pattie Maes, Robert H. Guttman and Alexandros G. Moukas, "Agents That Buy and Sell", Communications of ACM, vol. 42, no. 3, pp. 81–91, March 1999.
12. Stavros Papastavrou, George Samaras and Evaggelia Pitoura, "Mobile Agents for WWW Distributed Database Access", Proceedings of IEEE International Conference on Data Engineering (ICDE99), 1999.
13. Gian Pietro Picco and Mario Baldi, "Evaluating Tradeoffs of Mobile Code Design Paradigms in Network Management Applications", Proceedings of 20th International Conference on Software Engineering, ICSE98, Kyoto, Japan, IEEE CS Press, 1998.
14. David Wong, Noemi Paciorek, Tom Walsh, Joe DiCelie, Mike Young, Bill Peet, "Concordia: An infrastructure for collaborating mobile agents", In K. Rothermel & R. Popescu-Zeletin (Eds.), Lecture Notes in Computer Science: 1219, Mobile Agents, Proceedings of the First International Workshop (pp. 86–97), Berlin, Springer-Verlag, 1997.

Performance Evaluation of Real-Time Communication Services on High-Speed LANs under Topology Changes

Juan Fernández[1], José M. García[1], and José Duato[2]

[1] Dpto. Ingeniería y Tecnología de Computadores
Universidad de Murcia, 30071 Murcia (Spain)
{peinador, jmgarcia}@ditec.um.es
[2] Dpto. Informática de Sistemas y Computadores
Universidad Politécnica de Valencia, 46071 Valencia (Spain)
jduato@gap.upv.es

Abstract. Topology changes, such as switches being turned on/off, hot expansion, hot replacement or link re-mapping, are very likely to occur in NOWs and clusters. Moreover, topology changes are much more frequent than faults. However, their impact on real-time communications has not been considered a major problem up to now, mostly because they are not feasible in traditional environments, such as massive parallel processors (MPPs), which have fixed topologies. Topology changes are supported and handled by some current and future interconnects, such as Myrinet or Infiniband. Unfortunately, they do not include support for real-time communications in the presence of topology changes.

In this paper, we evaluate a previously proposed protocol, called *Dynamically Re-established Real-Time Channels (DRRTC) protocol*, that provides topology change- and fault-tolerant real-time communication services on NOWs. We present and analyze the performance evaluation results when a single switch or a single link is deactivated/activated for different topologies and workloads. The simulation results suggest that topology change tolerance is only limited by the resources available to establish real-time channels as well as by the topology connectivity.

1 Introduction

In the past few years, networks of workstations (NOWs) and clusters, based on off-the-shelf commodity components like workstations, PCs and high-speed local area networks (LANs), have emerged as a serious alternative to massive parallel processors (MPPs) and are the most cost-effective platform for high-performance servers. In fact, they are becoming the main infrastructure for science and engineering distributed processing. However, distributed real-time processing on NOWs and clusters is still a pending issue. This is because of the lack of schemes that are able to provide dependable real-time services on NOWs.

Distributed real-time applications impose strict conditions on traffic such as bounded delivery time (deadline) or guaranteed bandwidth [1]. In order to

B. Monien, V.K. Prasanna, S. Vajapeyam (Eds.): HiPC 2001, LNCS 2228, pp. 341–350, 2001.

provide real-time communications, real-time channels [2] establish unidirectional connections among source and destination hosts. Once a real-time channel has been established, that is, resource reservation has finished, maximum delivery time and bandwidth are guaranteed.

Faults have been traditionally considered a major problem in distributed real-time processing since they may interrupt real-time communications. Thus, several researchers have proposed efficient solutions to this problem. The backup channel protocol (BCP), developed by Shin et al. [3] and based on real-time channels [2], performs recovery from faults by means of additional resources (backup channels).

On the other hand, topology changes, such as switches being turned on/off, hot expansion, hot replacement or link re-mapping, are very likely to occur in NOWs and clusters. Moreover, topology changes are much more frequent than faults. However, their impact on real-time communications has not been considered a major problem up to now, mostly because they are not feasible in traditional environments, such as massive parallel processors (MPPs), which have fixed topologies. Topology changes degrade network ability to establish new real-time channels because the routing tables are not up-to-date. Therefore, not all the available resources can be fully exploited. To make the best use of resources, every time a topology change or fault occurs, routing tables must be updated to reflect the new configuration. Dynamic reconfiguration, recently proposed by Duato et al. [4,5], assimilates topology changes by updating routing tables without stopping traffic. Note that dynamic reconfiguration by itself provides neither quality of service nor real-time services, but it provides support for an additional mechanism designed to meet real-time requirements. Finally, topology changes are supported and handled by some current and future interconnects, such as Myrinet [6] or Infiniband [7]. Unfortunately, they do not include support for real-time communications in the presence of topology changes.

In this paper, we evaluate a previously proposed protocol [8], called Dynamically Re-established Real-Time Channels (DRRTC) protocol, to provide topology change- and fault-tolerant real-time communication services on NOWs. The novelty of this protocol primarily relies on the ability to assimilate hot topology changes and faults while still providing real-time communication services as well as best-effort ones. To do this, our protocol is based on real-time channels with single backup channels and dynamic reconfiguration. Real-time channels provide real-time communications, single backup channels provide single-fault tolerance, and dynamic reconfiguration provides topology change tolerance and tolerance to additional faults. We present and analyze the performance evaluation results when a single switch or a single link is deactivated/activated for different topologies and workloads. All interrupted real-time channels are re-established after a topology change when two real-time connections per host are established. As the workload increases, channel recovery guarantees decrease. In this way, the simulation results suggest that topology change tolerance is only limited by the resources available to establish channels as well as by topology connectivity.

The rest of this paper is organized as follows. The next section presents our protocol. Network model is depicted in Sect. 3. In Sect. 4, the performance evaluation results are shown. Sect. 5 describes related work. Finally, we present our conclusions and feasible ways of future work.

2 Dynamically Re-established Real-Time Channels

This section describes the protocol previously proposed in [8] in an informal way. In order to support real-time communications, we set up a primary channel and a single secondary channel[1] for each real-time channel. Once a real-time channel has been established, real-time messages flow through the primary channel from source host to destination host until the real-time channel is closed or the primary channel is broken down. When either a hot topology change or a fault breaks down a primary channel, real-time messages are redirected through its secondary channel. At the same time, the dynamic reconfiguration algorithm described in [4,5] is triggered. The dynamic reconfiguration process updates routing tables in such a way that secondary channels could be re-established for the affected real-time channels as long as the network topology still provides an alternative physical path. The dynamic reconfiguration process does not affect the deadline of real-time messages flowing through real-time channels because traffic is allowed during reconfiguration. After reconfiguration, a new secondary channel is allocated for each affected real-time channel regardless of whether the old secondary channel has become the new primary channel or the old secondary channel was broken down. The procedure for interrupted secondary channels is the same as for primary ones. However, in this case, real-time traffic is not affected. On the other hand, if several topology changes or faults concurrently occur, the dynamic reconfiguration protocol combines them in a single process [5], and so, our protocol remains valid. Next, we are going to describe the different stages of our protocol.

2.1 Real-Time Channel Establishment

A *real-time channel* is a unidirectional connection between a pair of hosts H_1 and H_2. A real-time channel consists of a primary channel and a single secondary one. Each of them is a set $\{H_1, H_2, B, D, T\}$ where H_1 is the source host, H_2 is the destination host, B is the required bandwidth, D is the maximum admissible latency or deadline for real-time messages, and T is the type of channel, that is, primary or secondary. Enough resources (a virtual channel and enough bandwidth) are assigned to each channel in each switch along its path to meet its bandwidth and deadline requirements before real-time messages are transmitted. Real-time messages flow through the primary channel from H_1 to H_2 until the real-time channel is closed or the primary channel is broken down. In the meantime, the secondary channel remains idle and does not consume link bandwidth but just a virtual channel in each switch. In order to maximize topology change tolerance, the primary and the secondary channels must not share physical resources. In this way, disjoint physical paths are essential.

---

[1] The terms secondary and backup are used interchangeably throughout this paper.

Before transmitting real-time messages, H_1 must reserve the necessary resources for both the primary and the secondary channels. A best-effort message, called RTC_REQUEST, is sent from H_1 to H_2 to set up the primary channel. In each switch, the request message is processed to check for the availability of resources and to reserve them as we will see later. If there are not enough resources, a best-effort message, called RTC_RESPONSE(False), is returned to H_1. When a request message arrives at H_2, the primary channel is accepted if, and only if, latency of request message is shorter than channel deadline. Then, H_2 sends a best-effort message, called RTC_RESPONSE(True), back to H_1. The path followed by the response messages may not be the same used by the request to establish the channel. If H_1 receives a false response message or channel timeout expires, resources are released by means of a RTC_MSG(Release) message that flows through the channel. Channel timeout allows H_1 to release resources when no response message is received. After a few cycles, H_1 will try to establish the primary channel again until it is established or a maximum number of attempts is reached. If H_1 receives a true response message, the secondary channel will be established likewise. A request is sent from H_1 to H_2 to set up the secondary channel. In each switch, the request message is processed to check for the availability of resources, to reserve them, and also to verify that the primary channel does not go through that switch. This is because the primary and the secondary channels must not share resources in order to maximize fault tolerance.[2] Once both the primary and the secondary channels have been successfully established, the real-time channel establishment process has finished. Otherwise, resources are released, and the real-time channel is rejected. Note that several channels could be concurrently established.

Finally, we are going to analyze the processing of requests in detail. First, for each possible output port provided by the routing function for a request message, we check for the availability of resources, that is, a free virtual channel and enough bandwidth. Output ports without enough resources are no longer considered. If the request corresponds to the secondary channel, it must also be verified that the primary channel does not go through the switch. To do this, each switch has a channel table that keeps track of all channels that go through it. If there are not enough resources or the primary channel goes through the switch, a false response message is returned to H_1. Once an appropriate output port has been found, the channel is added to the channel table, resources are reserved (a virtual channel and enough bandwidth), and the request is forwarded to the next switch. Note that selection of channel routes is distributed among all switches, that is, global information is not necessary to establish channels.

2.2   Real-Time Channel Operation

After a real-time channel has been successfully established, H_1 begins to inject real-time messages, called RTC_MSG messages, through the primary channel.

---

[2] Initially, two NICs per host are assumed so that the primary and the secondary channels have to share neither switches nor links.

Real-time messages flow from H_1 to H_2 through the primary channel until the real-time channel is closed or the primary channel is broken down. In the former case, resources are released by means of two release messages. In the latter case, real-time messages are redirected through the secondary channel and a new secondary channel will be allocated if possible. Note that if the secondary channel is broken down, real-time traffic is not affected.

2.3   Real-Time Channel Recovery

Once a real-time channel has been set up and is transmitting real-time messages, we have to deal with the problem of channel recovery while still satisfying real-time requirements. Let us assume that a single link is turned off or fails. Next, the two adjacent switches detect the fault and determine the broken channels looking up their real-time channel tables. For each channel whose output port matches with the broken link, a best-effort message, called RTC_REPORT, is sent to its corresponding source host. The switch remains in the releasing state until the corresponding release message is received. For each channel whose input port matches with the broken link, a release message is sent to the destination host through the channel. Every time a report arrives at a source host for the primary or the secondary channels, a release message releases resources from that source host up to the previous switch. In the former case, real-time traffic is redirected through the secondary channel, that is, the secondary channel becomes the new primary channel. In the latter case, real-time messages continue flowing through the primary channel. In any case, after reconfiguration, the secondary channel will be re-established if possible.

At the same time that the adjacent switches detect the fault, the dynamic reconfiguration protocol described in [4,5] is triggered. This process performs sequences of partial routing table updates to avoid stopping traffic, and trying to update routing tables in such a way that a secondary channel could be re-established for each affected real-time channel. After reconfiguration, a new secondary channel is allocated, if possible, for each affected real-time channel regardless of whether the old secondary channel has become the new primary channel or the old secondary channel was broken down.

2.4   Modified Switch Architecture

Although a detailed hardware design is out of the scope of this paper, switch architecture is depicted to help readers to understand our proposal (see [8]). As shown in Fig. 1, our approach uses an input-buffered switch with virtual channels. Virtual cut-through is used because it may replace wormhole in the near future in NOWs [9]. Physical link bandwidth is 1.24 Gbps, and links are 8-bit wide. Each output port has sixteen virtual channels: thirteen RTC virtual channels that can be reserved for real-time channels, Min and UD (Up*/Down* routing) virtual channels that are used by best-effort traffic, and C/RTC that is used by DRRTC control messages and all control messages generated by reconfiguration. To build a deadlock-free routing function we use the methodology

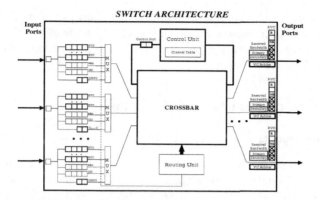

Fig. 1. Modified Switch Architecture

described in [10]. Primary and Secondary store the amount of bandwidth reserved by primary and secondary channels respectively. RVC stores the reserved virtual channels. Virtual Channel Arbiter implements the link scheduling algorithm used to forward messages. The scheduling algorithm is based on that of Infiniband [7]. The control unit processes requests and all control messages generated by reconfiguration. The Channel Table keeps track of all channels that go through the switch. The Control Port allows switch to inject messages.

3   Network Model

Network is composed of a set of switches connected by point-to-point links and hosts attached to switches through two network interface cards (NICs). In this way, we eliminate single points of failure. Network topologies were randomly generated and are completely irregular. However, some restrictions were applied for the sake of simplicity. First, all switches have eight ports, four ports connected to other switches and four ports connected to hosts. Therefore, the number of hosts is two times the number of switches. Second, two switches are not connected by more than one link. Finally, the assignment of hosts to switches was made so that the intersection of the sets of hosts connected to two switches is one host maximum. This is because we want to avoid bottlenecks when re-establishing channels after reconfiguration.

Simulation was used instead of analytical modeling. Our simulator models the network at the packet level. Simulation results were generated from three irregular topologies (T1, T2 and T3) consisting of 64 switches and 128 hosts each other (2 NICs per host). Workload is composed of CBR real-time channels of 1 Mbps and deadline is equal to inter-arrival time (IAT). Packet length of real-time messages is 16 Bytes and packet length of best-effort messages is 256 Bytes. All hosts try to establish the same number of real-time channels. Nevertheless, the number of real-time channels per host varies between two and four (2RTC, 3RTC and 4RTC). Therefore, the total number of real-time channels is 256, 384 or 512 according to the number of real-time channels per host. Destinations of channels are randomly chosen among all hosts in the network. Real-time channels are initially established during an interval of time proportional to the number of real-time channels per host.

4   Simulation Results and Analysis

4.1   Switch Deactivation and Activation

In Fig. 2(a), we show the evolution of real-time channels for different configurations when a single switch is turned off. Each column represents the total number of real-time channels that hosts are trying to establish, that is, 256, 384 and 512 for 2RTC, 3RTC and 4RTC respectively. NE corresponds to initially non-established real-time channels and the rest corresponds to successfully established real-time channels. As illustrated in Fig. 2(a), for all topologies, as the number of channels per host increases, the number of initially non-established channels increases too. Whereas 100% of channels are established for 2RTC per host, the percentage of non-established channels varies from 1.3% (T1) to 1.8% (T3) for 3RTC per host, and from 16.2% (T1) to 20.3% (T3) for 4RTC per host. For each topology, we successively simulate the deactivation of eight switches, one by one, that is, each switch is deactivated while the rest remain activated. Switches are randomly chosen among all switches in the network. Next, mean values are represented. NA corresponds to the average of non-affected real-time channels, and the rest corresponds to the average of interrupted real-time channels. RE is the average of re-established real-time channels after reconfiguration, and NRE is the average of non re-established real-time channels because a new secondary channel could not be allocated after reconfiguration. The average of affected real-time channels by switch deactivation is very similar for all configurations (it varies from 13% for T1/3RTC to 15% for T3/4RTC). However, the averages of re-established channels differ considerably from each other according to the number of real-time channels that hosts are trying to establish. As shown in Fig. 2(a), while all interrupted channels are re-established for 2RTC, the average of re-established channels varies from 81% (T3) to 94% (T1) for 3RTC per host, and from 61% (T1) to 66% (T2) for 4RTC per host. Finally, note that switch activation does not affect any already established real-time channel. Consequently, no further analysis is needed.

Now, we analyze why results get worse when increasing the number of real-time channels per host. In Fig. 3(a) and Fig. 3(b) we show the reserved virtual channels for all switch-to-switch links before reconfiguration for 2RTC and 4RTC per host respectively. For each switch, four bars represent the reserved virtual channels of its four ports connected to other switches. As we can observe, for 2RTC per host, only a few ports have no free virtual channels (only 47% of virtual channels are reserved). However, for 4RTC per host, most ports have consumed all its virtual channels (87% of virtual channels are reserved). Hence, the average of established channels decreases as the channels per host increase because free virtual channels are used up in most ports. After reconfiguration,

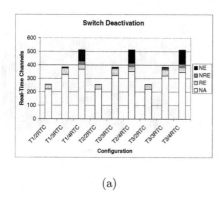
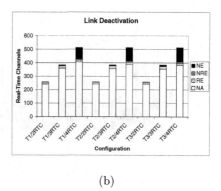

(a)   (b)

Fig. 2. Evolution of real-time channels for different configurations when a single switch/link is turned off. Configurations correspond to topologies T1, T2 and T3 when hosts try to establish 2RTC, 3RTC and 4RTC per host

the reasoning is same as the previous one (49% versus 87% of reserved virtual channels for 2RTC and 4RTC respectively).[3]

4.2   Link Deactivation and Activation

In Fig. 2(b), we show the evolution of real-time channels for different configurations when a single link is turned off. For each topology, we simulate the deactivation of the four switch-to-switch links, one by one, of the same switches used to generate the results in Fig. 2(a). As expected, the average of affected real-time channels by link deactivation is very similar for all configurations (it varies from 4% for T1/3RTC to 5% for T1/3RTC) but lower than the one by switch deactivation. Apart from that, results keep the same proportions as in the case of switch deactivation. Finally, note that link activation does not affect any already established real-time channel so that no further analysis is needed.

5   Related Work

Shin *et al.* have proposed the *Backup Channel Protocol* (BCP) [3] to achieve fault-tolerant real-time communications. In this approach, the maximum number of admissible faults depends on the maximum number of alternative paths provided by the routing function to establish the backup channels. Moreover, topology change tolerance is not provided.

Static reconfiguration techniques (Autonet [11] and Myrinet with GM [12]) stop user traffic to update routing tables. Although reconfigurations are not frequent, they can considerably degrade performance [11] and real-time constraints can not be met because of traffic disruption. Duato *et al.* [4,5] have proposed a

---

[3] These figures have been omitted for the sake of brevity.

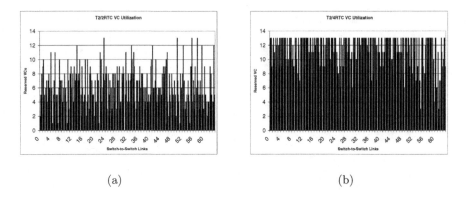

Fig. 3. VC usage before switch deactivation when establishing 2RTC/4RTC

dynamic reconfiguration algorithm, called *Partial Progressive Reconfiguration*, to minimize the negative effects of static reconfiguration on network performance. The protocol guarantees that the global routing algorithm remains deadlock-free at any time. Pinkston *et al.* [13] have developed a simple but effective strategy for dynamic reconfiguration in networks with virtual channels. Lysne *et al.* [14] aim at reducing the scope of reconfiguration identifying a restricted part of the network, the *skyline*, as the only part where a full reconfiguration is necessary. Avresky *et al.* have recently presented a new dynamic reconfiguration protocol, called *NetRec*, for high-speed LANs using wormhole routing [15].

6   Conclusions and Future Work

In this paper, the DRRTC protocol has been evaluated. The novelty of this protocol primarily relies on the ability to assimilate hot topology changes and faults while still providing real-time communication services. We have evaluated its behavior when a single switch or a single link is turned on/off for different topologies and workloads. Simulation results suggest that the DRRTC protocol provides topology change tolerance that is only limited by the resources available to establish real-time channels as well as by the topology connectivity.

Using the ideas presented in this paper, future work involves: a quantitative characterization of the DRRTC protocol under multiple topology changes, an analysis of the optimal assignment of hosts to switches within a bounded distance, and identifying of an upper bound to the protocol in order to ensure channel recovery. We are also planning to develop an InfiniBand [7] version of DRRTC.

Acknowledgments. The authors thank *Francisco J. Alfaro* at the University of Castilla La Mancha for his comments during the course of this work. This work has been supported in part by the Spanish CICYT under grant TIC2000-1151-C07-03.

References

[1] ATM Forum. *ATM Forum Traffic Management Specification. Version 4.1*, May 1995.

[2] D. D. Kandlur, K. G. Shin, and D. Ferrari. Real-Time Communications in Multi-hop Networks. *IEEE Transactions on Parallel and Distributed Systems*, 5(10):1044–1056, October 1994.

[3] Seungjae Han and Kang G. Shin. A Primary-Backup Channel Approach to Dependable Real-Time Communication in Multi-hop Networks. *IEEE Transactions on Computers*, 47(1):46–61, 1998.

[4] Francisco J. Alfaro, Aurelio Bermudez, Rafael Casado, Francisco J. Quiles, Jose L. Sanchez, and Jose Duato. Extending Dynamic Reconfiguration to NOWs with Adaptive Routing. In *Workshop on Communication, Architecture, and Applications for Network-based Parallel Computing*, January 2000. Held in conjunction with HPCA-6, Toulouse, France.

[5] Dimiter Avresky, editor. *Dependable Network Computing*, chapter 10, Dynamic Reconfiguration in High Speed Local Area Networks. José Duato, Rafael Casado, Francisco J. Quiles and José L. Sánchez. Kluwer Academic Press, 1999.

[6] Nanette J. Boden, Danny Cohen, Robert E. Felderman, Alan E. Kulawik, Charles L. Seitz, Jakov N. Seizovic, and Wen-King Su. Myrinet: A Gigabit-per-Second Local-Area Network. *IEEE Micro*, 15(1):29–36, February 1995.

[7] InfiniBand Trade Association. *InfiniBand Architecture Specification Volume 1. Release 1.0*, October 2000.

[8] Juan Fernéndez, José M. García, and José Duato. A New Approach to Provide Real-Time Services in High-Speed Local Area Networks. In *Workshop on Fault-Tolerant Parallel and Distributed Systems*, April 2001. Held in conjunction with IPDPS'01, San Francisco, CA.

[9] José Duato, Antonio Robles, Federico Silla, and Ramón Beivide. A Comparison of Router Architectures for Virtual Cut-Through and Wormhole Switching in a NOW Environment. In *Proceedings of International Parallel Processing Symposium*, 1998.

[10] Federico Silla and José Duato. Improving the Efficiency of Adaptive Routing in Networks with Irregular Topology. In *Proceedings of International Conference on High Performance Computing*, December 1997.

[11] Thomas Rodeheffer and Michael D. Schroeder. Automatic Reconfiguration in Autonet. In *Proceedings of ACM Symposium on Operating System Principles*, 1991.

[12] Myricom, Inc. *Myrinet GM documentation*, May 1999. http://www.myri.com/GM.

[13] R. Pang, T. Pinkston, and J. Duato. The Double Scheme: Deadlock-free Dynamic Reconfiguration of CutThrough Networks. In *Proceedings of International Conference on Parallel Processing*, August 2000.

[14] Olav Lysne and Jose Duato. Fast Dynamic Reconfiguration in Irregular Networks. In *Proceedings of International Conference on Parallel Processing*, August 2000.

[15] D. R. Avresky and Y. Varoglu. Dynamically Scaling Computer Networks. In *Proceedings of Workshop on Fault-Tolerant Parallel and Distributed Systems*, April 2001. Held in conjunction with IPDPS'01, San Francisco, CA.

Wavelength Conversion Placement and Wavelength Assignment in WDM Optical Networks*

Mahesh Sivakumar[1] and Suresh Subramaniam[2]

[1] Jasmine Networks, San Jose, CA 95134,
`msivakumar@jasminenetworks.com`
[2] Department of ECE, George Washington University, Washington, DC 20052,
`suresh@seas.gwu.edu`

Abstract. Wavelength conversion and wavelength assignment are key determinants of the blocking performance of wavelength-routing WDM networks. In this paper, we investigate the relative performance advantage offered by wavelength converter placement and wavelength assignment. A heuristic for conversion placement that considers constraints imposed by four different converter node architectures is proposed. Also, two heuristics for wavelength assignment are proposed, and the performance is evaluated through simulations. Results indicate that the share-per-link architecture provides a good balance between cost and performance, and that wavelength assignment algorithms are more important than converter placement algorithms in determining the blocking performance.

1 Introduction

Wavelength routing, together with wavelength division multiplexing (WDM), is a promising information transport mechanism for future all-optical networks. In a wavelength-routing network, network nodes route signals based on their wavelength and point of origin. WDM wavelength routing networks enable the formation of lightpaths which are envisioned to be the transport links for a variety of networks that may be overlaid on the physical optical infrastructure. Lightpaths are all-optical circuit-switched paths which are formed by choosing one wavelength on each link from the source to the destination of the lightpath and concatenating them. An important functionality of a wavelength-routing node is its capability or otherwise of converting a signal on one wavelength to a signal on another wavelength. This capability, called as wavelength conversion, along with routing and wavelength assignment (RWA) plays a key role in determining the performance of wavelength-routing networks.

A network can be provided with varying degrees of wavelength conversion capability depending on how wavelength conversion is implemented. For example, wavelength conversion may be unlimited at all nodes, or the number of

---

* This work was supported in part by NSF Grant #ANI-9973111 and DARPA Grant #N66001-00-1-8949.

B. Monien, V.K. Prasanna, S. Vajapeyam (Eds.): HiPC 2001, LNCS 2228, pp. 351–360, 2001.

conversions per node or link may be limited to some given constant, etc. Furthermore, the performance of a network for a given number of conversions may also depend on the locations of the converters. For example, a node handling a higher traffic load may utilize converters better than one that handles a lower load. Thus, network performance is a function of the number of wavelength converters, the locations of wavelength converters (which may be constrained by the architecture of the wavelength-convertible routing node), and the RWA algorithm. Our main objective in this paper is to investigate the interplay between these performance-affecting factors. We consider a wavelength-routing network in which lightpath requests arrive and depart dynamically, and use blocking probability as the network performance metric.

1.1 Wavelength Assignment

For simplicity, we assume throughout that the route for a lightpath request between a given pair of nodes is fixed. Then, wavelength assignment refers to the selection of a wavelength on each link of the path for a call (lightpath) arrival. Let a segment of a lightpath be defined as a sub-path with converters available (at the time of call arrival) only at the end-nodes of the sub-path and at no intermediate nodes. Wavelength assignment must obey the following constraints: (a) two lightpaths must not be assigned the same wavelength on a given link, and (b) the same wavelength must be assigned within each segment of the lightpath (wavelength continuity constraint).

Given the complexity of the wavelength assignment problem for any reasonable performance measure, heuristics have to be used in practice. Several heuristics for wavelength assignment under dynamic traffic have been proposed before [1,2,3,4,5]. We briefly refer to two that are commonly used for comparison, namely, the Random (R) and the First-Fit (FF) WA algorithms. In Random WA, an available wavelength is chosen randomly within each segment, whereas in FF WA, wavelengths are ordered, and the first wavelength that is available on each segment is assigned.

1.2 Conversion Placement

The performance of a wavelength-routing network depends not only on the amount of wavelength conversion but also on the locations of wavelength converters [6]. The amount of conversion and locations of wavelength converters may be constrained by the architecture of the wavelength-routing nodes at which wavelength conversion is done. We consider the following four wavelength-convertible node architectures that vary in their implementation complexities, first proposed by [7], in this paper.

- Node converter: In this architecture, the amount of wavelength conversion is unlimited at the node, and any lightpath passing through the node may use a converter if necessary. This provides the most flexibility in terms of wavelength assignment, but is also the most expensive.

- Link converter: Here, unlimited conversion is provided to lightpaths that use a limited number of outgoing links of the routing node.
- Share-per-link converter: Here, a limited number of conversions is allowed at each of a limited number of outgoing links of the routing node, and the lightpaths using those outgoing links share the limited number of conversions. Thus, lightpaths using some outgoing links do not enjoy wavelength conversion at all, whereas some, but not all, lightpaths that use one of the remaining outgoing links may use wavelength conversion.
- Share-per-node converter: Here, a limited number of conversions is allowed at a routing node, and these conversions are shared by all lightpaths that pass through the routing node.

Previous work on conversion placement with the objective of minimizing blocking probability has been reported in [6,8,9,10,11]. These efforts, however, did not consider the effect of the wavelength assignment algorithm on conversion placement. Some hybrid models for converter node architectures were presented in [5]. In the next section, we present some heuristics for WA and converter placement. Simulation results are presented in Section 3 and the paper is concluded in Section 4.

2 Wavelength Assignment and Converter Placement Heuristics

2.1 Minimized Conversion Wavelength Assignment

We first present two WA algorithms called Minimized Random (MinR) and Minimized First-Fit (MinFF). The basic idea behind MinR and MinFF is to minimize the use of converters. When a call arrives, we find the candidate set of wavelengths which are available on the *maximum* number of segments of the lightpath, starting from the source node. Within these segments, let t be the node closest to the destination. In MinR, one of the wavelengths from the candidate set is chosen randomly, whereas in MinFF, the lowest indexed wavelength from the candidate set is chosen. The chosen wavelength is assigned to the lightpath from the source node to t.

We then find another candidate set of wavelengths that are available on the maximum number of segments from t to the destination node of the call, and repeat the process until a wavelength has been assigned on all segments of the lightpath. It is easy to see that this greedy WA algorithm minimizes the number of converters that are used in setting up the call.

We next propose a heuristic for the placement of wavelength converters.

2.2 A Heuristic for Converter Placement

In general, determining the optimal placement of a given number of converters[1] depends on the wavelength assignment algorithm used. Given the possibly large

---

[1] In our model here, each converter is capable of performing exactly one wavelength conversion.

search space (especially for share-per-node and share-per-link architectures) involved in determining the optimal converter placement through a brute force approach, one has to turn to heuristics. Here, we propose one that iteratively determines the converter locations based on computed blocking probabilities. We compare the performance of this heuristic with some other simpler heuristics that do not rely on blocking probability computations, in the next section.

The algorithm begins with a rank-based placement and follows that by reshuffling the locations until there is no improvement in the blocking performance. It is based upon the particular wavelength assignment algorithm to be used (R, FF, MinR or MinFF) and proceeds as follows.

Step 1: Assign a rank to each node (link if a link-based converter architecture is used) based on the blocking performance using the given wavelength assignment algorithm. First, each node (link) is considered in turn, and given unlimited conversion capability, and the blocking probability is noted. The node (link) that gives the lowest blocking probability is given the highest rank.

Step 2: Distribute the given converters among the nodes (links) in an approximately uniform fashion, but giving preference to the higher-ranked nodes. For example, if node converters are used, then uniform placement would result in the highest ranked nodes having full conversion and the remaining nodes having no conversion at all. Stop if the converter architecture is node or link.

Step 3: Arrange nodes[2] in order of increasing rank. Consider the lowest ranked node as the candidate node.

Step 4: This step consists of several iterations. In each iteration, move a converter from the candidate node to each of the higher-ranked nodes in turn, and record the placement giving the best performance. If none of the placements gives a better performance than at the start of the iteration or if no converters remain at the candidate node, go to Step 5. Otherwise, continue with the next iteration.

Step 5: Repeat Step 4, each time picking the next higher-ranked node as the candidate node. The algorithm stops when the highest-ranked node becomes the candidate node.

3   Performance Evaluation

In this section, we present the results of our performance evaluation and some discussion. All results are presented for the 14-node NSFNET T-1 network topology that has 14 nodes and 21 links [12].

---

[2] While in the share-per-node architecture converters are assigned to nodes, converters are assigned to links in the share-per-link architecture. We use the term "node" henceforth with the implicit understanding that "link" applies for the share-per-link architecture.

In the simulation model, call requests arrive according to a Poisson process and have exponentially distributed holding times. Fixed shortest routes are used for a given node pair. The Erlang loads on these routes are chosen in proportion to the packet traffic measured between nodes and reported in [12].[3] Each converter is assumed to be capable of converting any input wavelength to any output wavelength. In the figures, each data point was obtained using 1 million call arrivals.

3.1   Simulation Results

This section is divided into three subsections based on the individual entities whose performance is analyzed. The proposed heuristic for wavelength converter placement was used to obtain all the results except in the comparison of converter placement heuristics presented later.

Convertible Node Architecture Analysis. As mentioned earlier, we evaluate the performance benefits of four different kinds of convertible-node architectures, namely, the node and link converter architectures, the share-per-link, and the share-per-node architecture, for wavelength converter placement. We first plot the blocking probability (P_b) against the number of wavelength converters in the network for cases with 10 and 20 wavelengths per link in Figure 1. The MinFF algorithm is used for wavelength assignment in both these cases.

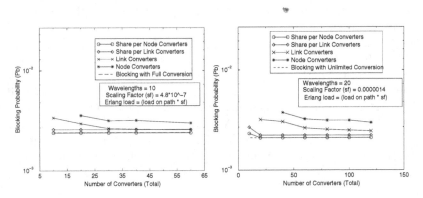

Fig. 1. Blocking probability vs. number of converters for the MinFF WA algorithm with $F = 10$ and $F = 20$.

As seen in the graphs, the share-per-node and the share-per-link architectures outperform the node and link converter architectures in terms of P_b for a

---

[3] Self-traffic and traffic to nodes in Canada which were reported in [12] are ignored. Actual traffic loads are obtained by using a scaling factor such that average blocking probabilities of the order of 10^{-3}.

given number of wavelength converters. In fact, one can see an improvement by a factor of 3 when share-per-node and node converter architectures are compared. Furthermore, very few converters are needed to achieve a performance similar to that obtained with full wavelength conversion (wherein all nodes have node converters). In the figure, it can be seen that MinFF requires just 10 converters. In Figure 2, the number of converters is fixed and P_b is plotted against the number of wavelengths per link for the MinR algorithm. The performance benefits of sharing wavelength converters are once again noticeable.

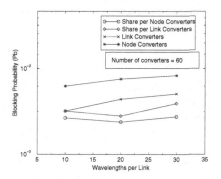

Fig. 2. Blocking probability vs. wavelengths per link for the MinR WA algorithm.

In our experiments, we found that between the share-per-node and share-per-link architectures, share-per-node performed better in almost all cases except in some. In Figure 3, P_b is plotted against the number of converters when there are 10 wavelengths per link. It can be seen that the curves for share-per-node and share-per-link cross over when the number of converters is increased to 30. However, the difference in blocking between the two is negligible.

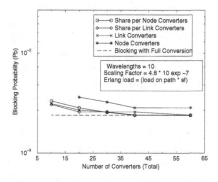

Fig. 3. Blocking probability vs. number of converters for the FF algorithm with $F = 10$.

Our results indicate that the performance improvement with share-per-node converters compared to share-per-link is only marginal. Considering the extra cost of deploying share-per-node converters, the utility of share-per-node converters appears questionable.

Performance of Wavelength Assignment Algorithms. We now evaluate the performance of the proposed wavelength assignment algorithms MinR and MinFF. These algorithms are compared with the FF and R algorithms. We first plot P_b against the number of converters in the network for 10 and 20 wavelengths per link in Figure 4. The share-per-node architecture is used in both the cases.

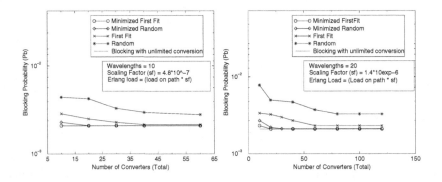

Fig. 4. Blocking probability vs. number of converters for share-per-node architecture with $F = 10$ and $F = 20$.

The MinFF algorithm outperforms all other algorithms. While there is significant blocking improvement with MinFF and MinR compared to Random, there is only a marginal difference between MinFF, MinR and FF algorithms. In general, as mentioned earlier, MinR and MinFF were designed to minimize the number of conversions for a connection. They only utilize converters when required and try to preserve them for future use. Therefore, as the number of converters available increases, the curves tend to converge.

One would be interested to know the actual number of converters used by these algorithms to get a desired P_b relative to the FF and R algorithms. In order to compute this, we allow unlimited conversions at all nodes, and record the number of converters that are actually utilized (i.e., the maximum number of converters used at some point of time) during a simulation run at each node. The total number of converters that are sufficient to give the same performance as unlimited conversion is obtained by adding the number of converters utilized at each node. Based on this, we plot the number of converters required to obtain the P_b with unlimited conversion against the number of wavelengths available per link in Figure 5. The advantage of minimizing the number of conversions in wavelength assignment is clearly seen. Furthermore, whereas the number of conversions required for MinFF and MinR remain more or less constant as the number of wavelengths per link increases, the number increases almost linearly for the FF and R algorithms.[4]

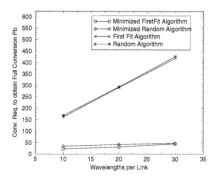

Fig. 5. Converters required to obtain the same P_b as unlimited conversion vs. wavelengths per link (F).

Performance of Conversion Placement Heuristics. In this subsection, we analyze the performance of the heuristic algorithm for wavelength converter placement proposed in Section 2.2. We compare the proposed heuristic with three other simple heuristics. The performance benefits with wavelength conversion depends on the topology of the network, traffic demand and wavelengths available among other factors [13]. Motivated by that observation, we also consider the following placement heuristics.

- Load-based heuristic: Here, converters are placed in accordance with the transit traffic at the nodes, with nodes having more transit traffic being preferred to those with less transit traffic.
- Converter utilization-based heuristic: As described earlier, the number of converters that are utilized at each node when unlimited conversions are allowed can be obtained from simulation. This heuristic places the given converters in proportion to the observed converter utilizations.
- Random placement heuristic: In this case, the wavelength converters are placed randomly at all the nodes (links) for the share-per-node (share-per-link) architecture. In case of the node and link converter architectures, the nodes (links) are selected randomly in such a way as to accommodate the given number of converters.

---

[4] This does not mean that FF and R would perform worse if fewer converters are available. As Figure 4 shows, about 40 converters suffice to give the same performance as full conversion when FF is used. However, if more converters are available, more may be used because FF and R do not attempt to minimize the use of converters.

We plot P_b against the number of converters with 10 wavelengths per link in Figure 6 for the FF and MinFF algorithms. The share-per-node architecture is used here.

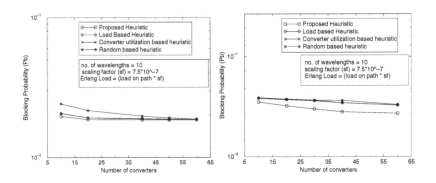

Fig. 6. Blocking probability vs. number of converters for FF and MinFF with $F = 10$.

As can be seen from the figure, the proposed heuristic performs slightly better than the other heuristics. For a small number of converters, the heuristic runs in very little time. Even though the performance improvement due to better converter placement is only marginal, it must be kept in mind that this improvement comes at almost no extra cost. However, considering the results in Figures 4 and 6, we conclude that the wavelength assignment algorithm plays a greater role in determining blocking performance than converter placement.

4   Conclusions

In this paper, we investigated the role of wavelength conversion architectures, placement, and wavelength assignment algorithms, in determining the blocking performance of all-optical wavelength routing networks. Towards this end, we proposed two heuristics for wavelength assignment, namely, Minimized First-Fit (MinFF) and Minimized Random (MinR) and a heuristic for converter placement. Finally, we evaluated their performance on the NSFNET for four different routing node architectures with wavelength conversion – link, node, share-per-link, and share-per-node converters.

The MinR and MinFF algorithms were found to be very useful in reducing the number of converters required to obtain a desired blocking performance. Besides utilizing converters more effectively, these algorithms provide a better blocking performance by preserving converters for more needy connections in the future. The results also substantiate previous studies that suggested that unlimited conversion is not necessary to obtain the best blocking performance.

The proposed heuristic for wavelength converter placement obtains a near optimal placement for a given number of converters and works better than the load-based, converter utilization-based, and random placement heuristics. An important conclusion of our study is that the wavelength assignment algorithm is more important than the converter locations when it comes to obtaining a desired blocking performance. We note that these conclusions were reached by using the NSFNET topologies as the test topology. Future work may consider the impact of adaptive routing and the performance advantage that it offers relative to wavelength assignment and conversion placement. Finally, we have considered centralized WA schemes only. It would be interesting to see if similar conclusions hold under distributed wavelength assignment.

References

1. I. Chlamtac, A. Ganz, and G. Karmi, "Purely optical networks for terabit commincation," in *Proc. INFOCOM '89*, April 1989, pp. 887–896.
2. A. Mokhtar and M. Azizoğlu, "Adaptive wavelength routing in all-optical networks," *IEEE/ACM Trans. Networking*, vol. 6, no. 2, pp. 197–206, April 1998.
3. E. Karasan and E. Ayanoglu, "Effects of wavelength routing and selection algorithms on wavelength conversion gain in WDM optical networks," in *LEOS 1996 Summer Topical Meeting on Broadband Optical Networks*, Aug. 1996.
4. S. Subramaniam and R. A. Barry, "Wavelength assignment in fixed routing WDM networks," in *Proc. ICC*, June 1997, pp. 406–410.
5. J. T. Lee and B. Ramamurthy, "A novel hybrid wavelength converter node architecture for WDM wavelength-routed networks," in *Proc. Optical Networking Workshop*, Jan. 2000.
6. S. Subramaniam, M. Azizoğlu, and A. K. Somani, "On the optimal placement of wavelength converters in wavelength-routed networks," in *Proc. INFOCOM*, April 1998, pp. 902–909.
7. K. C. Lee and V. O. K. Li, "A wavelength-convertible optical network," *IEEE/OSA J. Lightwave Tech.*, vol. 11, no. 5/6, pp. 962–970, May/June 1993.
8. J. Iness, *Efficient use of optical components in WDM networks*, Ph.D. thesis, University of California, Davis, 1997.
9. K. R. Venugopal, M. Shivakumar, and P. Sreenivasa Kumar, "A heuristic for placement of limited range wavelength converters in all-optical networks," in *Proc. INFOCOM*, March 1999, pp. 908–915.
10. S. Thiagarajan and A. K. Somani, "An efficient algorithm for optimal wavelength converter placement on wavelength-routed networks with arbitrary topologies," in *Proc. INFOCOM '99*, March 1999, pp. 916–923.
11. A. S. Arora and S. Subramaniam, "Converter placement in wavelength-routing mesh topologies," in *Proc. ICC*, June 2000, pp. 1282–1288.
12. B. Mukherjee, D. Banerjee, S. Ramamurthy, and A. Mukherjee, "Some principles for designing a wide-area WDM optical network," *IEEE/ACM Trans. Networking*, vol. 4, no. 5, pp. 684–696, Oct. 1996.
13. S. Subramaniam, M. Azizoğlu, and A. K. Somani, "Connectivity and sparse wavelength conversion in wavelength-routing networks," in *Proc. INFOCOM*, March 1996, pp. 148–155.

Identifying Long-Term High-Bandwidth Flows at a Router[*]

Smitha[**], Inkoo Kim, and A.L. Narasimha Reddy

Texas A & M University, reddy@ee.tamu.edu

Abstract. In this paper, we propose a novel approach for identifying long-term high-rate flows at a router. Identifying the long-term high-rate flows allows a router to regulate flows intelligently in times of congestion. We employ a limited amount of state to record the arrivals of packets of individual flows. The state is managed as an LRU cache. The LRU cache self converges to hold only the flows that are high rate. The size of the cache need not depend on the number of flows at the router. We analyze a number of internet packet traces to show the effectiveness of this LRU-based cache. It is shown that this scheme is highly scalable since a few flows contribute a significant fraction of the traffic at a router.

1 Background & Motivation

Recently, there has been much interest in developing resource management techniques that can effectively control nonresponsive applications. It has been shown that nonresponsive applications can effectively claim most of the bandwidth at a router while starving other applications that respond to congestion [1]. This has motivated a number of recent proposals at novel buffer management techniques that allow different droprates for different flows. In order to provide different treatment to different flows at the time of congestion, it is necessary to identify flows that are significant contributors to traffic at the router. Similarly, to prevent nonresponsive applications from monopolizing the resources, it is first necessary to identify the flows that are not responding to congestion. In this paper, we propose a simple scheme to identify long-term high-rate flows and nonresponsive flows. The proposed scheme is independent of the actual resource management technique that may be applied to control the traffic.

Current Internet traffic is heterogeneous. Most of the bytes, typically, are transferred by a small number of flows (like ftp) while a large number of flows (like HTTP) do not contribute much traffic in bytes. In such an environment, flow based schemes tend to be inefficient as the work done to establish state may not be useful for most short-lived flows. RED[2] and CHOKe[4], even though increase drop rates for high bandwidth flows, do not work well as the number of high bandwidth flows increases. With the growing use of multimedia (audio & video) applications, it is expected that traffic due to unresponsive flows will

---

[*] This work was supported by an NSF Career Award.
[**] Now with Intel, India

B. Monien, V.K. Prasanna, S. Vajapeyam (Eds.): HiPC 2001, LNCS 2228, pp. 361–371, 2001.

increase in the future. Hence, it is important to find mechanisms that do not employ per-flow state, yet are effective in controlling several high bandwidth (unresponsive) flows at the router.

Most schemes try to achieve fairness by estimating the number of flows that are present in the system. This is a difficult thing to do. With a great number of short lived flows in the system that are idle for longer periods than active, it is hard to estimate the number of flows accurately. There could be several short-lived flows that would require only a few KB of the bandwidth, but a few long lived flows that would need more bandwidth than their fair share. It would be inappropriate in this context to base the evaluation criterion on fairness. Achieving max-min fairness would not be possible since the demand of the flows is not known apriori.

Ideally, one would want to keep track of the long-lived flows and drop packets from these flows in situations of congestion, rather than drop packets from the short-lived flows. The rationale behind this is that dropping packets from short-lived flows may not reduce congestion as they are typically not the source of the congestion. As a result even if they reduce their rate, it would not make a significant difference to the network status. More importantly these flows may belong to HTTP traffic or Telnet like interactive traffic that are sensitive to delays. Dropping a packet from these flows would significantly affect the perceived network QOS without improving the network congestion. Whereas, dropping packets from flows that are causing the congestion to happen, viz., the high bandwidth flows, would make a great deal of difference. These high bandwidth flows may reduce their rate when they notice congestion and therefore experience less drops or they may not respond and continue experiencing high drops.

In this paper, we propose a simple method to identify high bandwidth flows at a router. The proposed method is totally decoupled from the underlying buffer management scheme in the routers i.e., it can be employed irrespective of the kind of the buffer management scheme used. The proposed method can be coupled with a resource management scheme to achieve the specific goals at the router.

2   Overview of the Scheme

We consider various types of flows in this scheme viz., long-term high bandwidth flows (referred to as high-BW flows), short-lived flows, and low bandwidth flows. Flows that pump data at a rate that is greater than acceptable to the network (this is typically decided by the ISP) over a period of time are high-BW flows. Flows that pump bursts of data over a short period and stay idle for some period and continue this process are short-lived flows. The others are classified as low bandwidth flows simply because they do not violate the rate limit. Among the high-BW flows we identify two classes, one that reduces its rate when congestion is indicated and the second class that are non-responsive to congestion. Ftp applications are typical examples of responsive high-BW flows. UDP sources pumping data at high rates with no congestion control mechanism built into them can be classified as nonresponsive high-BW flows. HTTP transfers over the Internet can be classified as short-lived flows. UDP sources that send at a low rate and telnet type interactive applications can be classified as low bandwidth flows.

We propose a scheme that can be used by a router to recognize high-BW flows. Also, our scheme can distinguish between responsive and non-responsive high-BW flows.

2.1 Identifying High Bandwidth Flows

Packets from high-BW flows will be seen at the router more often than the other flows. Short-lived flows, characterized by HTTP transfers, are typically the ON-OFF type and send data intermittently. Thus, packets from such flows are not seen at a constant rate at the router. When they are seen, the data is much less than that of high bandwidth flows. So, by observing the arrivals of packets for a period of time, the router can distinguish between high-BW and low-BW flows.

In order to identify high-BW flows at the router, we employ an LRU (Least Recently Used) cache. This cache is of a fixed pre-determined size, 'S'. In an LRU cache every new entry is placed at the topmost (front) position in the cache. The entry that was the least recently used is at the bottom. This is chosen to be replaced when a new entry has to be added and there is not enough space in the cache. This mechanism ensures that the recently used entries remain in the cache. The objective is to store state information for only long-term high bandwidth flows in the LRU cache.

With a cache of limited size, a flow has to arrive at the router frequently enough to remain in the cache. Short-term flows or low-BW flows are likely to be replaced by other flows fairly soon. These flows do not pump packets fast enough to keep their cache entries at the top of the LRU list and hence become candidates for replacement. High-BW flows are expected to retain their entries in the LRU cache for long periods of time.

Every time the router sees a packet, it searches the cache to check if that flow's entry exists in the cache. If yes (no), we say that a *hit* (*miss*) for that flow has occurred. On a miss, the flow is added to the cache if there is space in the cache. If there is no space in the cache, it replaces the least recently seen entry. It adds this entry in the topmost position in the cache. On a hit, the router updates the position of the entry in the cache (brings it to the topmost position). The scheme employed in SRED [5] based on "zombie list" is similar to this approach, but it does not work well in the presence of many short-lived http flows.

The use of LRU can be justified by the following data obtained from a network trace. A complete distribution of packet counts across the flows in one network trace is given in Table 1. This distribution shows that most of the flows send very few packets and that most of the bytes are contributed by very few flows. This seems to suggest that limited state caches may be successful in identifying the high-BW flows.

Table 1. Packet distribution per Flow for 12NCS-953135966 Trace

n (number of packets)	number of flows having less than n packets
2	1610
5	719
10	408
20	239
50	171
100	94
200	70
300	62
400	29
500	25
700	24
1000	18

To allow for burstiness of flows, we employ a 'threshold' below which a flow is not considered high bandwidth, even if its entry is in the cache. For each flow in the LRU, we keep track of its 'packet count' seen at the router. This count is updated on each packet arrival. Only when this count exceeds the 'threshold', a flow is regarded as a high-BW flow. Short-term flows and low-BW flows are likely to be replaced from the cache before they accumulate a count of 'threshold'. Packet sizes can be taken into account in determining the probability with which a flow is admitted into the cache. In order to keep the discussion simple, in the rest of the paper, we consider packets of the same size.

The LRU is implemented as a doubly linked list. Each node contains an entry for the flow id and the packet count. In order to make the search into the linked list easy, it is indexed by a hash table.

Cost Analysis The LRU when implemented as a doubly linked list, insertion and deletion of a flow takes $O(1)$ time. Searching for a flow in the linked list would take linear time if it were a simple doubly linked list. A hash table is used to make the search $O(1)$. Every time a new flow is added to the LRU, a hash table entry is made corresponding to this so that a search would take $O(1)$ time. The memory cost is proportional to the size 'S' of the cache.

2.2 Performance Measures

Typically, hit ratio is used as a performance measure of a cache. However, in our case, we need a different measure, because recording more hits is not our goal. Our goal is to identify the long term high rate flows. We ran the algorithm on the traces from NLANR, and obtained the list of flows which were identified by the LRU cache as exceeding a threshold value of hit count. We also collected the statistics about all the flows in the trace, and sorted all the flows in descending order of number of packets. We will call it 'sorted list'. We define 2 performance measures. **Measure 1:** If the cache identified m flows, ideally, the m flows should exactly match the m flows at the top of the sorted list. However, this was not always the case. We let the number of matching flows in the sorted list with m top entries be n. Then we define measure 1 to be

$$\text{measure1} = \frac{n}{m} \tag{1}$$

Measure 2: If the cache identified m flows, and not all of the m flows are in the top m of the sorted list, then some flows have escaped the filter mechanism. Assuming the least high packet count flow identified by the filter is the kth element on the sorted list, we define measure 2 to be

$$\text{measure2} = \frac{m}{k} \tag{2}$$

Based on these definitions, we present the results of employing an LRU cache in identifying the high-BW flows.

Table 2. Performance Measures for different traces, before Decimation

Trace	cache size	threshold	total identified	correct	measure1 %	eligible	measure2
Trace1	50	500	4	2	50.00	36	11.11
Trace2	50	500	8	4	50.00	31	25.80
Trace3	50	500	6	6	100.00	6.00	100.00
Trace4	50	500	14	8	57.14	24	58.33
Trace1	100	500	19	7	36.84	43	44.18
Trace2	100	500	31	25	80.64	38	81.57
Trace3	100	500	8	8	100.00	8.00	100.00
Trace4	100	500	21	19	90.47	24	87.50
Trace1	200	500	42	38	90.47	50	84.00
Trace2	200	500	40	39	97.50	42	95.23
Trace3	200	500	9	9	100.00	9.00	100.00
Trace4	200	500	23	22	95.65	24	95.83

3   Trace Analysis & LRU Performance

Fig. 1 shows the plots for measures 1 and 2, grouped by cache sizes and by threshold values. We can clearly see that larger caches improve the accuracy of identifying the high-BW flows for all threshold values. We can also see that small values for the threshold parameter increase inaccuracy because flows with small number of packets (but still exceeding the threshold) will be classified as high-BW flows. It is to be noted that we have used the entire length of the trace (roughly 95 seconds) as a measurement window. The identification accuracy is likely to be higher over shorter periods of time.

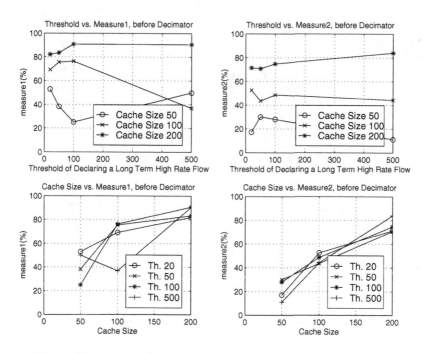

Fig. 1. Measure 1 and Measure 2 Plots for 12SDC-953145115 Trace

Table 2 shows the performance by measures 1 and 2 across four different traces. Whenever a flow's hit count reaches the threshold value, we declare the flow 'identified', and record it in a list. The third and fourth columns in the table indicate the number of such identified flows and those that correctly belong to so many top flows. For example, with a cache size of 200 and threshold of 500, there were 42 flows that recorded 500 or more hits. Among them, 38 flows correctly belonged to the top 42 flows (as sorted by the number of packets). So by measure 1, the accuracy is 90.47% which is shown in the fifth column. The sixth column is the rank of the identified flow which sent the least number of packets. That is, among the 42 identified flows, the one with the least number of packets is found at the 50*th* position in the sorted flow list. So ideally, a total of 50 flows would have been eligible for identification by this cache, but $50 - 42 = 8$ flows escaped it. We identified 42 from the 50 flows, so measure 2 is 84.00%.

As can be observed, the success rate depended on the cache size. At a cache size of 200, the identification is quite successful across all the traces. Since the traces consist of different mix of flows, in terms of life spans, pause periods or burstiness, or phases of flow arrivals, the performance is expected to be different across the traces. By both measures, the LRU based identification was quite successful. We discuss an improved algorithm in the next section.

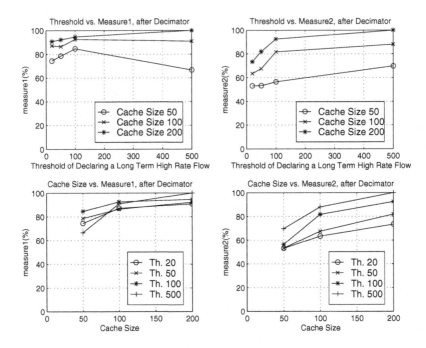

Fig. 2. Measure 1 and Measure 2 Plots for 12SDC-953145115 Trace

4 Revised Algorithm – Random Decimator

One factor that decreases the accuracy of the above-described algorithm in holding the high-BW flows in the cache is the effect of single-packet, or short-lived flows passing through the LRU filter. For example, when a single packet flow arrives, it is put at the top of the filter. Even though there's no other packet from this flow, this entry pushes an already existing entry (possibly high-BW flow's) out of the cache. For our purpose, state information for low-BW flows is not needed, we want to filter them out. They just act as noise. In order to admit a single packet flow into the filter (since we do not know if it's a single packet flow at the time of the first packet's arrival), we may have to drop a flow whose hit counts are several hundreds if it was somehow the least recently seen. We need to correct this problem.

The solution we adopted is putting a random decimator before the filter. A flow is admitted into the cache with a probability p. Hence, on an average, $1/p$ packets of a flow have to arrive at the router before that flow is given space in the cache. As a result, most low-BW packets do not disturb the state of the cache. By adjusting p, we can control the rate at which we allocate space in the cache. Smaller the p, the larger number of packets it takes to admit a flow. This is expected to help retain more high-BW flows in the cache.

4.1   Performance – After Random Decimator

Fig. 2 shows the performance of Trace 1 at different cache sizes and different thresholds after decimation with $p = 5\%$. The accuracy of identifying high-BW flows has improved significantly compared to the results in 1. The performance at the lower cache sizes has improved more significantly than at higher cache sizes. This is to be expected since the noisy affect of the low-BW flows on the cache state is felt more at smaller cache sizes. The results also indicate that at fairly small cache sizes, 100-200 entries, it is possible to achieve very high accuracies.

Table 3 show the performance of the LRU caches after decimation across four different traces, where we used $p=5\%$. The performance has improved considerably, compared to the results before decimation. For example, with a cache size of 100 and threshold of 500, measure 2 performance for Trace 1 was 11.11% before decimation and is now 69.76% after decimation. For the same cache size and threshold, the measure 1 performance for Trace 4 increases from 57.14% to 100.00%. All the traces show similar improvement.

Table 3. Performance Measures across different traces, after Decimation

Trace	cache size	threshold	total identified	correct	measure1 %	eligible	measure2
Trace1	50	500	30	20	66.66	43	69.76
Trace2	50	500	31	26	83.87	38	81.57
Trace3	50	500	23	22	95.65	24	95.83
Trace4	50	500	8	8	100.00	8	100.00
Trace1	100	500	44	40	90.90	50	88.00
Trace2	100	500	41	40	97.56	42	97.61
Trace3	100	500	23	22	95.65	24	95.83
Trace4	100	500	8	8	100.00	8	100.00
Trace1	200	500	50	50	100.00	50	100.00
Trace2	200	500	42	42	100.00	42	100.00
Trace3	200	500	25	25	100.00	25	100.00
Trace4	200	500	8	8	100.00	8	100.00

4.2   Responsive versus Nonresponsive Flows

In order to demonstrate the effectiveness of the LRU policy in identifying nonresponsive flows, we carred out an ns-2 [7] based simulation experiment consisting of 20 TCP, 20 UDP and 300 HTTP flows. The cache occupancy times of different flows with a 30-entry cache are shown in Fig. 3. It is evident from figure 3 that the LRU cache was able to hold the UDP flows for a longer period (500 seconds -length of the simulation) than the TCP flows (less than a second). This data shows that nonresponsive flows are likely to be retained longer and hence can be clearly identified.

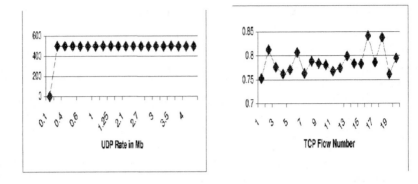

Fig. 3. Cache Occupancy

5   Scalability

To answer the scalability questions, we have looked at different traces from NLANR [8]. We collected six traces representing traffic at different times on different days. We reran our simulation with the traces to see the effectiveness of a small amount of state.

The results of caching with a 100-flow cache are shown in Table 4. The results show that a small cache of 100 entries can monitor slightly larger number of flows (189 to 327) over the simulation time since the captured flows are not active during the entire 95 second period of the trace. Even though this represents only about 0.66% to 7.70% of the flows, the packet monitoring rate (or packet hit rate) is much higher, ranging from 26.42% to 52.52%. Similarly the byte monitoring rate (or byte hit rate) is much higher, ranging from 23.45% to 56.25%. This shows that the cache policy resulted in most of the higher-rate flows being captured and monitored in the cache. This is a direct result of the nature of the network trace where a few flows contribute to most of the packets and bytes transferred through the link. This heavy-tailed property has been extensively reported in the literature [10].

We ran a second experiment with Trace5 to see the impact of the cache size. The results are shown in Table 5. As the cache is varied from 50 flows to 1000 flows, the packet hit ratio in the cache goes up from 21.63% to 54.31% and the byte hit ratio goes up from 31.16% to 59.04%. This shows that with modest size caches (1000 flows), it is possible to observe and monitor a significant (about 60%) fraction of the traffic. As the hardware improves, more state may be deployed, and thus allowing greater service from this approach.

The results show that as the cache size increases, more traffic can be monitored and controlled without requiring the network elements to establish complete state for all the flows through the link.

Table 4. Performance of a 100-flow cache on different traces

Trace	Flows			Packets			Bytes		
	Total Flows	Cached Flows	Hit Ratio	Total Packets	Cached Packets	Hit Ratio	Total Bytes(MB)	Cached Bytes(MB)	Hit Ratio
Trace1	4,245	327	7.70	131,044	61,794	47.16	43.58	10.22	23.45
Trace2	5,870	218	3.71	442,310	167,174	37.80	174.79	64.72	37.03
Trace3	9,059	207	2.29	458,691	240,922	52.52	185.86	104.54	56.25
Trace4	16,112	222	1.38	573,382	224,851	39.21	287.31	127.40	44.34
Trace5	21,348	257	1.20	557,001	147,179	26.42	236.19	80.71	34.17
Trace6	28,585	189	0.66	835,526	365,505	43.75	407.36	215.10	52.81

Table 5. Impact of cache size on Trace5

Cache Size	Flows			Packets			Bytes		
	Total Flows	Cached Flows	Hit Ratio	Total Packets	Cached Packets	Hit Ratio	Total Bytes(MB)	Cached Bytes(MB)	Hit Ratio
50	21,348	120	0.56	557,001	120,491	21.63	236.19	73.60	31.16
100	21,348	257	1.20	557,001	147,179	26.42	236.19	80.71	34.17
200	21,348	557	2.61	557,001	204,740	36.76	236.19	107.53	45.53
500	21,348	1,290	6.04	557,001	248,742	44.66	236.19	119.00	50.38
1000	21,348	2,175	10.19	557,001	302,496	54.31	236.19	139.44	59.04

6   Conclusion

In this paper, we have analyzed the Internet traffic traces and proposed an efficient method to identify long-term high-rate flows. We argued that it is effective and efficient to identify a few high rate flows which can significantly contribute to congestion resolution. We proposed an LRU cache method to identify the important flows. The proposed method is self-driven and employs fixed amount of partial state that does not depend on the number of flows at the router. In order to reduce the noise from short-lived flows, we introduced a random decimator and accepted a new flow probabilistically. We simulated the algorithm on actual traces from NLANR and demonstrated that the algorithm is effective in identifying the long term high rate flows. Possible improvements to our algorithm include preferential decimation based on port/protocol information in the packet header, and decimation of short-lived flows by delay.

References

1. S.Floyd and K.Fall, *Promoting the use of end-to-end congestion control in the internet*, IEEE-ACM Transactions on Networking, pp. 458-472, August 1999.
2. S.Floyd and V.Jacobson, *Random early detection gateways for congestion avoidance*, IEEE-ACM Transactions on Networking, pp. 397-413, August 1993.
3. B. Suter, T.V. Lakshman, D. Stiliadis and A K. Choudhary, *Design Considerations for supporting TCP with per-flow queueing*, INFOCOMM'98.

4. Rong Pan, Balaji Prabhakar and Konstantinos Psounis. *CHOKe, A Stateless Active Queue Management Scheme for Approximating Fair Bandwidth Allocation* in IEEE INFOCOMM, March 2000.
5. T.J. Ott, T.V.Lakshman, and L.H. Wong, *SRED: Stabilized RED* in Proceedings of IEEE INFOCOM, pp. 1346 -1355, March 1999.
6. Inkoo Kim *Analyzing Network Traces to Identify Long-Term High-Rate Flows*, Master's thesis, Technical Report, Texas A & M University, May 2001.
7. S. Floyd, *NS network simulator*, www.isi.edu/nsnam.
8. National Laboratory for Applied Network Research (NLANR), measurement and analysis team (MOAT), http://moat.nlanr.net/, Sept. 2000.
9. K. Claffy, G. Polyzos and H. W. Braun *A parameterizable methodology for Internet traffic flow profiling* in IEEE JSAC, pp. 1481-1494, March 1995.
10. M. Crovella and A. Bestavros *Self-Similarity in World Wide Web traffic: Evidence and possible causes* in IEEE/ACM Trans. on Networking, 1997.

Variable Length Packet Switches: Input Queued Fabrics with Finite Buffers, Speedup, and Parallelism

D. Manjunath[1] and Biplab Sikdar[2]

[1] Department of Electrical Engineering, Indian Institute of Technology,
Powai Mumbai 400 076 INDIA
[2] Department of ECSE, Rensselaer Polytechnic Institute,
Troy NY 12180 USA

Abstract. We investigate non blocking, variable length packet switches by focusing on performance evaluation and architectures to increase the throughput of such switches. With TCP/IP becoming the dominant protocol suite in the Internet, the analysis of variable length packet switches is necessary to understand the performance of the core routers and switches. We first present analytic models for delays and overflow probabilities in a variable length packet switch with finite buffers for both Poisson and self-similar packet arrival processes. The second part of the paper investigates various means to increase the throughput of these switches. As an alternative to VOQ-CIOQ switches that are known to be necessary to provide practical 100% throughput and QoS we consider a FIFO-CIOQ switch with speedup and multiple parallel planes of switches to minimise the delay in the input queue. We present analytic models for evaluating the impact of speedup and parallelism on increasing the throughput of the switch and show that with a parallelism of 4, it is possible to achieve 99.9% throughput.

1 Introduction

Fixed length packet switches have been studied extensively in the context of ATM switching fabrics. However, the Internet is primarily TCP/IP with variable length packets and arrivals not slotted in time and it is become important to analyse switching architectures for variable length packets. Also, while output queued (OQ) switches can provide 100% throughput and arbitrary QoS provisions efficiently, they are deemed infeasible to implement at high speeds and high port densities due to the switch and memory speed requirements. Thus there is considerable interest in architecture and performance of input queued (IQ) switches. Much of switching literature, be it on performance analysis, design and architectures or QoS issues, assume fixed length packets. Variable length packets can be switched using by breaking them up into fixed length cells, switch the cells and then reassemble the cells at the output but they have overheads from cell padding and headers and have additional circuit and control complexities for fragmentation and reassembly. This leads us to believe that for variable length

B. Monien, V.K. Prasanna, S. Vajapeyam (Eds.): HiPC 2001, LNCS 2228, pp. 372–382, 2001.

packet switches with non slotted arrivals, a FIFO-CIOQ switch with speedup will be architecturally simpler than a VOQ-CIOQ switch and will achieve low latencies in the input queue, which in turn will enable it to use output port QoS schedulers. Alternative to speedup, FIFO-CIOQ switch could have multiple parallel switching planes, such that more than one packet is switched to an output queue *simultaneously*. This latter feature will be called parallelism. In this paper we develop analytical models for the throughput, delay and loss performance of a variable length FIFO-CIOQ switch with speedup and parallelism.

There is surprisingly little study of non time-slotted variable length packet switches. The only architecture that does not breakup variable length packets into smaller cells is reported by Yoshigoe and Christensen [19]. Performance models for variable length packet switches are by Fuhrman [7] and Manjunath and Sikdar [11,12]. In [7], the packet delay in a non blocking $M \times N$ input queued packet is analysed with variable length packets arriving according to an i.i.d. Poisson processes at all the inputs and having uniform routing probabilities. In [11,12] we analysed these switches for arbitrary Poisson and self similar arrival and routing probabilities and infinite buffers. In this paper we continue that work with finite buffer analysis for Poisson and self-similar packet arrivals. We also analyse speedup and parallelism. The rest of the paper is organised as follows. In the next section we describe the model and the analysis method. In Section 3 we present the analysis and results from the analysis for a finite buffered input queued switch with Poisson and self-similar arrivals. In Section 4 we discuss the comparative merits of speedup and parallelism and present some results from our analysis for CIOQ switches.

2   Analytic Models: Solution Method

We consider a single stage unslotted, internally nonblocking $M \times N$ input queued packet switch with packet arrivals of rate λ_i to input port i. Packet lengths are exponential with unit mean and a packet input to port i chooses a destination j with probability p_{ij}. The line rate on output port j is μ_j. Input queue is FIFO. When a packet moves to the head of its input queue, if its destination is busy, the packet will wait at the head of the input queue till the destination output port is free and chooses to evacuate it. When an output port finishes service, the first blocked HOL packet for that output is served. From above, arrival rate to output port j, Λ_j, and its utilisation, η_j, are

$$\Lambda_j = \sum_{i=1}^{M} \lambda_i p_{ij} \qquad \eta_j = \frac{\Lambda_j}{\mu_j} \qquad (1)$$

There are two components to the sojourn time of an input packet in the switch - waiting time till it moves to the head of the line (HOL) and the time spent at the HOL till the HOL packets blocked earlier at other input queues finish their service and the packet is evacuated. The time spent at the HOL of the input queue corresponds to the "service time" in the input queue. For Poisson packet arrivals, each input queue is a M/G/1/K queue with service time equal to the time spent by a packet at its HOL. Thus, we now need the distribution of the time spent at the HOL of the queue, which is obtained using techniques similar to the analysis of blocking queueing networks [18].

The analysis is similar to the technique outlined in [12]. We summarise that technique here. Consider output port j. Output port j evacuates a packet from the HOL of the inputs at rate μ_j. The throughput from output j will be $\Lambda_j(1 - PB_j)$, (PB_j is the blocking probability of packets destined for output j) because of finite input buffers. Now, we approximate the arrival process to the virtual queue of output j by a Poisson process of throughput $\Lambda_j(1 - PB_j)$ and model the output queue as a M/M/1/M queue. In fact, we can show that as $M \to \infty$, the arrival process to the output queue is indeed Poisson following the technique in [9]. Therefore the "arrival rate" corresponding to this throughput, let us call this the effective arrival rate Λ'_j, will be obtained by solving for Λ'_j in the equation

$$\Lambda_j(1 - PB_j) = \Lambda'_j \left[\frac{1 - \eta'^M_j}{1 - \eta'^{M+1}_j} \right] \tag{2}$$

where $\eta'_j = \Lambda'_j/\mu_j$. The probability that a packet arriving to the head of an input queue wanting to go to output j sees k packets ahead of it, $\pi_j(k)$, using M/M/1/M queueing theory is given by

$$\pi_j(k) = \left[\frac{(1 - \eta'_j)(\eta'_j)^k}{1 - (\eta'_j)^M} \right] \qquad \text{for } k = 0, 1, \cdots M - 1 \tag{3}$$

In the virtual queue of output port j if there are k packets ahead of it, the packet has to wait for the evacuation of these packets before it can begin its service and its waiting time is a k stage Erlangian distribution (sum of the k independent, exponentially distributed evacuation times). In addition to the blocking delay there is the evacuation time that has an exponential distribution of mean $1/\mu_j$. The Laplace-Stieltjes Transform (LST) of the unconditional distribution of the sojourn time at the head of input i, $\mathcal{X}_i(s)$, can be seen to be

$$\mathcal{X}_i(s) = \sum_{j=1}^{N} p_{ij} \left[\sum_{k=1}^{M-1} \pi_i(k) \left(\frac{\mu_j}{\mu_j + s} \right)^k \right] \left[\frac{\mu_j}{\mu_j + s} \right] \tag{4}$$

Here the term in the first square brackets corresponds to the blocking delay and that in the second corresponds to the evacuation time given that the packet wants to go to output j. The first three moments of the blocking delay at input queue i, $\overline{B_i}$, $\overline{B_i^2}$ and $\overline{B_i^3}$ respectively, are

$$\overline{B_i} = \sum_{j=1}^{N} p_{ij} \sum_{k=1}^{M-1} \pi_j(k) \frac{k}{\mu_j} \qquad \overline{B_i^2} = \sum_{j=1}^{N} p_{ij} \sum_{k=1}^{M-1} \pi_j(k) \frac{k(k+1)}{\mu_j^2}$$

$$\overline{B_i^3} = \sum_{j=1}^{N} p_{ij} \sum_{k=1}^{M-1} \pi_j(k) \frac{k(k+1)(k+2)}{\mu_j^3} \tag{5}$$

Likewise, the first three moments of the service time for the input queue, $\overline{X_i}$, $\overline{X_i^2}$, $\overline{X_i^3}$ respectively, are

$$\overline{X_i} = \overline{B_i} + \sum_{j=1}^{N} \frac{p_{ij}}{\mu_j} \qquad \overline{X_i^2} = \overline{B_i^2} + 2\overline{B_i}\sum_{j=1}^{N} \frac{p_{ij}}{\mu_j} + 2\sum_{j=1}^{N} \frac{p_{ij}}{\mu_j^2}$$

$$\overline{X_i^3} = \overline{B_i^3} + 3\overline{B_i^2}\sum_{j=1}^{N} \frac{p_{ij}}{\mu_j} + 6\overline{B_i}\sum_{j=1}^{N} \frac{p_{ij}}{\mu_j^2} + 6\sum_{j=1}^{N} \frac{p_{ij}}{\mu_j^3} \tag{6}$$

3   Finite Buffer Analysis for Poisson and Self-Similar Arrivals

Along with the approximate distribution of the input queue service time, we use well known results from M/G/1/K and MMPP/G/1/K queueing theory to analyse the input queued switch for these packet arrival models. We first derive the expressions for the Poisson arrivals and then for self-similar arrivals.

3.1   Poisson Arrivals

The queue at input port i is modeled as a $M/G/1/K_i$ queue with Poisson arrivals of rate λ_i and service time distribution given by Eqn. 4. We analyse this queue using the diffusion approximation method of [8]. From [8], the probability that there are n packets in input queue i, $\zeta_i(n)$, is given by

$$\zeta_i(n) = \begin{cases} c_i\hat{\zeta}_i(n), & 0 \leq n < K_i \\ 1 - \frac{1-c_i(1-\rho_i)}{\rho_i}, & n = K_i \end{cases} \tag{7}$$

$$\text{where} \quad c_i = \left\{ 1 - \rho_i\left[1 - \sum_{n=0}^{K_i-1} \hat{\zeta}_i(n) \right] \right\}^{-1} \tag{8}$$

Here $\rho_i = \lambda_i\overline{X_i}$, $\hat{\zeta}_i(n)$ is the probability that there are n packets in an M/G/1/∞ with arrival rate λ_i and service time distribution of Eqn. 4. Using the diffusion approximation, the probabilities $\hat{\zeta}_i(n)$ can be written as

$$\hat{\zeta}_i(n) = \begin{cases} 1 - \rho_i, & n = 0 \\ \rho_i(1 - \hat{\rho}_i)(\hat{\rho}_i)^{n-1} & n \geq 1 \end{cases} \tag{9}$$

$$\text{where} \quad \hat{\rho}_i = \exp\left(\frac{2\overline{X_i}^2(\lambda_i\overline{X_i} - 1)}{\lambda_i\overline{X_i}^3 + \overline{X_i^2} - \overline{X_i}^2} \right) \tag{10}$$

The mean packet delay from input i, D_i, is obtained from Little's theorem.

$$D_i = \frac{1}{\lambda_i(1 - \zeta_i(K_i))} \sum_{n=0}^{K_i} n\zeta_i n \tag{11}$$

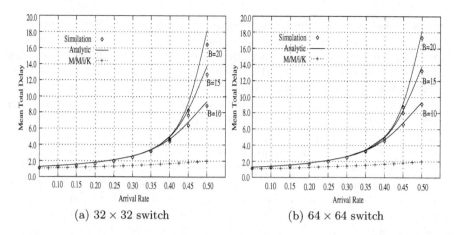

Fig. 1. Mean delay for a 32 × 32 and 64 × 64 switch from analysis and simulation

Since arrivals are Poisson, blocking probability at input i will be $\zeta_i(K_i)$. Note that the effective arrival rate for a virtual queue of an output was derived using the blocking probability for packets destined for output port i and

$$PB_j = \sum_{i=1}^{M} \zeta_i(K_i) \frac{p_{ij}\lambda_i}{\Lambda_j} \qquad (12)$$

The analytical model is solved by iterating on Eqns 1-12.

Numerical results are obtained for 32 × 32 and 64 × 64 switches with i.i.d. input arrivals and uniform routing. Fig. 1 shows the mean queuing delay. The good agreement between the simulation and the analytical results justifies our approximations. The significant "delay-penalty" for input queued switch can be seen by comparing with that output queued switch, modeled as a M/M/1/K queue. Table 1 shows the loss probabilities. Once again, the close match between analytical and simulation results indicate the goodness of our approximations. Also observe that the "blocking-penalty" of input queueing is significant.

3.2   Self-Similar Arrivals

Having modeled switch behavior for a "simplistic" Poisson arrivals, we now analyse under a more realistic model of self similar packet arrival processes. Among

Table 1. Loss probabilities in a 32 × 32 and 64 × 64 switch

λ	32 × 32, B=15			32 × 32, B=20			64 × 64, B=15			64 × 64, B=20		
	Sim.	Ana.	o/p Q	Sim.	Ana.	o/p Q	Sim.	Ana.	o/p Q	Sim.	Ana.	o/p Q
0.4	0.0006	0.0008	6.4E-7	0.0001	0.0001	6.6E-9	0.0007	0.0008	6.4E-7	0.0001	0.0001	6.6E-9
0.5	0.0363	0.0443	1.5E-5	0.0256	0.0335	4.7E-7	0.0401	0.0443	1.5E-5	0.0295	0.0336	4.8E-7
0.6	0.1595	0.1718	0.0002	0.1560	0.1687	1.5E-5	0.1657	0.1718	0.0002	0.1622	0.1687	1.5E-5

the many models for simulation of exact and approximate self similar processes, we will choose the one by Andersen and Nielsen [1]. Here a Markovian approach in which a self-similar process is obtained by superposing a number of two state Interrupted Poisson Processes (IPPs) with the resultant arrival process being an MMPP. This method, in addition to allowing the burstiness over a number of time scales with the desired covariance structure, allows us to use the well developed MMPP queuing theory for our analysis.

The distribution of the time spent by a packet at head of the input queue, its "service time" distribution, is approximated by a phase type distribution whose LST is given by Eqn. 4. Let the generator and the rate matrix of the MMPP process corresponding to the arrivals at input port i be denoted by $\mathbf{Q_i}$ and $\mathbf{R_i}$ respectively. We only report the loss analysis for the MMPP arrival process. Having obtained the service time distribution and described the arrival process we use the MMPP/G/1/K analysis of [2] and the efficient evaluation techniques for evaluating the loss probabilities from [3] to solve our model. The following notation will be used for each input i and to simplify the notation we will omit the subscript corresponding to the input port.

$\mathbf{U}$: $m \times m$ matrix given by $(\mathbf{R} - \mathbf{Q})^{-1}\mathbf{R}$

$\mathbf{P}(n, t)$: $m \times m$ matrix whose (p, q)th element denotes the conditional probability of reaching phase q of the MMPP and having n arrivals during a time interval of length t, given that we start with phase p at time $t = 0$.

$\mathbf{A_n}$: $m \times m$ matrix given by $\int_0^\infty \mathbf{P}(n, t)dX(t)$, $n \geq 0$ where $X(t)$ is the distribution of the service time as obtained previously.

$\mathbf{A}$: $m \times m$ matrix given by $\sum_{n=0}^\infty \mathbf{A_n} = \int_0^\infty e^{\mathbf{R}t}dX(t)$

$\mathbf{\Pi}(i)$: m dimensional vector whose p^{th} element is the limiting probability at the embedded epochs of having i packets in the queue and being in phase p of the MMPP, $i = 0, 1, \cdots, K - 1$.

$\mathbf{I}$: $m \times m$ identity matrix

The steady-state probability distribution of the queue length of the embedded Markov chain at the departure instants can be calculated using the following approach. The matrices $\mathbf{A_n}$ are first calculated using the technique described in [6]. From $\mathbf{P}$, we then find the matrix sequence $\{\mathbf{C_i}\}$, independent of the buffer size K, such that $\mathbf{\Pi}(i) = \mathbf{\Pi}(0)\mathbf{C_i}$ for $i = 0, 1, \cdots, K - 1$. The matrices $\mathbf{C_i}$ are calculated using the following equation

$$\mathbf{C_{i+1}} = \left[\mathbf{C_i} - \mathbf{U}\mathbf{A_i} - \sum_{v=1}^{i} \mathbf{C_v}\mathbf{A_{i-v+1}}\right]\mathbf{A_0^{-1}} \quad i = 1, \cdots, K - 2 \quad (13)$$

beginning with $\mathbf{C_0} = \mathbf{I}$. The vector $\mathbf{\Pi}(0)$ is then determined using

$$\mathbf{\Pi}(0)\left[\sum_{v=0}^{K-1} \mathbf{C_v} + (\mathbf{I} - \mathbf{U})\mathbf{A}(\mathbf{I} - \mathbf{A} + \mathbf{e\Phi})^{-1}\right] = \mathbf{\Phi} \quad (14)$$

The loss probability can then be found using the following expression

$$P_{\text{loss}} = 1 - (\overline{X}\mathbf{\Phi}\mathbf{Re})^{-1}\left[1 + \mathbf{\Pi}(0)(\mathbf{R} - \mathbf{Q})^{-1}\overline{X}^{-1}\mathbf{e}\right]^{-1} \quad (15)$$

Table 2. Loss probabilities in 8×8 switch for Bellcore traces pAug.TL and pOct.TL.

λ	pAug.TL B=500			pAug.TL, B=1000			pOct.TL, B=500			pOct.TL, B=1000		
	Sim.	Ana.	o/p Q	Sim.	Ana.	o/p Q	Sim.	Ana.	o/p Q	Sim.	Ana.	o/p Q
0.1	0.057	0.065	0.049	0.054	0.061	0.045	0.022	0.030	0.0003	0.014	0.020	6.9E-5
0.2	0.076	0.099	0.057	0.075	0.094	0.054	0.086	0.110	0.027	0.074	0.095	0.018

Numerical results are obtained as follows. The Hurst parameter, correlation at lag 1 (ρ), arrival rate λ and the time scales over which burstiness occurs in the Bellcore traces pAug.TL and pOCT.TL [10] are used to fit the parameters c_i^{1j}, c_i^{2j} and r_i for $j = 1, \cdots, 4$ of the MMPP model described in [1]. The analytical results are obtained for the MMPP/G/1 queue as described earlier. We also develop a simulation model in which the arrivals are MMPP with parameters derived above. The arrival process generator is validated by simulating a single server queue and comparing with the results given in [5]. A separate and independent MMPP arrival process generator is used for each input port with the traces generated by each of the sources having identical statistical properties. Thus, statistically identical self-similar traces but with different sample paths are used as the input processes to the simulation model. We use the Bellcore traces pAug.TL ($H = 0.82$ and $\rho = 0.582$) and pOct.TL ($H = 0.92$ and $\rho = 0.356$) with burstiness over 4 time scales.

Table 2 shows the overflow probabilities for a 8×8 switch fed with Bellcore traces pAug.TL and pOct.TL. Packet loss is significant for low loads and the drop in loss rates does not scale with increase buffer sizes. Such behavior of queues for long-range dependent arrivals are well known [5]. A large number of buffers are required with long-range dependent traffic to support even moderate loss probabilities. The tables also show the loss probabilities for the MMPP/M/1/K queue for different values of K. We only use it as an approximation to get an indication of the "input queueing penalty" for the more realistic model of self-similar packet arrivals. (Bernoulli splitting of packet arrivals into N streams corresponding to splitting the packet arrivals to the N outputs and combining of M such streams at output queue will make the characteristics of the arrival process to the output queue different from that to the input queue.)

4 Increasing Throughput in an Input Queued Switch

For Bernoulli packet arrivals and uniform routing, it is well known that the maximum throughput of an input queued switch is 0.586 [9] and it is 0.5 for variable length packet switches with Poisson arrivals and exponential packet lengths [12]. Output queued switches can provide 100% throughput, but with higher implementation complexity, typically of $\mathcal{O}(N)$ more than the input queued switch. The following two mechanisms can approximate the output queued switch but with reduced complexity in fixed length packet switches. The first method is a virtual output queued (VOQ) switch was shown to be able to achieve 100% throughput in [13]. Switch complexity is traded for scheduling complexity. The scheduling algorithms of [13] cannot be used in practical switches and many "practical" algorithms that have been reported in literature [14,15,16]. The algorithms of [14] impractical to be implemented in hardware [15] and the "practical algorithm" of [15] has a complexity of $\mathcal{O}(N^{2.5})$. Thus, although VOQ reduces the effect of HOL blocking, they are complex and do not scale well enough to offset the throughput disadvantage of input queued switches for large N. Further, note that the result of [13] is for iid arrivals and uniform destinations. In [4] it has been shown that if in each slot there are two scheduling cycles, then 100% throughput is achieved under any arrival process and destination distributions. However, this will require the use of queues at the output! Further, for variable length packets, an obvious implementation will be to break the packets into fixed length cells, switch the cells and reassemble them at the output. This method, in addition to requiring disassembly and reassembly circuits, will also mean that the switching hardware has to operate at a higher rate than the line rate to account for padding and cell header overheads.

The second method is the combined input-output queued (CIOQ) switch in which either the switch fabric operates at a higher rate than the line rate on the output ports or there are multiple parallel switching fabrics or both. Given that to achieve high throughput in a practical VOQ switch we need to operate the switch at a rate higher than the line rate, a CIOQ switch with speedup or parallel fabrics seems appealing. At this time we would like to separate the concepts of parallelism and speedup that have been used interchangeably in switching literature. In the former multiple packets can be evacuated to the same output and/or from the same input *simultaneously* or in parallel possibly through multiple parallel switching fabrics. In the case of speedup, the evacuation rate from the input queue HOL is higher than the line rate. Oie et al [17] use the term speedup to connote parallelism. In our analysis speedup is modeled by using a higher μ for the evacuation rate, or equivalently a lower λ for the arrival rate. In our analysis speedup is modeled by using a higher μ for the evacuation rate. We present only the analysis for the case of parallelism.

4.1 Analysis for Parallelism Factor m, $m > 1$

The delay analysis is similar to the previous analyses. First, consider the infinite buffer case. If there are more than m HOL packets at the inputs destined for a particular output port, m of them are served simultaneously while the others are blocked. We approximate the input process to the queue to be Poisson. Thus the virtual queue of each output port will be modeled as an $M/M/m/M$ queue and the effective arrival rate to output port j corresponding to a throughput of Λ_j is obtained by solving for Λ'_j in

$$\Lambda_j(i - PB_j) = \Lambda'_j \left[1 - \frac{\left[\frac{(\eta'_j)^M m^m}{m!} \right]}{\left[\sum_{k=0}^{m-1} \frac{(m\eta'_j)^k}{k!} + \sum_{k=m}^{M} \frac{(\eta'_j)^k m^m}{m!} \right]} \right] \tag{16}$$

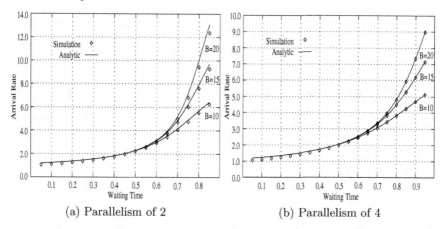

(a) Parallelism of 2   (b) Parallelism of 4

Fig. 2. Mean delay for a 32×32 and 64×64 switch from analysis and simulation

where $\eta'_j = \frac{\Lambda'_j}{m\mu_j}$. $\theta_j(k)$, $\pi_j(k)$ and $\mathcal{X}_i(s)$ are obtained as before by considering an M/M/m/M queue at the outputs. Similarly the blocking and total delay moments are also obtained like before and are given by,

$$\overline{B_i} = \sum_{j=1}^{N} p_{ij} \sum_{k=m}^{M-1} \pi_j(k) \frac{k-m+1}{\mu_j}$$

$$\overline{B_i^2} = \sum_{j=1}^{N} p_{ij} \sum_{k=m}^{M-1} \pi_j(k) \frac{(k-m+1)(k-m+2)}{\mu_j^2} \quad (17)$$

$$\overline{B_i^3} = \sum_{j=1}^{N} p_{ij} \sum_{k=m}^{M-1} \pi_j(k) \frac{(k-m+1)(k-m+2)(k-m+3)}{\mu_j^3}$$

We now use these moments of the blocking delay in Equation 6 to obtain the first three moments of the total delay. The summations over the index k for the blocking delay is from $k = m$ to $k = M - 1$ because only when there are $\geq m$ waiting in the virtual queue will the packet at the HOL of an input queue have to wait. Mean delay and loss probabilities at the input ports are obtained as in the previous section by modeling each input port as a M/G/1/K and MMPP/G/1/K queue for Poisson and self-similar packet arrivals respectively.

Fig. 2 plots the throughput-mean delay characteristics for a CIOQ switch with Poisson arrivals, and parallelism of 2 and 4 for various buffer sizes and arrival rates normalised to the switch rate. Note the excellent agreement between analytic and simulation results. Tables 3 and 4 show the blocking probabilities for parallelism of 2 and 4 for a 64×64 switch with Poisson arrivals and an 8×8 switch self similar arrivals respectively. The maximum throughput for a given parallelism is obtained by solving for λ in $\lambda \overline{X} = 1.0$. We find that for a parallelism of 4 the achievable throughput is 99.3% while it is 82.8% for a parallelism of 2.

Table 3. Loss rates in a 64×64 switch with parallelism of 2 and 4 for Poisson arrivals.

	Parallelism=2								
	K=10			K=15			K=20		
λ	Sim.	Ana.	o/p Q	Sim.	Ana.	o/p Q	Sim.	Ana.	o/p Q
0.6	0.0044	0.0048	0.0024	0.0005	0.0006	0.0002	0.0001	0.0001	1.46E-5
0.7	0.0196	0.0202	0.0086	0.0057	0.0060	0.0014	0.0017	0.0018	0.0002
0.8	0.0571	0.0590	0.0235	0.0333	0.0352	0.0072	0.0214	0.0233	0.0023
	Parallelism=4								
0.60	0.0024	0.0026	0.0024	0.0002	0.0002	0.0002	1.40E-5	1.80E-5	1.46E-5
0.70	0.0087	0.0090	0.0086	0.0015	0.0015	0.0014	0.0003	0.0003	0.0002
0.80	0.0238	0.0240	0.0235	0.0074	0.0075	0.0072	0.0024	0.0024	0.0023

Table 4. Loss rates in a 8×8 switch with parallelism of 2 and 4 for self-similar arrivals corresponding to the Bellcore trace pAug.TL

	Parallelism=2								
	K=250			K=500			K=1000		
λ	Sim.	Ana.	o/p Q	Sim.	Ana.	o/p Q	Sim.	Ana.	o/p Q
0.10	0.0573	0.0505	0.0506	0.0516	0.0487	0.0488	0.0444	0.0453	0.0454
0.15	0.0643	0.0549	0.0549	0.0523	0.0531	0.0531	0.0513	0.0497	0.0498
0.20	0.0686	0.0596	0.0591	0.0595	0.0579	0.0573	0.0559	0.0545	0.0540
	Parallelism=4								
0.10	0.0544	0.0502	0.0506	0.0504	0.0484	0.0488	0.0439	0.0450	0.0454
0.15	0.0631	0.0541	0.0549	0.0513	0.0523	0.0531	0.0510	0.0489	0.0498
0.20	0.0681	0.0581	0.0591	0.0603	0.0563	0.0573	0.0546	0.0530	0.0540

5   Conclusion

We presented a general analytical model for finite buffer analysis of input queued, variable length packet switches. This analytical model can also be used with other arrival process and for random order of service and priority queues at the input.

Numerical results show that the buffer requirements to provide acceptable blocking probabilities for self-similar arrivals are considerably higher than those for Poisson arrivals, even for low utilisations. These results emphasise the fact that realistic assumptions about the packet arrival process needs to be made to resonably estimate the performance characteristics.

We also discussed two alternatives to increase the throughput of input queued switches, VOQ and FIFO-CIOQ, and analysed FIFO-CIOQ switches based on parallelism. Our analysis suggests that FIFO-CIOQ can be a viable alternative to time slotted, fixed packet length VOQ-CIOQ switches especially if the latter have speedup ≥ 1. A complex centralised scheduler that has to collect information about N^2 queues and fragmentation and reassembly overhead is traded for multiple switching planes. Although a VOQ-CIOQ switch can suppor arbitrary QoS, technology limitations will make the centralised QoS scheduler infeasible in the Tbps region. Instead, a FIFO-CIOQ switch with speedup and parallelism can minimise input queueing delay and an output QoS scheduler can be used.

References

1. A. T. Andersen and B. F. Nielsen, "A Markovian approach for modeling packet traffic with long-range dependence," *IEEE Jour on Sel Areas in Commun,* vol. 16, no. 5, pp. 719-732, June 1998.
2. A. Baiocchi and N. Bléfari-Melazzi, "Steady-State Analysis of the MMPP/G/1/K Queue," *IEEE Trans on Commun,* vol. 41, no. 4, pp. 531-534, Apr 1993.
3. A. Baiocchi and N. Bléfari-Melazzi, "Analysis of the Loss Probability of the MAP/G/1/K Queue Part II: Approximations and Numerical Results," *Stochastic Models,* vol. 10, no. 4, pp. 895-925, 1994.
4. J.G. Dai and Balaji Prabhakar, "The Throughput of Data Switches with and without Speedup," *Proc of IEEE Infocom-2000,* Mar 2000.
5. A. Erramilli, O. Narayan and W. Wilinger, "Experimental Queuing Analysis with LRD Packet Traffic," *IEEE/ACM Trans on Networking,* vol 4, no 2, Apr 1996.
6. W. Fischer and K. Meier-Hellstern, "The Markov-modulated Poisson process (MMPP) cookbook," *Performance Evaluation,* vol. 18, no. 2, pp. 149-171, 1993.
7. S. W. Fuhrman, "Performance of a Packet Switch with a Crossbar Architecture," *IEEE Trans. on Commun.,* vol COM-41, pp. 486-491, 1993.
8. E. Glenbe and G. Pujolle, "Introduction to Queueing Networks," *John Wiley and Sons,* 1987.
9. M. J. Karol, M. G. Hluchyj and S. P. Morgan, "Input Versus Output Queuing on a Space-Division Packet Switch," *IEEE Trans on Commun,* vol. COM-35, no. 12, pp. 1347-1356, Dec 1987.
10. W. E. Leland, M. S. Taqqu, W. Willinger and D. V. Wilson, "On the self-similar nature of Ethernet traffic (Extended Version)," *IEEE/ACM Transactions on Networking,* vol. 2, no. 1, pp. 1-15, Feb 1994.
11. D. Manjunath and B. Sikdar, "Input Queued Packet Switches for Variable Length Packets : A Continuous Time Analysis," *Proc of IEEE Broadband Switching Systems,* Kingston, ON, pp. 65-69, June 1999.
12. D. Manjunath and B. Sikdar, "Variable Length Packet Switches: Delay Analysis of Crossbar Switching under Poisson and Self Similar Traffic," *Proc of IEEE Infocom-2000,* Tel Aviv, Israel, Mar 2000.
13. N. McKeown, V. Anantharam and J. Walrand, "Achieving 100% throughput in an Input Queued Switch," *Proc of IEEE Infocom-96,* pp.296–302.
14. N. Mckeown and A. Mekkittikul, "A Starvation Free Algorithm for Achieving 100% Throughput in an Input Queued Switch ", *Proc of ICCCN 96,* Oct.1996.
15. N. McKeown and A. Mekkittikul, "A Pratical Scheduling Algorithm to Achieve 100% Throughput in Input-Queued Switches", *Proc of IEEE INFOCOM-98,* Apr.1998.
16. N. McKeown, "iSLIP: A Scheduling Algorithm for Input Queued Switches," *IEEE/ACM Trans on Networking,* vol 7, pp 188-201, 1999.
17. Y. Oie, M. Murata, K. Kubota and H. Miyahara, "Effect of Speedup in Nonblocking Packet Switch," *Proc of IEEE ICC, 1989,* pp 410-414.
18. H. Perros, "Queuing Networks with Blocking," *Oxford University Press,* 1994.
19. K. Yoshigoe and K. Christensen, "A Parallel Polled Virtual Output Queued Switch with a Buffered Crossbar," *Proc of IEEE HPSR,* pp. 271-275, May 2001.

Web Mining Is Parallel

Masaru Kitsuregawa, Iko Pramudiono, Katsumi Takahashi*, and
Bowo Prasetyo

Institute of Industrial Science, The University of Tokyo
4-6-1 Komaba, Meguro-ku, Tokyo 153-8505, Japan
{kitsure,iko,katsumi,praz}@tkl.iis.u-tokyo.ac.jp

Abstract. The emerging WWW poses new technological challenges for
information processing. The scale of WWW is expected to keep growing
as more devices, such as mobile phones and PDAs are equipped with the
ability to access internet. Here we report the application of data mining
techniques on large scale web data of a directory service for users of i-
Mode, a major mobile phone internet access in Japan. We develop tool
to visualize the behavior of web site visitors. We also report experiments
on PC cluster as promising platform for large scale web mining. Parallel algorithms for generalized association rules are implemented on PC
cluster with 100 PCs.

1 Introduction

The phenomenon of i-Mode, mobile phone internet access service from NTT
Docomo, has fueled the great expectation for next generation mobile phone.
Mobile phone is not only a device to make phone call from any place but it is
also proven to be a tool to be "connected" anywhere. The number of i-Mode
subscribers has outnumbered the largest internet provider in Japan. In Japan
alone, there are 39.4 million mobile internet users. The coming of third generation
mobile phone that starts this year certainly will accelerate the importance of
internet access of this kind.

However, the browsing behavior of mobile internet users is largely unknown.
There are already many techniques, some are derived from existing data mining
techniques, to analyze browsing behavior that is available in web server access
logs. However as far as we know, there are no report on this yet. The problem
lies on the scale of data that has to be processed. This is also true for most large
portals.

Recently commodity based PC cluster system is regarded as one of the most
promising platforms for data intensive applications such as decision support
query processing and data mining. The power of PC is superior to the workstation for integer performance and the price of PC is also much lower.

Data mining has attracted a lot of attention for discovering useful information such as rules and previously unknown patterns existing between data items

---

* NTT Information Sharing Platform Laboratories, Midori-cho 3-9-11, Musashino-shi,
Tokyo 180-8585, Japan

B. Monien, V.K. Prasanna, S. Vajapeyam (Eds.): HiPC 2001, LNCS 2228, pp. 385–396, 2001.
© Springer-Verlag Berlin Heidelberg 2001

embedded in large databases, which allows effective utilization of large amount of accumulated data like web access logs.

Association rule mining is one of the most important problems in data mining. Example of association rule is the rule about what items are bought together within the transaction. Usually, the classification hierarchy over the data items is available. Users are interested in generalized association rules that span different levels of the hierarchy, since sometimes more interesting rules can be derived by taking the hierarchy into account[7].

We implemented the parallel algorithms for mining generalized association rules proposed in [9] on a cluster of 100 PCs interconnected with an ATM network, and analyzed the performance of our algorithms using a large amount of transaction dataset such as web access logs. We showed that our system can handle large amount of transactions [10].

We employ generalized association rule mining and sequential pattern mining to analyze web logs from an online directory service namely i-Townpage. Approximately 40% of i-Townpage users are from mobile internet users, mostly i-Mode subscribers. The version of i-Townpage for i-Mode users is also known as Mobile Townpage.

Here we report the development of a tool to visualize user paths. The importance of visualization tool to help human users interpret the result of web mining has been pointed out by many researchers[5,11]. Combining the visualizer with web mining system on web access logs of Mobile Townpage gives us clear picture of the behavior of mobile phone users that visit the site.

2   Association Rule Mining

An example of an association rule is "if a customer buys A and B then 90% of them buy also C". Here 90% is called the *confidence* of the rule. Another measure of a rule is called the *support* of the rule that represents the percentage of transactions that contain the rule.

The problem of mining association rules is to find all the rules that satisfy a user-specified minimum support and minimum confidence, which can be decomposed into two subproblems:

1. Find all combinations of items, called large itemsets, whose support is greater than minimum support.
2. Use the large itemsets to generate the rules.

Here we briefly explain the Apriori algorithm for finding all large itemsets, proposed in [1].

In the first pass, support count for each item is incremented by scanning the transaction database. All items that satisfy the minimum support are picked out. These items are called large 1-itemset. Here k-itemset is defined as a set of k items. In the k-th pass, the candidate k-itemsets are generated using set of large $k-1$-itemsets. Then the support count of the candidate k-itemsets is incremented by scanning the transaction database. At the end of scanning the transaction data, the large k-itemsets which satisfy minimum support are determined. The process is repeated until no candidate itemsets generated.

2.1 Generalized Association Rule with Taxonomy

In most cases, items can be classified according to some kind of "is a" hierarchies. [7] For example "Sushi is a Japanese Food" and also "Sushi is a Food" can be expressed as taxonomy. Here we categorize sushi as descendant and Japanese food and food are its ancestors. By including taxonomy as application specific knowledge more interesting rules can be discovered.

Cumulate algorithm [7] is the first algorithm to mine generalized association rule mining. It is based on Apriori algorithm, and it is extended with optimizations that make use of the characteristics of generalized association rule such as pruning itemsets containing an item and its ancestors at second pass and pre-computing the ancestors of each item.

3 Highly Parallel Data Mining

3.1 Parallel Association Rule Mining

J.S.Park, et.al proposed bit vector filtering for association rule mining and naive parallelization of Apriori [2,6], where every node keeps the whole candidate itemsets and scans the database independently. Communication is necessary only at the end of each pass. Although this method is very simple and communication overhead is very small, memory utilization efficiency is terribly bad. Since all the nodes have the copy of all the candidate itemsets, it wastes memory space a lot.

In [8] Hash Partitioned Apriori(HPA) was proposed. The candidate itemsets are not copied over all the nodes but are partitioned using hash function. Then each node builds hash table of candidate itemsets. The number of itemsets at second pass is usually extremely high, sometimes three orders of magnitude larger than the first pass in a certain retail transaction database which we examined. When the user-specified support is low, the candidate itemsets overflow the memory space and incur a lot of disk I/O.

While reading transaction data for support counting, HPA applies the same hash function to decide where to send the transactions and then probe the hash table of candidate itemsets to increase the count. Although it has to exchange transaction data among nodes, utilization whole memory space through partitioning the candidates over nodes instead of duplication results in better parallelization gain.

Hybrid approach between candidate duplication and candidate partitioning is proposed at [4]. The processors are divided into some number of groups. Within each group, all the candidates are duplicated and among groups, candidates are partitioned.

3.2   Parallel Algorithms for Generalized Association Rule Mining.

In this subsection, we describe our parallel algorithms for finding all large itemsets on shared-nothing environment proposed in [9].

Non Partitioned Generalized association rule Mining: NPGM. NPGM copies the candidate itemsets over all the nodes. Each node can work independently.

Hash Partitioned Generalized association rule Mining: HPGM. HPGM partitions the candidate itemsets among the nodes using a hash function like in the hash join, which eliminate broadcasting.

Hierarchical HPGM: H-HPGM. H-HPGM partitions the candidate itemsets among the nodes taking the classification hierarchy into account so that all the candidate itemsets whose root items are identical be allocated to the identical node, which eliminates communication of the ancestor items. Thus the communication overhead can be reduced significantly compared with original HPGM.

H-HPGM with Fine Grain Duplicate: H-HPGM-FGD. In the case the size of the candidate itemsets is smaller than available system memory, H-HPGM-FGD utilizes the remaining free space. H-HPGM-FGD detects the frequently occurring itemsets which consists of the any level items. It duplicates them and their all ancestor itemsets over all the nodes and counts the support count locally for those itemsets like in NPGM.

4   Performance Evaluation

We implemented all the above algorithms in PC cluster system. Currently the PC cluster consists of total 189 PCs :

- 100 PCs with Pentium Pro 200 MHz
- 19 PCs with Pentium II 333 MHz
- 8 PCs with Pentium II 450 MHz
- 62 PCs with Pentium III 800 MHz

Those PCs are interconnected with 128 ports 100 Mbps ATM switch, 192 ports Gigabit Ethernet and Fibre Channel. The configuration of the PC cluster is depicted in Figure 1

For performance evaluation here we used 100 PCs, connected with ATM switch[13]. Each PC has Intel Pentium Pro 200MHz CPU, 4.3GB SCSI hard disk and 64MB main memory.

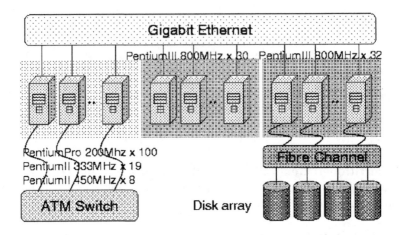

Fig. 1. PC cluster

4.1 Transaction Dataset

We use synthetic transaction data generated using procedure in For large scale experiments of generalized association rules we use the followin g parameters : (1)the number of items is 50,000, the number of roots is 100, the number of levels is 4–5, fanout is 5, (2)the total number of transactions is 20,000,000(1GBytes), the average size of transactions is 5, and (3)the number of potentially large itemsets is 10,000.

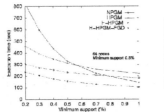

Fig. 2. Execution time

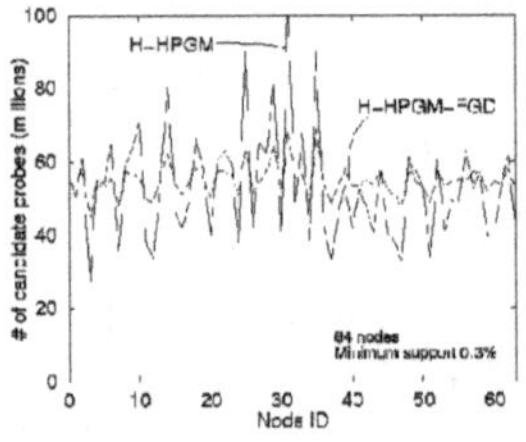

Fig. 3. Candidate probes

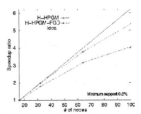

Fig. 4. Speedup ratio

4.2 Performance Evaluation Results

We show the execution time at pass 2 of all parallel algorithms varying the minimum support in Figure 1. The execution time of all the algorithms increases when the minimum support becomes small. When the minimum support is small, the candidate partitioned methods can attain good performance. H-HPGM-FGD significantly outperforms other algorithms.

Next, the workload distribution of H-HPGM and H-HPGM-FGD is examined. Figure 2 shows the number of candidate probes to increment the support count in each node at pass 2. In H-HPGM, the distribution of the number of probes is largely fractured, since the candidate itemsets are partitioned in the unit of hierarchy of the candidate itemsets. H-HPGM-FGD detects the frequently occurring candidate itemsets and duplicate them. The support counting process for these duplicated candidate itemsets can be locally processed, which can effectively balance the load among the nodes.

Figure 3 shows the speedup ratio with varying the number of nodes used 16, 32, 64 and 100. The curves are normalized by the execution time of 16 nodes system. H-HPGM-FGD attains higher linearity than H-HPGM. Since H-HPGM duplicates no candidate itemsets, the workload skew degrades the linearity. The skew handling methods detect the frequently occurring candidate itemsets and duplicate them so that the remaining free memory space can be utilized as much as possible. In Figure 3, H-HPGM-FGD achieves good performance on one hundred nodes system.

5   i-Townpage

Townpage is the name of "yellow pages", a directory service of phone numbers in Japan. It consists 11 million listings under 2000 categories. At 1995 it started internet version namely i-Townpage whose URL *http://itp.ne.jp/*. The visitors of i-Townpage can specify the location and some other search conditions such as industry category or any free keywords and get the list of companies or shops that matched as well as their phone number and address. Visitors can input the location by browsing the address hierarchy or from the nearest station or landmark. Currently i-Townpage records about 40 million page views monthly.

At the moment i-Townpage has four versions :

- standard version
  For access with ordinary web browser. A snapshot of its English version is shown in Figure 5. [1] It is also equipped with some features like online maps.
- lite version
  Simplified version for device with limited display capability such as PDA.
- Mobile Townpage
  A version for i-Mode users. An illustration of its usage is shown in Figure 6. I-mode users can directly make a call from the search results.
- L-mode version
  A version for L-mode access, a new service from NTT that enables internet access from stationary/fixed phone.

---

[1] The page is shown in English at *http://english.itp.ne.jp/*

Fig. 5. Standard English version of i-Townpage

5.1   Mobile Townpage

Because the limited display and communication ability of mobile phones, a web site for mobile phone has to be carefully designed so that the visitors can reach their goal with least clicks.

The Mobile Townpage is not simply a reduced version of standard one but it is completely redesigned to meet the demand of i-Mode users. Nearly 40% visitors of i-Townpage accesses from mobile phones.

Figure 6 gives a simple illustration of Mobile Townpage typical usage. A user first inputs industry category by choosing from prearranged category list or entering free keywords. Then he/she decides the location. Afterward he/she can begin the search or browse more detailed location.

5.2   Mining Web Access Logs of Mobile Townpage

We employed some data mining techniques such as sequential pattern mining and clustering to the web access logs of i-Townpage. We also mined generalized association rules from web access log combined with application server logs. We have implemented parallel database kernel on our PC cluster system so we can also manage large scale data efficiently[13].

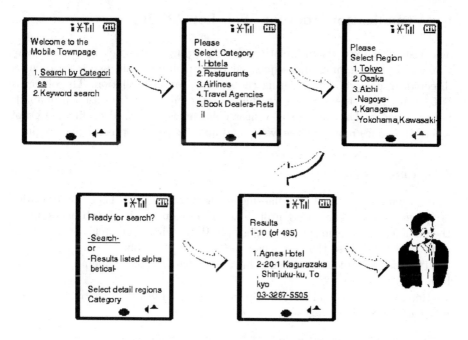

Fig. 6. i-Mode version of i-Townpage

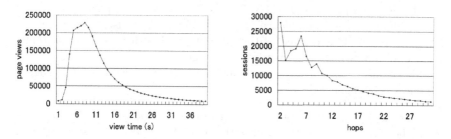

Fig. 7. View time distribution(left) Visitors' hops distribution(right)

Mobile phone internet users behave differently than their counterparts that use ordinary web browsers. Figure 7(left) shows the distribution of view time. The visitors spend less than 9 seconds on 32% of their page views. Even 67% of page views only require less than 19 seconds. The standard in preprocessing web logs often states that subsequent accesses with more than 30 minutes interval between them are considered as different sessions. However we employ a much shorter 5 minutes interval as the threshold to determine the end of a session.

Figure 7(right) derives the distribution of visitors' hops. The average hops is 7.5, 33% of the sessions are less than 4 hops length and the curve drops drastically after 6 hops.

6   Visualization Tool for Web Log Mining

The visualizer extracts the relations between pages from web access log using association rule mining and sequential pattern mining and expresses them as directed graphs while considering the strength of each relation. The objective is to make clear the visitor flow that is represented by transition probability between pages. The patterns of visitor behavior is visualized in directed graph where the nodes represent the web pages and the edges represent page request.

6.1   Implementation

The visualizer is based on placement algorithm from AT&T Laboratories' *Graphviz* tools[3]. The strength of relation between a pair of web pages affects the distance of nodes in the directed graph. Pages with strong relation are placed as near as possible so that users can easily figure them out.

The tool is implemented using Java Servlet so it can be operated with web browser.

Here we give the explanation of parameters used to represent the strength of relation between web pages :

- Support: The support is defined as the percentage of page request between a pair of pages from overall number of page requests We can think "Support" of edge from page1 to page2 as "Traffic" from page1 to page2.
- Confidence: The "Confidence" is "Probability" that visitor from page1 will go to page2. It is defined as the percentage of such page request from the overall number of page requests from page1.
- Minimum/Maximum Support/Confidence: limitation Support/Confidence of edge to be displayed.

Pages with stronger support value are also connected with bolder edges. Here the higher the confidence, the color of the edge that represents it shifts from blue to red.

6.2   Interpretation of the Graph

Based on the relation strength between any pair of pages, we can derive the following knowledge from the graph :

1. Number of visitors
   a) Pages with many visitors
   b) Pages with few visitors
2. Flow of the transition between pages
3. Transition probability between pages

On the other hand, the visualization of sequential patterns helps us to figure out:

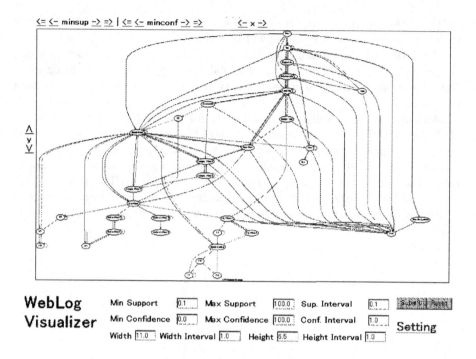

Fig. 8. Snapshot of weblog visualizer

1. Contiguous visitors' flow
2. Visitor behavior pattern until they succesfully reach destination page :
   a) Longest visitor paths
   b) Shortest visitor paths
3. Visitor behavior pattern that can not reach destination page :
   a) Longest visitor paths
   b) Shortest visitor paths

6.3   The Usage of Visualizer

The snapshot of the tool and short description of its usage is given below :
   Symbol Explanation:

 − ⇐ ← minsup → ⇒ : decrease/increase minimum support
 − ⇐ ← minconf → ⇒ : decrease/increase minimum confidence
 − ⇐ ⇒ : change minimum support/confidence while creating new graph layout.
 − ← → : change minimum support/confidence without creating new graph layout.
 − ← x → : change width of the image.
 − ← y → : change height of the image.

When the minimum support or minimum confidence is changed above certain threshold, the placement of nodes will be adjusted.

We extracted the relation between any pair of pages using association rule mining to reconstruct the sitemap of Mobile Townpage, as shown in figure 8. Notice that pages with similar functionality are well grouped with the placement algorithm of *Graphviz*.

Although we can not give the detail of the mining results, we give an example of visitor path that reach the destination page (Search Result) as a visualization of mined sequential patterns. In figure 9, we show a path of visitors that use "Input Form" to get "Search Result" and then quit the site. However we notice that those visitors get lost once in the "Category List" page. This result suggest the possibility to enhance the design of the site. Recommendation system also might work well to give shortcut link from "Category List" page to "Input Form" page.

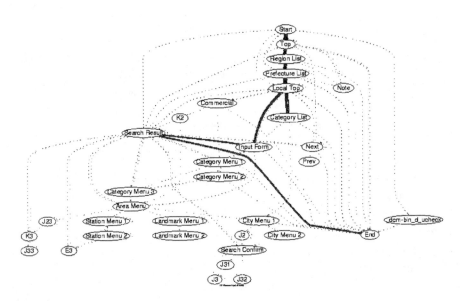

Fig. 9. Sequential pattern mining

7   Summary

We have reported our challenge to implement large scale data mining system on PC cluster, particularly with the problem of memory utilization within shared-nothing environment.

On the other hand, we also reported one prospective domain for large scale data mining applications that is the analysis of mobile internet user behavior. We have developed a tool to visualize the results of web mining. In the future we are planning to integrate the visualization tool and parallel web mining system.

The configuration of PC cluster is also evolving rapidly because of technological advance of the PC hardware. The mix of PCs from different generations to compose PC cluster is inevitable. Our PC cluster has been enhanced with new Pentium III PCs. Such heterogenous configuration needs special load balancing mechanism that will be a challenging research task [12].

References

1. R. Agrawal and R. Srikant. "Fast Algorithms for Mining Association Rules". In *Proc. of VLDB* , pp. 487–499, Sep. 1994.
2. R. Agrawal and J. C. Shafer. "Parallel Mining of Associaton Rules". In *IEEE TKDE*, Vol. 8,
3. http://www.research.att.com/sw/tools/graphviz/
4. E.-H.Han and G.Karypis and Vipin Kumar "Scalable Parallel Data Mining for Association Rules." In *Proc. of SIGMOD*, pp. 277–288, May. 1997
5. J.I. Hong and J.A. Landay "WebQuilt: A Framework for Capturing and Visualizing the Web Experience" In *Proc. of WWW10*, 2001.
6. J.S.Park, M.-S.Chen, P.S.Yu "Efficient Parallel Algorithms for Mining Association Rules" In *Proc. of CIKM*, pp. 31–36, Nov. 1995
7. R. Srikant, R. Agrawal. "Mining Generalized Association Rules". In *Proc. of VLDB*, 1995.
8. T. Shintani and M. Kitsuregawa "Hash Based Parallel Algorithms for Mining Association Rules". In *Proc. of PDIS*, pp. 19–30, Dec. 1996.
9. T. Shintani, M. Kitsuregawa "Parallel Mining Algorithms for Generalized Association Rules with Classification Hierarchy." In *Proc. of SIGMOD*, pp. 25–36, 1998.
10. T.Shintani, M. Oguchi, M.Kitsuregawa. "Performance Analysis for Parallel Generalized Association Rule Mining on a Large Scale PC Cluster". In *Proc. of Euro-par*, 1999.
11. M. Spiliopoulou and L.C. Faulstich "WUM:A tool for Web Utilization Analysis" In *Proc. of EDBT Workshop WebDB'98*, 1998.
12. M. Tamura, M.Kitsuregawa. "Dynamic Load Balancing for Parallel Association Rule Mining on Heterogeneous PC Cluster System". In *Proc. of VLDB*, 1999.
13. T. Tamura, M. Oguchi, M. Kitsuregawa "Parallel Database Processing on a 100 Node PC Cluster: Cases for Decision Support Query Processing and Data Mining." In *Proc of Super Computing 97::High Performance Networking and Computing*, 1997

An Optical Booster for Internet Routers

Joe Bannister, Joe Touch, Purushotham Kamath, and Aatash Patel

University of Southern California
Information Sciences Institute
Los Angeles, California, U.S.A.
{joseph, touch, pkamath}@isi.edu, aatashpa@usc.edu

Abstract. Although optical technologies have been effectively employed to increase the capacity of communication links, it has proven difficult to apply these technologies towards increasing the capacity of Internet routers, which implement the central forwarding and routing functions of the Internet. Motivated by the need for future routers that will forward packets among several high-speed links, this work considers the design of an Internet router that can forward packets from a terabit-per-second link without internal congestion or packet loss. The router consists of an optical booster integrated with a conventional (mostly electronic) Internet router. The optical booster processes Internet Protocol packets analogously to the hosting router, but it can avoid the time-consuming lookup function and keep packets in an entirely optical format. An optically boosted router is an inexpensive, straightforward upgrade that can be deployed readily in a backbone IP network, and provides optical processing throughput even when not deployed ubiquitously.

1 Introduction

Because of steady progress in the development of optical technology, the rate of data transfer over long distances has grown continuously and rapidly over the past several years. Data rates of 1 terabit/second (Tb/s) are attainable in the not-so-distant future. This paper describes the design of a future Internet router that forwards packets at the line rate of 1 Tb/s and can be built with optical and electronic components that will evolve easily from today's technology. The router is constructed by integrating an optical booster element with a conventional router.

With the advent of low-loss, dispersion-compensated optical fibers, high-quality lasers, optical amplifiers, and efficient modulation systems, link speeds have scaled rapidly upwards. A favored approach has been to apply wavelength-division multiplexing (WDM) on a single optical fiber to create multiple high-speed, independent channels that serve as virtual links between routers or other communication equipment. Often associated with WDM is the idea of an "all-optical network," which offers ingress-to-egress transport of data without intervening optoelectronic conversions, promising reductions in end-to-end latency and data loss as well as improvements in throughput. Routers are sometimes enhanced by wavelength-selective optical crossconnect switches and label-switching software [1, 2, 3, 4], which maps flows of Internet Protocol (IP) packets to lightpaths of WDM channels created dynamically between specific packet-forwarding end-points. Label

B. Monien, V.K. Prasanna, S. Vajapeyam (Eds.): HiPC 2001, LNCS 2228, pp. 399–413, 2001.
© Springer-Verlag Berlin Heidelberg 2001

switching for optical networks is being developed by the Internet Engineering Task Force in its Multiprotocol Label/Lambda Switching (MPLS) [5] and (MP S) [6] standards.   Although MPLS/MP S has proven valuable as a tool for traffic engineering, it has not made significant inroads as a method for dynamically managing label-switched paths. Label switching in a WDM optical network suffers from the disadvantage that it can be difficult to achieve acceptable channel utilization unless flows are sufficiently aggregated and can be mapped to the small set of WDM channels available in the network. Switching gains are achieved only for paths through contiguous sets of label-capable routers. Because WDM label-switched paths are scarce, they are used only for heavy flows. Detecting such a flow takes time, establishing a switched path requires several round-trip times, and deploying the flow on a label-switched path introduces further delay [7]. These combined latencies reduce the amount of time over which a flow can be switched.

To overcome the high latency of lightpath establishment in optical label switching, the alternative of optical burst switching [8, 9] has been considered as a way to exploit WDM. The first of a stream of packets destined for one address blazes a path through the network by setting up segments of the lightpath as it is being routed, the succeeding packets in the stream use the lightpath, and the final packet of the stream tears down the lightpath. Although this eliminates the wait for the signaling protocol to establish a lightpath, it also incurs the undesirable possibility that the burst will encounter a failure to complete the lightpath, because the search for a lightpath must be conducted without the benefit of backtracking. Streams of packets also must be buffered at the border routers so that the stream can be shaped by timing gaps between packet releases from the border router into the core network. Again, as with label switching, gains for burst switching are achieved only through contiguous paths of burst-capable routers.

Rather than continuing to force-fit connectionless IP onto these circuit-switching models, it is useful to consider routers that connect to other routers using a single high-speed channel on an optical fiber, i.e. each router pair may be connected by an unmultiplexed fiber. Router-to-router connections are largely provisioned on a link-by-link basis, e.g., SONET links and long-haul (gigabit) Ethernet links. This style of network deployment will likely persist into the foreseeable future, which implies that a network operator would derive greater benefit from the use of a conventional router that handles very-high-speed links than from a label- or burst-switching router that relies on dynamic mappings of IP flows to lower-speed WDM channels. The Internet performs best when the connectionless model at the network (IP) layer is replicated at the link layer; attempts to combine connectionless mode data transfer with connection-oriented data transfer have not been generally successful. A network of conventional (connectionless) routers is considerably easier to deploy, because these are the kind of routers the Internet already uses — the kind that network operators understand and appreciate.

This paper describes the architecture of an optically boosted Internet router that adds a fast, simple optical switch to the conventional router. The level of integration between these two components is moderate, so that the booster switches can be added to a network of conventional routers easily, modularly, and inexpensively. Minor redesign of router control interfaces would be necessary so that the boosting router and the base router can exchange control and routing information. A network may be upgraded router by router, rather than as a whole (as required by label and burst switching). A proper subset of router links may be boosted, at the discretion of the network operator. No new protocols are required, and interoperability is not affected in any way. The resulting enhanced router operates like the base router, except that the forwarding speed is increased substantially.

The remainder of this paper is organized into five sections. Section 2 discusses the challenges that router designers face in scaling up to Tb/s speeds. Section 3 presents the architecture of the optically boosted Internet router. Section 4 provides a simple analysis of the router's performance. Section 5 presents the results of a detailed simulation of the router, using real traffic traces taken from a production network. Section 6 offers conclusions to be drawn from this study.

2 Background

Consider a wide-area backbone network composed of IP routers connected by 1-Tb/s links. Clearly, the connecting links would be optical fibers. Although IP's soft-state properties and proven flexibility are strong attractions, IP forwarding is definitely expensive. Forwarding speed is key to the router's overall performance and is often its bottleneck. In a router with 1-Tb/s links, the worst-case scenario is to forward a continuous stream of minimum-size IPv4 packets, each consisting of 20 bytes (i.e., IP headers only). This scenario requires that a packet be forwarded every 160 picoseconds (ps), which equates to 6.25 billion packets per second per interface. Even if the interface were receiving a stream of back-to-back maximum-size packets of 1500 bytes each, then the forwarding rate would still be over 83 million packets per second per interface. These performance levels are considerably beyond the capabilities of today's routers, which today achieve about 10 million packets per second. The central task of forwarding is to look up the packet's destination IP address in the routing table and dispatch it to the correct next hop. Routers maintain a table of IP address prefixes, and the lookup of the address is actually a longest-prefix match among all routing-table entries and the packet's destination address. Given the rapid growth of routing tables in the core of the Internet, longest-prefix matching requires a significant — and growing — amount of time; to illustrate the magnitude of the problem, the size of a typical routing table in early 2001 was nearly 110,000 entries, up from about 70,000 entries at the start of 2000 [10].

Finding the longest matching prefix of an address every 160 ps is difficult, because it requires several memory accesses to be completed during the 160-ps interval. Routing-table memory is relatively large (needing to store 100,000 entries or more at core routers), and it is normally implemented with commodity, high-density random-access memory (RAM) parts, which have access times well above the subnanosecond times required by the worst-case routing scenario. Moreover, RAM access times have historically decreased at a slow pace; it is unlikely that high-density RAMs will be available with the required memory-access times in the near future.

The high cost of searching the routing table for the longest-prefix match with a destination IP address has motivated researchers to seek innovative algorithms and data structures to perform this search efficiently [11, 12, 13, 14]. These schemes have sought to exploit the fact that RAM cost and density are steadily improving. It is, however, difficult to extend these techniques to very high-speed links, because their performance is limited by RAM access times. Growth of the routing table, which is exacerbating lookup times, will undoubtedly continue. The joint growth in routing table size and link speed means that it will be difficult to keep lookup times low enough to handle packets at the line rate.

Concerns for the future of routing surround the planned transition from version 4 of IP (IPv4) to version 6 (IPv6). The 32-bit addresses of IPv4 are conceptually represented in the routing table by a binary tree of 2^{32} leaves, which must be matched to the lookup IP address and referred back to the smallest subtree that corresponds to an entry in the routing table (i.e. matched to the longest prefix of the address). The use of 128-bit addresses in IPv6 strains the lookup process even more, as there are, *prima facie*, 2^{128} distinct leaves in the lookup tree. Although the prefixes in the IPv6 routing tree are significantly shorter than 128 bits, and some schemes even route in two phases based on a 64-bit provider part initially, there is, nevertheless, still a heavy burden in searching the tree for prefixes that match the lookup address. If the Internet widely adopts IPv6 as the replacement for IPv4, then matters will become only grimmer for router design.

To summarize the current trends: (1) routing table sizes are growing rapidly because of the addition of new customers to the Internet; (2) for technological reasons the access times of the RAM in which routing tables are stored are decreasing very slowly; (3) as optical technology spreads, the link speeds and packet arrival rates are growing extremely rapidly; and (4) the gradual acceptance of longer IPv6 address formats is making routing tables larger and prefix matching more complicated. These trends make it more and more difficult to construct a cost-effective router that can operate at line speeds. The next section describes a router architecture designed to meet the challenges described above.

3   The Architecture of an Optically Boosted Router

The optically boosted Internet router is designed to enhance an existing conventional router by coupling with the base router to accelerate packet forwarding. If the base router has a number of high-speed optical (Tb/s) interfaces, then the boosting router is inserted into the path of these interfaces, as pictured in Fig. 1. Whereas the boosting router is an optical device that maintains incoming packets in an entirely optical format, the base router is normally a conventional router implemented in digital electronics. It is therefore possible to boost a router with heterogeneous links, because only a subset of the base router's links might need to be boosted. In this way the boosting router may be added to any router, enabling incremental deployment and upgrading of a population of routers — it is not necessary for all links of the router to have the same speed and signal formats.

A central feature of the architecture is the use of routing caches in the boosting router. It is well established that destination IP addresses exhibit a high degree of temporal locality, often achieving hit rates above 90% with modest cache sizes and simple cache-replacement policies [15]. The routing cache reproduces entries of the full routing table. The cache is a commodity content-addressable memory (CAM), which looks up the destination address and returns the identity of the outgoing interface for that address, if it appears in the cache. If a packet's destination address does not appear in the cache, the boosting router ignores the packet and allows the

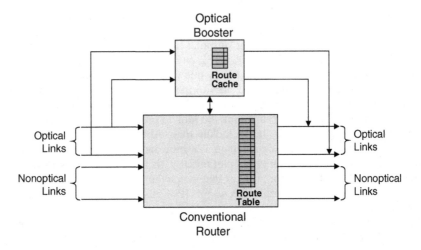

Fig. 1. A Base Router Enhanced by a Boosting Router

base router to handle the packet. When addresses hit in the cache, their packets are passed to a totally nonblocking optical space switch that attempts delivery to the correct output link. The most-likely candidate for the optical switch is a LiNbO$_3$ device. Compared to RAMs, CAMs are relatively fast, but smaller and more expensive. Today's CAMs accommodate a few thousand entries and can achieve access times of about a nanosecond. Although it is impossible to predict the future reliably, CAMs and optical space switches of the required performance will likely be available in the next few years. Specifically, to route a stream of back-to-back 20-byte IPv4 packets, the boosting router requires CAMs with access times of less than 160 ps and a totally nonblocking optical space switch that can switch at the rate of more than 6 gigahertz (GHz).

Because it is implemented optically, the boosting router supports a much higher forwarding rate than the base router. The boosting router incorporates an optical space switch that switches packets among the inlets and outlets of the boosting router. If the routing-cache hit rate of the incoming packet stream is high, then a large fraction of the input packet stream is offered to the optical switch. A packet whose destination address is not found in the routing cache will be passed on to the base router, which will buffer and switch the packet to its next hop.

Whenever a set of packets is submitted simultaneously to the optical (or any) space switch, more than one packet might need to be switched to the same outlet. Such contention for outlets demands that all but one of the contending packets be "dropped" from the switch; in the boosting router these packets are then submitted for forwarding via the base router. This packet loss reduces the number of packets that pass through the boosting router. These packets, however, are not truly lost, since they get passed to the base router. Contention for outlets of the optical space switch causes packets to be forwarded through the slower base router. The reason that contending packets are diverted through the base router is that the optical boosting router cannot buffer packets, given that there are no true optical buffers in existence at this time.

A function diagram of the optically boosted Internet router is shown in Fig. 2. Note the absence of packet buffers in the optical subsystems; all packet buffers are in the electronic subsystems of the base router, and hold packets dropped by the boosting router after contention. Fig. 2 shows the base router functionality distributed among the booster's line cards and switching fabric.

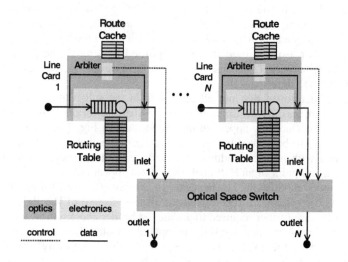

Fig. 2. The Architecture of the Optically Boosted Router

When a packet arrives on a link, it is replicated physically so that two identical copies are fed in parallel to the electronic and optical subsystems, until one of the copies is deleted or "killed" and the other is allowed to proceed and be forwarded to the next hop. First, the destination IP address is read by the booster, latched, and presented to the routing cache. This requires that the address be converted from its native optical format to an electrical format so that it can control the digital electronics of the booster; it means, also, that the initial segment of the packet header must be parsed at line rate to locate the destination IP address. (To enable line-rate parsing, parts of the packet header could be encoded at a lower bit rate than the main portion of the packet.) If the sought-after address is not in the cache, then the optical signal in the booster is deleted, and the copy of the packet in the base router is allowed to proceed through. If the address is found in the cache, then the booster's arbiter determines whether the desired output is in use or is sought by another input packet. If the arbiter finds no contention, the switch fabric is configured so that the packet's inlet and outlet are physically bridged; the packet is effectively forwarded to its next hop without optoelectronic conversion; the copy of the packet, however, must be deleted from the base router. If the booster's arbiter detects contention for the output, then all but one of the contending packets must be deleted from the booster and allowed to survive in the base router, where they will be forwarded to their next-hop routers.

The system incorporates some subtleties in how packets are handled and the switch controlled. To choreograph the movement of packets and to coordinate the control actions require very accurate and precise timing. Cache lookup and switch control must be completed in less than the critical minimum-size–packet time of 160 ps, so that a succeeding packet will be served without competing for these resources. Although the degree of integration between the base and boosting routers is low, there is a coupling so that the two subsystems can coordinate packet-deletion signals with each other as well as the exchange of routing-table information. Although this does not require a redesign of the base router, it would be necessary to introduce minimal modifications into the router that allow for routing-cache updates and packet deletions. Because packets from both subsystems merge onto the outgoing links of the router, the control of the base router is modified to defer to the boosting router when transmitting a packet, because — unlike the booster — the base router has buffering in which to defer packets.

It should be mentioned that not all addresses in the full routing table might be cached. For example, only destination addresses that use an outgoing link to which the booster is connected may appear in the cache (note in Fig. 1 that not all the base router's links need to be served by the booster, e.g. those that are not optical). It is difficult to cope with multicast addresses. Addresses that correspond to the router itself, such as an endpoint of an IP tunnel, should not appear either.

Internet routers modify packet headers, in addition to forwarding them. All Internet routers are required to decrement the IP time-to-live (TTL) field by at least one. IPv4 routers are further required to update the header checksum as well. Both of these operations present challenges to an optical booster, because they affect all packets processed by the router. Header option fields, though rarely used, require more complex processing, but those are likely to be processed outside the "fast-path"; even in conventional routers packets with options are relegated to slow-path processing in a separate processor. Fragmentation is to be avoided, because of the additional computational complexity it would require. The booster, however, should never face a fragmentation decision, because its input and output links are homogeneous and have the same maximum frame size.

The TTL can be decremented optically by zeroing the least-significant nonzero bit (the TTL needs to decrease by at least one). Alternatively, the TTL can be used as an index into a 256-entry, fast, read-only memory that stores the hard-wired ones' complement value of $k-1$ in location k, or copied to electronics and decremented by conventional means. The IPv4 checksum must be recomputed electronically; this is simple when only the modified TTL need be incorporated. IPv6 omits the header checksum, so it is simpler to process. All variations of these operations require that the electronics be pipelined and parallelized sufficiently to accommodate the header arrival rate, and also require optical bit, byte, or word replacement of header data. None of these is an insurmountable engineering challenge.

4 Performance Analysis

The optical booster increases throughput by sending some of the router's incoming packets through a low-latency optical booster capable of forwarding at very high speed. A certain portion of the incoming packets are not able to pass through the booster, because either their destination addresses are not in the fast-lookup routing cache or they experience contention at an outlet of the optical switch. Cache hit rate and output-port–contention probability are thus critical performance parameters. The higher the hit rate and the lower the contention probability, the better will be the boosted router's performance. The hit rate is determined by the traffic characteristics, the routing cache size, and the replacement policy for newly accessed addresses. The contention probability is influenced by the switch size and structure.

To gain insight into the performance of the booster, consider a simple model of the optical switch in the worst-case scenario mentioned above: each of the switch's N links has an average load of   minimum-size packets per 160-ps timeslot (where 0   1). Equivalently,   is the probability of a minimum-size packet arriving on an incoming link during a timeslot. Assume that packets are generated independently of each other, and that the next hop of a packet is chosen randomly from among the N outlets. For this analysis, suppose that all packet arrivals are synchronized to the timeslot.   is the rate of arrival after the cache hit rate has been factored in, i.e., if   is the arrival rate to an incoming link of the router and h is the hit rate, then h packets per timeslot.

Of the packets whose destination addresses hit in the cache, only a fraction of them will pass through the switch, on account of the contention that causes some of these packets to be deleted from the booster and sent back through the base router. Given the traffic load of   packets per timeslot per booster inlet, let   be the switch throughput efficiency, i.e., the fraction of offered packets that pass successfully through the optical switch. The throughput efficiency of the booster switch fabric is [16]

$$ 1 \quad 1 \quad /N \quad ^N $$

which has a limiting value of 1 $1/e$ 0.632 as N approaches   and   approaches 1. Thus, in a large, fully loaded router with a hit rate of 100%, about 37% of the packets presented to the switch will be turned away from the optical subsystem for handling by the base router. Another way of looking at this is to observe that 63% of a stream of back-to-back minimum-size IP packets presented to the router will take the all-optical booster path; this path has no loss and virtually no latency. The load on the base router is reduced dramatically.

Increasing port contention as the traffic load increases forces the value of   down to a fairly low level. A tactic to reduce contention is to allow the packets contending for the same outlet to choose a second, alternate outlet from the routing cache. These alternate outlets might indeed lead to longer routes than would the primary outlet, but the cost is often less than sending the packet through the base router. The risk of going through the base router, especially under high loads, is that congestion in the buffers will cause packets to be dropped, which carries a high penalty for many Internet end-to-end protocols.

Next assume that the base router is able to determine (and communicate to the cache) both a primary and an alternate route for every address in its routing table. Although routing protocols seldom compute alternate routes to a destination, it is straightforward to do so. The process of switching now becomes a two-stage event. In the first stage arriving packets are switched to their outlets. Some packets will have sole access to the outlet and some will contend with other packets for an outlet. Once the arbiter determines exactly which of the contending packets is granted access to the outlet and which of them is to be turned away, the second stage may begin. If the mean number of packets offered to the switch is k during the first stage, then the average number of packets granted access at the end of the first stage is $k_1 \le k$. Remaining packets must now be submitted to their alternate routes. After submission, some of the packets might actually find themselves tentatively switched to those outlets that already have been granted to the k_1 successful packets of stage 1; these packets are immediately deleted from the booster. In stage 2 the surviving k_4 packets are sent to their alternate, open outlets, and some of these packets might experience contention with each other. The contention is resolved just as it was in stage 1, but the new switch size is N' and the new traffic load is . After resolving contention there are $k_5 \le k_4$ packets that emerge unscathed from stage 2. The net throughput of k_6 packets consists of the successful packets from both stages, and the overall throughput efficiency is *. The derivation outlined above is depicted in Fig. 3, and the equations for the two stages are given below.

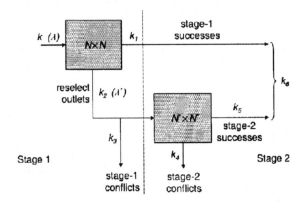

Fig. 3. Two-Stage Process Used in the Derivation of Throughput Efficiency for Dual-Entry Caches

$k \ N$ packets submitted to switch in stage 1

$1 \quad /N$

$1 \quad ^N/ \ = $ efficiency of stage-1 switch

$k_1 \quad k$ successful packets in stage 1

$k_2 \quad k \ k_1$ packets resubmitted in stage

$k_3 \quad k_1/N \ k_2$ immediately failed reroutes

$k_4 \quad k_2 \ k_3$ packets resubmitted in stage 2

$N \quad N \ k_1$ free outlets in stage 2

k_4/N load on stage-2 switch

$$1 \quad /N$$

$1 \quad ^N / \quad$ efficiency of stage-2 switch

$k_5 \quad k_4 \quad$ successful packets in stage 2

$k_6 \quad k_1 \quad k_5 \quad$ successful packets in stages 1 and 2

$* \quad k_6/k \quad$ overall switch efficiency

The equations above may be simplified to yield an expression for the overall switch efficiency in terms of the given parameters   and N. Combining and simplifying the equations, one gets the result

$$* \quad \frac{N}{N} \quad 1 \quad 1 \quad \frac{1}{N^{2N}} \quad N^{N}$$

where 1 . It is instructive to consider the limits of performance as the load and switch size grow. Taking the limit of $*$ as   approaches 1 and N approaches ', one gets

$$\underset{\text{lim}}{*} \quad \underset{1,N}{\text{lim}} \quad *$$

$$\underset{\substack{1 \\ N}}{\text{lim}} \quad \frac{N}{} \quad 1 \quad 1 \quad \frac{1}{N^{2N}} \quad N^{N}$$

$$1 \quad \frac{1}{e^{1 \ 1/e}} \quad 0.745$$

Shown in Figs. 4 through 6 are comparisons of the theoretical and simulated values of throughput efficiency   (and $*$) vs. traffic load   for different switch sizes N. In each graph one set of curves represents the results obtained when the routing cache has only one next-hop interface per destination address, and the second set of curves represents the results obtained when the routing cache has both a primary and an alternate next-hop interface per destination address. We assume that the primary and alternate next-hop interfaces are chosen uniformly from among the available outlets. In each of the graphs the efficiency for dual-entry caches exceeds single-entry caches by a comfortable margin. The asymptotic efficiency for single-entry caches is 63%, while for dual-entry caches it is 74%. The real difference may be seen, however, in the degree of roll-off in the curves: for example, in Fig. 6 (16 interfaces) the efficiency drops below 95% at a load of 0.1 packets per timeslot per interface with a single-entry cache, but the efficiency with a dual-entry cache drops below 95% only after a load of 0.4 packets per timeslot per interface. Thus, for a given level of performance, a booster with a dual-entry cache can carry four times as much traffic as one with a single-entry cache. Given the modest cost of extending the cache to hold two entries, this is clearly a winning strategy to reduce outlet contention in the booster. One notes that the analytical model produces reasonably accurate results compared to the simulator, especially as the size of the switch grows.

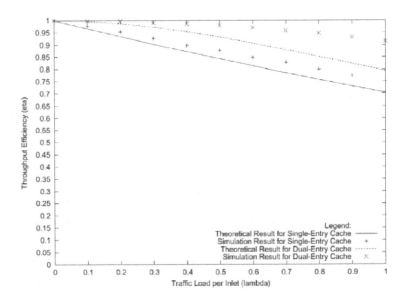

Fig. 4. Comparison of Theoretical and Simulation Results for a Booster with 3 Links.

5 Simulation Studies

Although the analytical results of the previous section are useful in characterizing the gross behavior of the optically boosted router, one cannot be sure how the system performs under real conditions unless a more-detailed and higher-fidelity model or prototype is developed and exercised. In this section the results from such a detailed simulation model driven by traffic traces collected from the Internet are presented.

The simulator models the base and boosting routers at the packet level. The packets are copied to both the electronic and optical subsystems, and lookups of their destination address are performed. If the routing cache misses, then the packet is deleted from the optical subsystem and the packet proceeds only through the base router. The base router normally queues the incoming packet until lookup in the full routing table is completed. The base router has a 10-megabyte packet buffer per interface. The routing cache can hold 100 addresses. The simulator uses 160 ps as the lookup time in the routing cache and 200 ns for the lookup time in the full routing table. Given today's technology, a lookup time of 200 ns is conservative for the base router, corresponding to a packet-forwarding rate of 5 million packets per second. In principle, such lookup times could be achieved by using a fast RAM large enough to contain all IPv4 address prefixes of 24 bits or fewer (16 megabytes of RAM with

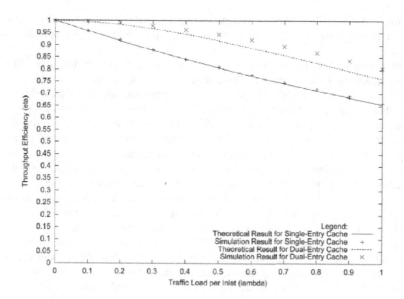

Fig. 5. Comparison of Theoretical and Simulation Results for a Booster with 8 Links

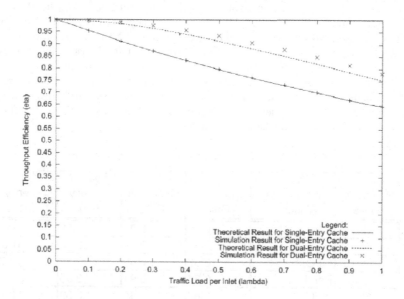

Fig. 6. Comparison of Theoretical and Simulation Results for a Booster with 16 Links

better than 200-ns cycle time) [13]. After address lookup the base router switches the packet to the appropriate output link; the simulation assigns a negligible delay to packet processing and switching. It is assumed that no slow-path options processing occurs. As described above, packets emerging from a booster outlet are granted higher priority over packets emerging from the base router's corresponding output link.

Providing real traffic to drive the simulator is not a trivial matter, given that there are no links today that operate at Tb/s speeds. Thus, all traces collected have inherent bit rates far less than 1 Tb/s. It is therefore necessary to scale the empirical traffic in a way that the apparent bit rate approaches 1 Tb/s. This is accomplished by upscaling times in the router rather than downscaling times in the packet stream. In this way, the relative timings of packets and router are such that the link utilization is moderate to high. The simulator does not implement a routing protocol. Instead it is assumed that all traffic is uniformly destined to one of the router's output links. Two different schemes for updating the routing cache are evaluated, first-in first-out (FIFO) and random replacement. In the FIFO scheme a missed entry is read from the full routing table into the routing cache such that the oldest entry in the cache is replaced. In the random scheme, the missed entry is placed into a random location of the cache. The packet traces are from the University of Auckland's Internet uplink (between New Zealand and the United States) and are provided by NLANR MOAT and the WAND research group [17]. Several traces, taken from November 1999 to June 2000, are used to feed the individual input ports of the router, and routing is assumed to be uniform among the N output ports (actually $N-1$ ports, since packets are never looped back).

The averaged results of four groups of simulation runs are displayed in Table 1. For each value of N (3, 8, and 16) and cache-replacement scheme (FIFO and random), the mean values of switching gain (defined as the fraction of traffic that passes completely through the booster), hit rate, and latency are measured. The latency measurements are for the overall packet delay (from first bit in to last bit out) through the entire router and for the packet delay through just the booster; the astonishing variance between them covers four orders of magnitude. The results further validate the design by demonstrating that high gain and hit rates are achieved under realistic traffic conditions.   The simulated switching gain is somewhat lower than the analytical results, because there is considerably more burstiness in the real traffic than in the artificial traffic of the theoretical model.

Table 1. Average Performance with Scaled Packet Traces

Number of Links	Cache Replacement Algorithm	Switching Gain	Cache Hit Rate	Booster Latency (ns)	Total Latency (us)
3	FIFO	0.89	0.95	3.23	41.02
3	Random	0.89	0.94	3.23	51.02
8	FIFO	0.87	0.95	3.34	39.53
8	Random	0.87	0.94	3.35	50.32
16	FIFO	0.87	0.95	3.39	44.96
16	Random	0.87	0.94	3.39	55.99

6 Conclusion

The optically boosted router is an approach to leveraging optics more effectively for packet switching. The simple ideas of a fast electronic cache and an all-optical path through the router come together to enable the implementation of an Internet router that could operate with high port counts and 1-Tb/s links, very low packet loss, and negligible latency. Analytical and detailed simulation models indicate that an optically boosted router under moderately heavy traffic can forward close to 90% of its packets through the all-optical switch path. The latency through the all-optical path is gauged in nanoseconds.

However, further studies of the optically boosted router need to be conducted, as important questions remain to be answered. The traffic model used is inadequate in two ways: (1) it must be scaled to mimic high data rates and (2) the routing is arbitrarily chosen to be uniform. The reason for the first shortcoming is that there appear to be no packet traces in the very high-speed regimes. The reason for the second shortcoming is that it is exceptionally difficult to collect packet traces along with associated routing-table information from commodity Internet providers. Our future approach will be to generate statistically aggregated packet traces that mimic real high-speed traffic and assign addresses that link back to actual routing tables.

The use of an electronic routing cache in the booster could eventually be replaced by an optical function to improve further the performance of the router. Although optical header recognition and switching are feasible [18], the capabilities of present implementations are limited. Similarly, dual-entry routing caches used for contention resolution might one day be supplanted by optical contention-resolution subsystems [19]. These technologies are maturing and might soon prove useful in the booster.

References

1. Y. Rekhter *et al.*, "Cisco Systems' Tag Switching Architecture Overview," IETF RFC 2105, Feb. 1997.
2. P. Newman *et al.*, "IP Switching — ATM Under IP," *IEEE/ACM Trans. Networking*, vol. 6, no. 2, pp. 117–129, Apr. 1998.
3. D. Blumenthal *et al.*, "WDM Optical IP Tag Switching with Packet-Rate Wavelength Conversion and Subcarrier Multiplexed Addressing," *Proc. OFC '99*, San Diego, pp. 162–164, Feb. 1999.
4. J. Bannister, J. Touch, A. Willner, and S. Suryaputra, "How Many Wavelengths Do We Really Need? A Study of the Performance Limits of Packet Over Wavelengths," *SPIE/Baltzer Optical Networks*, vol. 1, num. 2, pp. 1–12, Feb. 2000.
5. E. Rosen, A. Viswanathan, and R. Callon, "Multiprotocol Label Switching Architecture," IETF RFC 3031, Jan. 2001.
6. D. Awduche, Y. Rekhter, J. Drake, and R. Coltun, "Multi-Protocol Lambda Switching: Combining MPLS Traffic Engineering Control with Optical Crossconnects," IETF Internet Draft, Mar. 2001.
7. S. Suryaputra, J. Touch, and J. Bannister "Simple Wavelength Assignment Protocol," *Proc. SPIE Photonics East 2000*, vol. 4213, pp. 220–233, Boston, Nov. 2000.

8. J. Turner, "Terabit Burst Switching," Tech. Rep. WUCS-9817, Washington Univ., Dept. of Comp. Sci., July 1998.
9. C. Qiao and M. Yoo, "Optical Burst Switching — A New Paradigm for an Optical Internet," *J. High Speed Networks*, vol. 8, no. 1, pp. 69–84, 1999.
10. "AS1221 BGP Table Data," http://www.telstra.net/ops/bgp/bgp-active.html, Aug. 2000.
11. M. Degernark *et al.*, "Small Forwarding Tables for Fast Routing Lookups," *Proc. ACM Sigcomm '97*, pp. 3–14, Cannes, Sept. 1997.
12. M. Waldvogel *et al.*, "Scalable High Speed IP Lookups," *Proc. ACM Sigcomm '97*, pp. 25–36, Cannes, Sept. 1997.
13. P. Gupta, S. Lin, and N. McKeown, "Routing Lookups in Hardware at Memory Access Speeds," *Proc. IEEE INFOCOM '98*, pp. 1240–1247, San Francisco, Apr. 1998.
14. B. Lampson, V. Srinivasan, and G. Varghese, "IP Lookup Using Multiway and Multicolumn Binary Search," *Proc. IEEE INFOCOM '98*, pp. 1248–1256, San Francisco, Apr. 1998.
15. B. Talbot, T. Sherwood, and B. Lin, "IP Caching for Terabit Speed Routers," *Proc. IEEE GlobeCom '99*, pp. 1565–1569, Rio de Janeiro, Dec. 1999.
16. M. Karol, M. Hluchyj, and S. Morgan, "Input Versus Output Queueing on a Space-Division Packet Switch," *IEEE Trans. Commun.*, vol. COM-35, no. 12, pp. 1347–1356, Dec. 1987.
17. "Auckland Internet Packet Traces," http://moat.nlanr.net/Traces/Kiwitraces/auck2.html, Jan. 2001.
18. M. Cardakli *et al.*, "All-Optical Packet Header Recognition and Switching in a Reconfigurable Network Using Fiber Bragg Gratings for Time-to-Wavelength Mapping and Decoding," *Proc. OFC '99*, San Diego, Feb. 1999.
19. W. Shieh, E. Park, and A. Willner, "Demonstration of Output-Port Contention Resolution in a WDM Switching Node Based on All-Optical Wavelength Shifting and Subcarrier-Multiplexed Routing Control Headers," *IEEE Photonics Tech. Lett.*, vol. 9, pp. 1023–1025, 1997.

Intelligent Traffic Engineering of Internets: Towards a Model-Based Approach

Anurag Kumar

Dept. of Electrical Communication Engg.,
Indian Institute of Science, Bangalore 560 012, INDIA
anurag@ece.iisc.ernet.in

Abstract. Traffic engineering can be viewed as the mechanisms that are used to reoptimise network performance on traffic variation time scales. The traffic engineering process involves measurements, detection of performance problems, and the adjustment of routes and various quality of service mechanisms in order to restore the desired performance. Traffic engineering models are essential in this feedback loop. In this paper we provide an overview of our recent contributions to analytical models for traffic engineering of elastic traffic in internets, and discuss the role that such models could play in automated or intelligent traffic engineering.

1 Introduction

At the present time, the predominant use of the Internet is by so-called "elastic" traffic which is generated by applications whose basic objective is to move chunks of data between the disks of two computers connected to the network. In terms of the volume of data carried, web browsing, email, and file transfers are the main elastic applications. Elastic flows can be speeded up or slowed down depending on the availability of bandwidth in the network. Even for these non-real-time applications, however, the performance of the best-effort Internet is extremely sensitive to traffic conditions, often providing just a trickle of throughput, a source of much frustration among web users. In addition, indications are that the growth of elastic traffic in the Internet is very rapid (various reports have shown growth rates of at least 100% per year; see [15]). If the simple best-effort delivery paradigm is to be retained, the only remedy appears to be to continually increase the bandwidths of the links, such that the links operate at a low utilisation; i.e., overprovisioning. Bandwidth is expensive, however, and even though large amounts of it have been made available by fibre optics and high-speed modem technology, huge investments are required. Further, as Internet applications evolve they will require more and more bandwidth. In fact, it can be expected that it is from novel elastic applications that Internet users will have increasing expectations: e.g., diskless nomadic computing (see [17]), remote medical radiodiagnosis, movie editing over a wide-area internet, etc.

If we expect that, in most situations, overprovisioning and rapid deployment of excess bandwidth may not be a viable alternative, there will be a need

B. Monien, V.K. Prasanna, S. Vajapeyam (Eds.): HiPC 2001, LNCS 2228, pp. 414–424, 2001.

to manage the bandwidth in internets. At the present time, network operators make traffic engineering decisions in a fairly ad hoc manner, relying mainly on extensive experience to predict the possible consequences of a traffic engineering decision. In the recent years, however, considerable research has been done on performance models of internets, and it is conceivable that eventually these models will become powerful enough to be deployed in decision support tools, or even automatic control devices for traffic engineering. In this paper we outline a view of traffic engineering in terms of time scales of resource allocation controls in packet networks. Then we review some of our own work on developing traffic engineering models for TCP traffic. Finally, we provide a suggestion as to how such models may be used in the traffic engineering process, or even in an automatic network management device.

2   Control Time Scales and Traffic Engineering

A very important part of networking technology are the controls that coordinate the allocation of the network resources among the various competing traffic flows. There is a variety of such controls, and these are best understood in terms of the time scales over which they act. There are the following four time scales.

1. Packet time scale
2. Session or flow time scale
3. Busy hour (or traffic variation) time scale
4. Provisioning time scale

Packet time scale controls discriminate between the treatment that the network provides to individual packets. These controls include mechanisms such as priority queueing, fair scheduling, adaptive window based congestion control (as in TCP), and random packet dropping (for TCP performance optimisation).

When a session or flow (e.g., a packet phone call) arrives to the network, several decisions can be made. Should this call be accepted or not? The call may not be accepted if the network is unable to offer it the required quality of service (QoS). If the call is accepted the route should be such that it has enough resources to handle the call. Such decisions are made only when flows or calls arrive and hence can be viewed as operating over flow arrival and completion time scales, which are certainly longer than packet time scales.

The way that packet level and flow level controls must operate depends on the ambient traffic conditions. In telecommunication networks traffic is not stationary, but varies continually over the hours of the day. In most places there are typical busy hour patterns during which the traffic is at its peak; at other times the traffic is less. Given a particular pattern of traffic, the packet and flow level controls can be configured for optimal network performance. If the traffic changes, however, then a different set of configurations for these controls may be more efficient. Hence at traffic variation time scales it becomes necessary to adjust the packet and flow level controls in order to optimise network

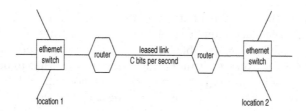

Fig. 1. A simple enterprise network connecting two locations.

performance. For example, the network routing may need to be adjusted to accommodate varying patterns of traffic at different times of the day. Thus it is at these time scales that MPLS (Multiprotocol Label Switched) paths are set up or torn down, or OSPF (Open Shortest Path First) weights are adjusted, in order to optimise the network performance. *Traffic Engineering can be thought of as all the mechanisms, manual or automatic, that work at this time scale* (see also [8]).

Provisioning is the longest time scale control. It entails increasing the bandwidth on some links, or adding new links and switches. Reprovisioning could be triggered when attempts to reroute traffic, or reconfigure the other traffic controls fail to achieve the desired results, and this happens frequently over several traffic engineering time scales. Observe that the overprovisioning approach, mentioned in the Introduction, basically uses only this control, keeping the packet and flow level controls very simple; with such an approach, traffic engineering would essentially involve ensuring that the routing plan is such that the utilisation of network links is uniformly small. Modern networks are being built on optical fibre links, with intelligent optical cross-connects. Such an infrastructure can be used to rapidly provision additional bandwidth, with little or no manual intervention.

3   Traffic Engineering Models

3.1   A Simple Network: Models with a Single Bottleneck Link

Consider the very simple network shown in Figure 1. This could represent the interlocation "network" of a small enterprise with two locations, one being a data center (say, Location 1) and the other a design center or call center (say, Location 2). Engineers or operators sitting at Location 2 need to access the data at Location 1. Now a problem faced by the network administrator is to size the leased link, so as to provide the users with some throughput performance. A more sophisticated problem would be to set up the system so that a more important category of users at Location 2 obtain a better quality of service. In addition to link sizing, queue management mechanisms in the routers will need to be configured; if differential services are required then a differential random drop based mechanism would need to be set up. Can performance modelling provide some guidance in such questions?

Let λ denote the total rate of arrival of requests for transfers from Location 1 to Location 2. Each such request is for the transfer of a file of random size. We denote the generic random variable for file size by V, with expectation EV bits. Hence $\lambda \times EV$ is the average bit rate being requested. This has to be transported on a link of rate C bits/sec. Define $\rho = \frac{\lambda EV}{C}$, the normalised load on the link. Clearly it is necessary that $\rho \leq 1$, and it may be expected that as ρ approaches 1 the average throughput would degrade. One would like a model that would fairly accurately provide an estimate of the average throughput as a function of the offered (normalised) load ρ. Such a model would need to capture the following.

1. The working of TCP's bandwidth sharing mechanism; i.e., adaptive window based packet transmission.
2. The pattern of arrival of downloads at the data center; i.e., a model of the arrival process (of rate λ) is needed.
3. The distribution of the file transfer sizes. Typically, each request may involve the transfer of several objects (e.g., web pages usually have a base document, in which are embedded several image objects), and the particular implementation of the application would decide in what order these transfers are made, using how many TCP connections, and in what manner (persistent or nonpersistent, pipelined or sequential).

In the recent past, several researchers have worked with the following traffic model. The transfer request arrival instants are assumed to constitute a Poisson process. The Poisson arrival process renders the mathematical models analytically tractable. A better model would be to take the arrival instants to constitute a Poisson process, but to model each arrival as bringing a batch of file transfers. The transfer sizes $V_1, V_2, \cdots$, are assumed to be independent and identically distributed (i.i.d). It is recognised that the distribution of V has to be heavy tailed, usually modeled as Pareto with a finite mean and infinite second moment. For analytical tractability, and in order to perform simulations with a finite variance distribution, the heavy tailed distribution is approximated with a mixture of several exponentials (or hyperexponential; see [7]). With the Poisson arrivals, and i.i.d. transfer volumes assumptions, several papers are now available (see [13,16, 18,10,3,11,4,9]); [9] permits a more complex session model, with which one can model "think" times.

Figure 2 shows the average bandwidth shares, normalised to the link capacity, obtained by randomly arriving file transfers from Location 1 to Location 2 in the network of Figure 1, plotted versus the normalised offered load ρ. Results for three sets of network parameters have been shown. The fixed round-trip propagation delays are 0.2ms, 3ms, and 40ms, corresponding to 20km, 300km, and 4000km between the locations, or in other words the spread of the network is metropolitan, statewide, or nationwide, respectively. The link speed in each case is 10Mbps, the mean transfer volume is 30Kbytes, the file size distribution is a mixture of three exponential distributions (hyperexponential) matched to a Pareto distribution with tail decay parameter 1.6, the link buffer size is 100 TCP packets, and buffer congestion results in arriving packets being dropped ("tail-drop"). Several important observations can be made from these plots: (i)

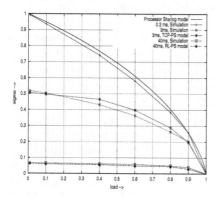

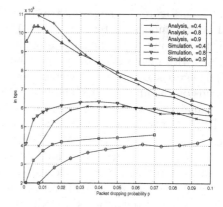

Fig. 2. Average bandwidth shares (normalised to link capacity) obtained by randomly arriving transfers from Location 1 to Location 2 in Figure 1, with tail-drop queue management, plotted against link load.

Fig. 3. Average flow throughputs (τ, in bps) obtained by randomly arriving transfers from Location 1 to Location 2 in Figure 1, with random drop queue management, plotted against the drop probability, for various link loads.

As the normalised load approaches 1, as expected, the flow throughputs reduce to zero; (ii) As the network spreads over a larger distance, for a fixed load, the throughput reduces. This happens because TCP's adaptive window mechanism is driven by acknowledgements (acks), or lack of them. Over a large network, acks take longer to return, resulting in a slower build-up of the window. When a buffer loss takes place, it takes longer to reestablish a window large enough to fill up the round-trip pipe.

For each set of parameters, Figure 2 shows two plots, one obtained from the analysis of a model, and the other from simulations. Notice that the analysis matches the simulation quite well; we have found that three different analytical models are needed to closely match the results for the three cases shown [12]. Such an analysis can be used by the network administrator to size the link bandwidth depending on the network parameters.

The results in Figure 2 are for tail drop queue management at the router. It is well known that TCP performance can be optimised by deliberately (randomly) dropping TCP packets at the router. This technique eliminates the detrimental effects of loss synchronisation that sometimes results from tail-drop, and can also be used to keep the router queue within reasonable limits without causing link starvation. The throughput of small file transfers depends mainly on the round trip delay in their path, and hence a small router queue will help improve the throughput of short files. On the other hand, excessive packet drops would result in the longer TCP transfers backing off excessively, resulting in link starvation. Thus it would seem that when a link is carrying a mix of short and long transfers the random drop probability should be tunable to optimise the average throughput performance. This is indeed the case, as is shown in Figure 3. The network is as in Figure 1. The round-trip propagation delay is 8.83ms, the link speed is 10Mbps, and the router at Location 1 drops packets from the ongoing TCP flows with probability p. The file size model is a mixture of two exponential distributions with means of 2Kbytes and 600Kbytes, and an overall mean of 25Kbytes. The figure shows a plot of average flow throughput versus p, for three normalised loads on the link, 0.4, 0.8, and 0.9. For each case, two plots are shown, one from an analytical model, and one from a simulation; the analysis is seen to capture the performance quite well, except for the high load case ([6]). Notice that, for each load, as p increases from 0, first the average throughput increases (owing to the queue length decreasing and the small file transfer throughput improving), and, after reaching a peak, the average throughput decreases for further increase in p. If the network administrator chooses to use random drop, then given the parameters of the network, the above analysis can be used to obtain the optimal value of p to operate at. Notice that as the load ρ increases the optimal p increases, and, as expected, the achievable average throughput decreases. We have also found that as the propagation delay increases, for a given ρ, the optimising p decreases.

3.2   Models for General Topologies

In Section 3.1 we considered an aggregate of randomly arriving TCP transfer sharing a single link, and showed that it is possible to come up with fairly accurate analytical estimates of bandwidth sharing performance. Such an analysis is immediately usable in a network in which the performance of each aggregate is limited by a single bottleneck link. Consider an enterprise that has a very high speed backbone, or that is interconnecting its branches over some service provider's high-speed backbone. Suppose that the branch offices are connected to the backbone by relatively low speed links (e.g., 10Mbps), and the primary traffic being generated by the branch offices is to access data at the enterprise's data-center, that is itself attached to the backbone by a high speed access link. Clearly, in this scenario, the network model shown in Figure 1 will characterise the performance between a branch location (Location 2) and the data-center (Location 1).

There is very little published work available for the situations in which the network topology is a general mesh, there are several TCP flow aggregates on arbitrary routes, and there are multiple bottlenecks. In [2] the authors present a computation technique for obtaining the bandwidth shares of long-lived TCP flows in an arbitrary network, with multiple bottlenecks. In practice, however, elastic transfers are finite, and hence short lived. In our work, we have studied the possibility of applying single link models to develop an analysis for arbitrary networks.

To this end we have explored the usefulness of this idea for computing bandwidth shares in a network with idealised max-min fair bandwidth sharing. The model is as defined in [18]. The network topology is an arbitrary mesh. There is a number of flow aggregates, on fixed routes, in which flows arrive in independent Poisson processes. For a given vector of active flows at time t, the network

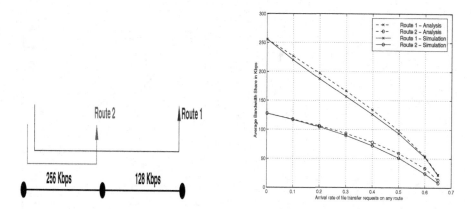

Fig. 4. Average bandwidth share vs. arrival rate; IMMF bandwidth sharing.

bandwidth is shared in an *instantaneous* max-min fair (IMMF) fashion; i.e., as the number of flows changes, the bandwidth shares adjust instantaneously so that max-min fair sharing prevails at all t. This can be expected to be a good approximation to the actual network behaviour when the delay-bandwidth product is small relative to the transfer volumes. Further, TCP can be viewed as a whole class of adaptive window based algorithms, and the bandwidth sharing that is achieved depends on the network feedback and source adaptation. With this point of view, it can be argued that with appropriate feedback from routers TCP can actually achieve max-min fairness (see [14], and [5]). For IMMF we have studied a heuristic based on an exact analysis of several flow aggregates sharing a single link. The heuristic works by considering all the network links, and the flows in them, in isolation. The bandwidth shares of the flows in each flow aggregate on each (isolated) link are calculated. The aggregate that gets the minimum bandwidth share over all the links is identified. It is assigned this minimum bandwidth share, and is then removed from the network; the link capacities are reduced by the aggregate's bit rate. The procedure is then repeated until all the flow aggregates are accounted for. Details are provided in [5].

Figures 4 and 5 show the results from the above procedure for two topologies. The file sizes are distributed as a mixture of 7 exponentials matched to a Pareto distribution. In Figure 4 we have a simple two link, low speed network, shared by two TCP aggregates. The bandwidth shares obtained by flows in each flow aggregate are shown in the right half of the figure, plotted against the flow arrival rate (the same for each aggregate). Notice that the analytical results match well with the simulation results, for both the routes. As would be expected, there is a an arrival rate at which the bandwidth shares go to zero. In Figure 5 we have a more complex network with 9 links and 18 routes. The right side of the figure shows the flow throughputs for routes 1, 2, 3, and 4, all single link routes. Results for other routes can be found in [5]. The analysis approach continues to do quite well even for this more complex network.

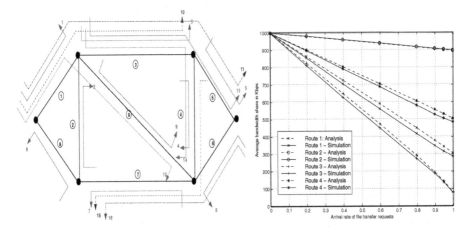

Fig. 5. Average bandwidth share vs. arrival rate for four routes in the mesh network shown; IMMF bandwidth sharing; 1 Mbps link speeds.

The performance of a TCP flow is sensitive to the drop probability along its path, and the round trip delay in the path. For small probabilities, the drop probability for a flow aggregate is the sum of the drop probabilities along its path. If the network uses the Differentiated Services approach the packets of the flows are coloured (green, yellow, or red) at the network edge. Each router through which a flow aggregate passes would have a drop probability for each colour. Thus, at a router, the drop probability for a flow aggregate depends on the fraction of packets of each colour in it. In order to apply the above analysis approach to a network of random drop routers it is necessary to develop an analysis of finite volume TCP flows sharing a single link, with each flow aggregate having its own drop probability and its own round trip delay. We have developed an analysis for such a model (in [6]). Figure 6 shows the results for an example. A 10Mbps link is being shared by two TCP aggregates, with fixed round-trip delays of 40.832ms and 60.832ms, with aggregate rates 4.5Mbps and 3.5Mbps, and with drop probabilities p_1 and $p_1 + 0.01$. This situation might arise if aggregate 2 traverses another link with propagation delay 20ms, and with drop probability 0.01. The average flow throughput for each class is plotted against p_1; analysis and simulation results are shown. Again it is seen that there is an optimal value of p_1. The analysis captures the performance quite well. We are in the process of attempting to use this approach to develop an analysis for a general network.

4   Towards Intelligent Traffic Engineering

An internet backbone operator is typically faced with the problem of carrying several aggregates of finite volume TCP transfers between its various attachment points. The problem faced by the operator is to configure the QoS mechanisms in the network, and to route the various flow aggregates so as to assure some

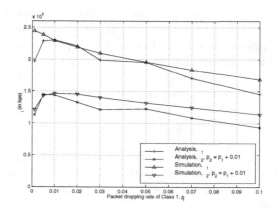

Fig. 6. Flow throughputs (τ_i) for two TCP aggregates at a bottleneck link. Arbitrary p_i and arbitrary d_i; $d_1 = 40.832$ ms, $d_2 = d_1 + 20$ ms, $r_1 = 4.5$ Mbps and $r_2 = 3.5$ Mbps; link speed 10Mbps.

level of performance to the flows in each aggregate. It is clear that if several flow aggregates share a link, and if the total bit rate from these is close to the capacity of the link, then the throughputs of the flows would become unacceptably small (see the results in Section 3.1). Such a situation would arise if all flows are allowed to follow the default OSPF routes. Since OSPF routing basically works on the administratively configured link metrics, it is unaware of link congestion, and will route all traffic between a pair of points (in a domain) on the route with the smallest total metric (see Figure 7). Measurements would then reveal that some link is getting congested, and the operator would need to move some traffic off the congested route. An MPLS LSP (Label Switched Path) can then be set up between the two end-point routers to carry the diverted traffic. In

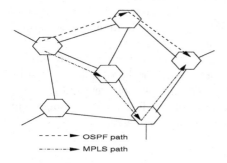

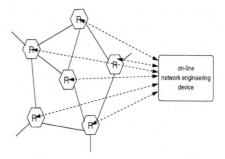

Fig. 7. An MPLS path can be set up to divert traffic from a congested OSPF (shortest) path.

Fig. 8. An on-line network engineering device receives measurements from the network, and makes or suggests traffic engineering decisions.

this process, however, some steps would be involved. First, it will need to be determined which traffic aggregates are sharing the congested link, the total bit rate each such aggregate is offering, and what the actual performance of the flows in the aggregates is. This process will require inferences to be made from the routing information bases in the routers (see [8]). Second, the operator will need to determine the consequence of moving a particular aggregate from the existing OSPF route to a proposed MPLS LSP. A "what-if" question will need to be asked: which aggregate should be moved, and to which alternate route. In addition, quality of service mechanisms may need to be reconfigured to achieve desired performance with the new routing. This may involve adjusting the packet colouring devices at the network edge, the queues, their WFQ weights, and random drop parameters at the network links. It is in this second process that traffic engineering models of the kind that we have been discussing in this paper can be useful.

In fact, it can be envisaged that with sophisticated enough models it may be possible to devise an automatic network engineering system (see Figure 8). Such a system will receive measurements from the routers, and perhaps other measurement probes. It will identify congestion points, and determine new MPLS paths to be created. It can then instruct the routers to set up the new MPLS paths, and configure the forwarding rules (see [1] for an example of such a system). It should even be able to calculate the new parameters for the QoS mechanisms, if necessary, and configure the routers with these new parameters. On a longer time scale, several hours or days, (after too many MPLS paths have been set up) the device could even suggest new OSPF metrics so that OSPF routes could then optimally carry all the traffic with the required QoS.

5   Conclusion

We have argued that unless a network can be overprovisioned, traffic engineering mechanisms need to work at the traffic variation time scale to maintain the quality of service provided to the flows using the network. Such traffic engineering would require analytical models so that "what-if" questions can be answered in the traffic engineering control loop. We have outlined our own contributions to analytical models for the evaluation of the throughput performance of aggregates of TCP controlled elastic file transfers (such as web transfers).

At present the state of the art of such modelling has not progressed to a point such that entire internets, with all their routing and traffic complexity, can be modelled accurately. The results obtained are promising, however, and it can be envisaged that such models can actually be deployed in network management devices for making automatic and intelligent traffic engineering decisions.

References

1. P. Aukia, M. Kodialam, P.V.N. Koppol, T.V. Lakshman, H. Sarin, B. Suter, "RATES: A server for MPLS traffic engineering," *IEEE Network*, Vol. 14, No. 2, pp 34-41, March/April 2000.

2. Lotfi M. Benmohamed, S. Dravida, P. Harshavardhana, W.C. Lau, and A. Mittal, "Designing IP networks with performance guarantees," *Bell Labs Technical Journal*, Vol. 3, No. 4, pp 273-296, Oct-Dec 1998.
3. Thomas Bonald and Laurent Massoulie, "Impact of fairness on internet performance," *Proc. of ACM SIGMETRICS, 2001*, 2001.
4. Tian Bu and Don Towsley, "Fixed point approximations for TCP behavior in an AQM network," *Proc of ACM SIGMETRICS, 2001*, 2001.
5. Pinaki Shankar Chanda, Anurag Kumar, and Arzad Alam Kherani, "An approximate calculation of max-min fair throughputs for non-persistent elastic sessions," In *IEEE Globecom 2001, Internet Performance Symposium.*
6. Ranjay John Cherian and Anurag Kumar, "An engineering model for assured differentiated services to aggregates of non-persistent TCP sessions," submitted, 2001.
7. Anja Feldmann and Ward Whitt, "Fitting mixtures of exponentials to long-tail distributions to analyze network performance models," *Proceedings IEEE INFOCOM 1997*, April 1997.
8. Anja Feldmann, Albert Greenberg, Carsten Lund, Nick Reingold, Jennifer Rexford, "NetScope: traffic engineering for IP networks," *IEEE Network*, Vol. 14, No. 2, pp 11-19, March/April 2000.
9. S. Ben Fredj, T. Bonald, A. Proutiere, G. Regnie, and J. W. Roberts, "Statistical bandwidth sharing: a study of congestion at flow level," in *Proc. ACM SIGCOMM 2001*, 2001.
10. Arzad A. Kherani, Anurag Kumar, Pinaki S. Chanda and Vijay H. Sankar, "On throughput of nonpersistent elastic sessions on an internet link with small propagation delay," Manuscript, available at http://ece.iisc.ernet.in/~ anurag/pubm.html, 2000
11. Arzad A. Kherani and Anurag Kumar, "On processor sharing as a model for TCP controlled elastic sessions," Manuscript, available at http://ece.iisc.ernet.in/~ anurag/pubm.html, 2001
12. Arzad A. Kherani and Anurag Kumar, "Stochastic models for throughput analysis of randomly arriving elastic flows in the Internet," submitted, 2001.
13. T.V.Lakshman, D.P.Heyman and A.L.Neidhardt, "New method for analyzing feedback protocols with applications to engineering web traffic over the Internet," *Proceedings of ACM SIGMETRICS*, 1997.
14. Jeonghoon Mo and Jean Walrand, "Fair end-to-end window-based congestion control," *IEEE/ACM Transactions on Networking*, vol. 8, no. 5, pp. 556–567, October 2000.
15. A. Odlyzko, "The current state and likely evolution of the Internet," *IEEE Globecom'99*, 1999.
16. J.W. Roberts and L. Massoulie, "Bandwidth sharing and admission control for elastic traffic," in *ITC Specialist Seminar*, October 1998.
17. J. Satran et al., "iSCSI," *IETF draft-ietf-ips-iscsi-02.html.*
18. Gustavo De Veciana, Tae-Jin Lee and Takis Konstantopoulos, "Stability and performance analysis of networks supporting services with rate control-could the Internet be unstable," *Proc. IEEE INFOCOM'1999*, April 1999.

Performance Analysis of Data Services over GPRS*

Marco Ajmone Marsan[1], Marco Gribaudo[2], Michela Meo[1], and Matteo Sereno[2]

[1] Dipartimento di Elettronica, Politecnico di Torino, 10129 Torino, Italy
{ajmone,michela}@polito.it
[2] Dipartimento di Informatica, Università di Torino, 10149 Torino, Italy
{marcog,matteo}@di.unito.it

Abstract. In this paper we describe a model for the performance analysis of mobile packet data services sharing a common GSM/GPRS (Global System for Mobile communications/General Packet Radio Service) cellular infrastructure with mobile telephony services. The performance model is developed using the DSPN (Deterministic and Stochastic Petri Net) paradigm. With limited computational cost, the DSPN solution allows the derivation of a number of interesting performance metrics, which can be instrumental for the development of accurate design and planning algorithms for cellular networks offering integrated services.

1 Introduction

Seven hundred millions mobile telephony service users are being offered (or will soon be offered) mobile packet data services for a variety of different applications, the most relevant of which is the mobile, ubiquitous access to the Internet.

In GSM (Global System for Mobile communications) systems [1,2], the technology that has been standardized for the packet transfer of data from/to mobile terminals is GPRS (General Packet Radio Service) [3,4,5,6,7], which is considered the forerunner of the Universal Mobile Telecommunication System (UMTS) [8,9], for whose licenses mobile telephony companies in Europe have made huge investments.

The fact that GPRS exploits the same resources used by mobile telephony raises a number of questions concerning the dimensioning and the management of the GSM/GPRS radio interface. In this paper we tackle this issue, and we develop a Petri net based model of the behavior of the radio interface in a GSM/GPRS cell, providing numerical performance results that give insight into the possible design tradeoffs.

The particular class of Petri nets used for the model construction is called DSPN (Deterministic and Stochastic Petri Net) [10,11]. These are Petri nets with timed transitions, where the transition firing delays can be either deterministic or exponentially distributed. While no restriction exists in the use of

---

* This work was supported in part by the Italian National Research Council, and by CERCOM.

B. Monien, V.K. Prasanna, S. Vajapeyam (Eds.): HiPC 2001, LNCS 2228, pp. 425–435, 2001.
© Springer-Verlag Berlin Heidelberg 2001

exponentially distributed transition firing delays, constant firing delays must be used for transitions that are enabled one at a time. In our case this is not a problem, since just one constant delay will be used to model the elementary time slot in our model.

The paper is organized as follows. In Section 2 we describe the characteristics of the GSM/GPRS cellular mobile communication network that we consider, together with the probabilistic assumptions that are needed to describe the system dynamics with a DSPN model. In Section 3 we describe the characteristics of the DSPN model and its solution. Results are shown and discussed in Section 4. Section 5 briefly comments on the complexity of the DSPN model solution. Section 6 concludes the paper.

2   System and Modeling Assumptions

GSM adopts a *cellular architecture*: the area served by a GSM network is covered by a number of non-overlapping *cells*. Each cell is controlled by a *base station* (called BTS – Base Transmitter Station, in the GSM jargon), which handles the transmissions from and to mobile terminals over the wireless link. Base stations are then connected through a (wired or wireless) point-to-point link to the fixed part of a GSM network.

A small number of frequency bands (resulting from a partition of the portion of spectrum allocated by international organizations to GSM services, around 900 MHz and 1.8 GHz), each one of 200 kHz, is activated in any cell (typical numbers are a few units). Each frequency band defines a FDMA *channel*, on which a TDMA *frame* of 60/13 ms is defined, comprising 8 *slots* of 15/26 ms. A circuit (or channel) is defined by a slot position in the frame, and by a frequency band; thus, the maximum number of telephone calls that can be simultaneously activated in a cell corresponds to 8 times the number of frequency bands. At least one channel must be allocated for signalling in each cell, so that a frequency band can use for the transmission of end user information (either voice or data) from 6 to 8 channels, depending on the cell configuration; we will assume that in cells with one active frequency band, the TDMA frame allocates 7 slots to end users and 1 slot to signalling. The pair (frequency,slot) defines a *control channel* if the channel is used for signalling, and defines a *traffic channel* if the channel is used for the transmission of end user information (either voice or data).

When an end user requests the activation of a new GSM telephone call, the system must transfer some signalling information to inform all the involved entities about the request and to distribute the necessary commands, and allocate a traffic channel on one of the frequency bands of the cell. If no slot is free among those that can be allocated to end-users, the new call cannot be established, and is *blocked*.

Similarly, when users move from one cell to another (we call the former origin cell, and the latter destination cell) during a telephone call, the system must transfer the call from the origin cell to the destination cell through a *handover* procedure. If no traffic channel can be allocated in the destination cell, the call handover fails (a *handover blocking* occurs) and the telephone call is forced to terminate.

As customary in models of telephone systems, we shall assume that the sequence of new call requests follows a Poisson process, and that the duration of calls is an exponentially distributed random variable. Moreover, as is normally done when modeling cellular telephony systems, we consider one cell at a time [12], we neglect the impact of signalling, and we include incoming handover requests into the Poisson arrival process, whose rate will be set to λ_v (thus, λ_v is the total rate of requests incoming to a cell, and each request should be satisfied with one of the slots on the frequency bands that are active within the cell, i.e., with a traffic channel). The total active time of a call within a cell (which is normally called the call *dwell time*) differs from the call duration, because a call may traverse several cells during its active time; we extend the exponential assumption to dwell times also, defining a rate μ_v for the time from the beginning of the life of a call in a cell (resulting from either the successful activation of a new call or a successful incoming handover) until the end of the call activity in the same cell (due to either the termination of the call by one of the end users or an outgoing handover). Note that exponential assumptions are generally considered not to be critical in telephony models: telephone systems have been dimensioned using exponential assumptions for almost a century. In the derivation of numerical results we shall assume $1/\mu_v = 120$ s. This value is quite often used in design and planning of mobile telephony systems; it is somewhat shorter than the traditional average call holding time of fixed telephony (180 s), in order to account for handovers.

GPRS was conceived for the transfer of packets over a GSM infrastructure, with a simplified allocation of resources over the wireless link, and an IP transport among additional elements of the wired GSM network, called SGSNs (Serving GPRS Support Nodes). In order to cross the wireless link, IP packets are fragmented in *radio blocks*, that are transmitted in 4 slots in identical positions within consecutive GSM frames of the same frequency band. Depending on the length of the IP packet, the number of radio blocks necessary for the transfer may vary. The allocation of the radio link to radio block transmissions can either use dedicated resources for signalling, or (more usually) the same signalling resources that are available for telephony.

In order to describe the GPRS traffic, we adopted the model of Internet traffic defined by the 3GPP (3rd Generation Partnership Project) in [13]; a sketch of the GPRS traffic model that we use is shown in Figure 1. Active users within a cell execute a *packet session*, which is an alternating sequence of *packet calls* and *reading times*. According to [13], the number of packet calls within a packet session should be a geometrically distributed random variable with average Npc; however, since we will study the system behavior for a fixed number of concurrently active packet sessions, we will assume that packet sessions remain active for an indefinite amount of time. The reading time between packet calls is an exponentially distributed random variable with rate μ_{Dpc}. Each packet call comprises a geometrically distributed number of packets with average Nd; the

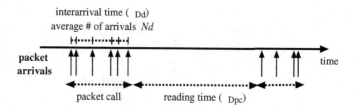

interarrival time (μ_{Dd})
average # of arrivals Nd

packet
arrivals

time

packet call   reading time (μ_{Dpc})

Fig. 1. Model of GPRS traffic: a packet session.

interarrival time between packets in a packet call is an exponentially distributed random variable with rate μ_{Dd}. According to [13], we shall assume $1/\mu_{Dpc} = 41.2$, $Nd = 25$ and $1/\mu_{Dd} = 0.5$, all times being expressed in seconds. The packet size in radio blocks can have a number of different distributions, some with heavy tail. In our model the packet size is equal to either 1 or 16 radio blocks, with the same probability. Other discrete distributions could be easily introduced in our models in order to approximate heavy tailed distributions of the packet size.

The transfer of radio blocks over the radio channel can either be successful, thus allowing the removal of the radio block from the buffer, or result in a failure due to noise, fading, or shadowing. These probabilities are modeled with a random choice: with probability c a radio block transfer is successful and with probability $1 - c$ it fails.

3   The DSPN Model

In this section we describe the DSPN model of one cell of a GSM system where just one frequency band is available to simultaneously offer telephony and GPRS services. We focus on transmissions from the base station to end user terminals (this is the so-called *downlink* direction).

The DSPN model is shown in Fig. 2. The only deterministic transition in the model is T_{4F}, whose delay corresponds to four consecutive GSM frames, i.e., $4 \cdot 60/13$ ms. The beginning and end of each quadruple of frames is modeled by the firing of immediate transitions t_b and t_e, respectively. Thus, the marking of place BEG-4F indicates that a quadruple of frames is starting, the marking of place END-4F indicates that a quadruple of frames is ending, and the marking of place 4F indicates that a quadruple of frames is in progress. One token is in place 4F in the initial marking.

We assume that a traffic channel can be allocated to new telephone call requests and incoming handover requests only at boundaries of quadruples of frames. Similarly, the scheduling of transmissions of radio blocks on traffic channels not used by voice can take place only at boundaries of quadruples of frames. Both assumptions appear to be quite realistic, at least for the first implementations of GPRS services.

At each firing of transition T_{4F}, one radio block can be transferred on each traffic channel that is not used by voice connections.

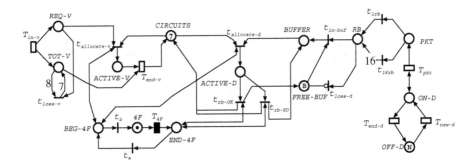

Fig. 2. The DSPN model.

The number of tokens in place CIRCUITS in the initial marking of the DSPN equals the number of traffic channels available in the cell (i.e., 7, according to our assumptions). Voice calls that await the allocation of a traffic channel are modeled by tokens in place REQ-V. Voice calls that are in progress are modeled by tokens in place ACTIVE-V. The total number of voice calls (that either await the allocation of a traffic channel, or are using a traffic channel) in the cell is modeled by tokens in place TOT-V.

Arrivals of new voice call requests and incoming voice call handover requests into the cell are modeled by the timed transition T_{in-v}, which must thus have rate λ_v. The arrival of a voice call increases the marking of places REQ-V and TOT-V; then, since no more than 7 calls can be simultaneously established in a cell, if the marking of place TOT-V exceeds 7, the immediate transition t_{loss-v} fires, modeling a voice call blocking (either a new call blocking or a handover blocking), and one token is removed from both REQ-V and TOT-V. At the beginning of a quadruple of GSM frames, voice calls that await the allocation of a traffic channel become active by firing the immediate transition $t_{allocate-v}$. This removes one token from REQ-V, and increases by one the marking of ACTIVE-V, but does not alter the marking of TOT-V.

The completion of voice calls during a quadruple of frames is modeled by the firing of the timed transition T_{end-v}, which must thus have rate μ_v and infinite server semantics. The firing of T_{end-v} decreases the marking of both ACTIVE-V and TOT-V.

The number N of ongoing packet sessions in the cell is assumed to be constant, and is modeled by the number of tokens in place OFF-D in the initial marking. The beginning and end of packet calls within a packet session is modeled by the firing of the timed transitions T_{new-d} and T_{end-d}, respectively, which must thus have infinite server semantics with rates μ_{Dpc} and μ_{Dd}/Nd. Tokens in place ON-D model active packet calls. The generation of a packet within an active packet call is modeled by the firing of the timed transition T_{pkt}, and the generation of one token in PKT; the firing rate is μ_{Dd} with infinite server semantics. The firing of one of the two immediate transitions t_{1rb}, t_{16rb}, models the generation of either 1 or 16 radio blocks from the packet segmentation, with equal probability. Note that these two immediate transitions describe the pdf of the number of radio blocks generated by segmenting a packet. We chose to describe a sort of bimodal packet length distribution with just two transitions, but the inclusion in the model of a larger number of transitions does not increase the model complexity (the model state space remains the same), so that more complex or realistic distributions can be very simply incorporated in the model. Tokens in place RB model radio blocks requesting access to the buffer where they can wait their turn for transmission. Radio blocks enter the buffer (modeled by place BUFFER) if free positions are available (some token marks place FREE-BUF), through the immediate transition t_{in-buf}. If the buffer is full (place FREE-BUF is empty), radio blocks are lost through transition t_{loss-d}. The initial marking of place FREE-BUF corresponds to the buffer capacity in radio blocks, which is denoted by B.

The allocation of traffic channels not used by voice calls to radio blocks is modeled by the immediate transitions $t_{allocate-d}$, which is enabled at the beginning of a quadruple of GSM frames, with lower priority than $t_{allocate-v}$. Transmitted radio blocks are modeled by tokens in place ACTIVE-D. If the radio block transmission is not successful, a copy of the radio block must remain in the buffer for retransmission. The immediate transitions t_{rb-OK} and t_{rb-KO}, with probabilities c and $1 - c$, model the successful and unsuccessful transmission of a radio block.

Table 1 reports weights and priorities for all immediate transitions in the DSPN. Table 2 reports the firing rates for all exponential transitions; Table 3 reports the initial marking for all places.

Table 1. Weights and priorities of immediate transitions in the DSPN model.

Transition	Weight	Priority
t_{loss-v}	1	4
t_{loss-d}	1	4
t_{in-buf}	1	4
$t_{allocate-v}$	1	3
$t_{allocate-d}$	1	2
t_{1rb}	1/2	4
t_{16rb}	1/2	4
t_b	1	1
t_e	1	1
t_{rb-OK}	c	2
t_{rb-KO}	$1 - c$	2

Table 2. Firing rates of exponential transitions in the DSPN model.

Transition	Firing rate	Semantics
T_{in-v}	λ_v	single server
T_{end-v}	μ_v	infinite server
T_{new-d}	μ_{Dpc}	infinite server
T_{end-d}	μ_{Dd}/Nd	infinite server
T_{pck}	μ_{Dd}	infinite server

Table 3. Initial marking of the DSPN model.

Place	Initial marking
CIRCUITS	7
4F	1
FREE-BUF	B
OFF-D	N

3.1 Performance Indices

The DSPN model allows the computation of many interesting performance indices. In this paper we shall mainly consider the following performance measures:

- voice call blocking probability (for both new call and handover requests); this is the ratio between the throughput of transition t_{loss-v} and the throughput of T_{in-v}
- average number of radio blocks in the base station buffer; this is the average number of tokens in place BUFFER
- average number of radio blocks in the base station buffer conditional on the number of active voice calls being equal to k; this is the average number of tokens in place BUFFER when the number of tokens in place ACTIVE-V equals k
- radio block loss probability; this is the ratio between the throughput of transition t_{loss-d} and the sum of the throughput of transitions t_{loss-d} and t_{in-buf}.

4   Numerical Results

We consider a cell with just one active frequency band, which allows the definition of one control channel and seven traffic channels. We assume that the voice and data traffic in this cell are as specified in Section 2. The traffic parameters values are summarized in Table 4, and are used for the derivation of all numerical results, unless otherwise specified.

The left plot of Figure 3 shows results for the voice call blocking probability and the radio block loss probability versus increasing values of the voice call

Table 4. Parameters of the basic scenario.

Parameter	Value
Number of traffic channels	7
Buffer size, B	100
Number of packet sessions, N	1, 3, 4, 5
$1/\mu_v$	120 s
Nd	25
$1/\mu_{Dd}$	0.5 s
$1/\mu_{Dpc}$	41.2 s
c	0.95

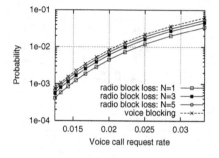 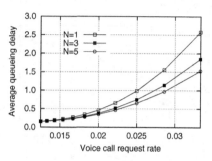

Fig. 3. Voice call blocking probability and radio block loss probability (left plot) and average queueing delay of radio blocks (right), with 1, 3 or 5 packet sessions.

request rate, λ_v. The radio block loss probability is plotted for different values of the number of active data sessions, N. The voice call blocking probability is insensitive to data traffic; instead, the radio block loss probability slightly increases with the number of data sources. This effect is mostly due to the increase in data traffic resulting from the higher number of data sources.

For the same scenario, the average queueing delay experienced by radio blocks is presented in the right plot of the same Figure. It can be observed that the delay suffered by radio blocks is quite large, due to the low bit rate of the radio channel and to the priority of voice traffic in accessing the radio resources.

The average number of radio blocks in the base station buffer is shown in the left plot of Figure 4, again as a function of the voice call request rate in the case of 1, 3 or 5 packet sessions. It is interesting to observe that the average number of radio blocks in the base station buffer remains quite small, even for rather large values of the radio block loss probability. The responsibility for the buffer overflow resides in the fact that voice has priority over data and in the different time scales of data and voice traffic. Indeed, the radio block service rate depends on the voice calls dynamics and, in particular, when all channels in the cell are busy with voice connections, no radio block can be transmitted until at least one traffic channel becomes free. During these intervals of service

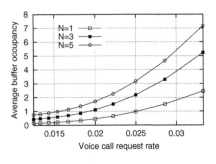
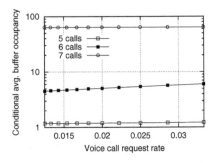

Fig. 4. Average buffer occupancy with 1, 3 or 5 packet sessions (left), and conditional average buffer occupancy, given that 5, 6 or 7 voice calls are active, with 3 packet sessions (right).

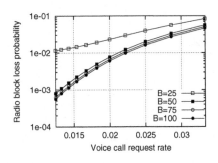
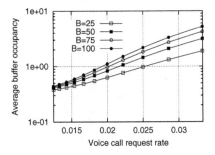

Fig. 5. Radio block loss probability (left) and average buffer occupancy (right) for different values of buffer size, with 3 packet sessions.

interruption, which are quite long compared to the data traffic dynamics, the radio block buffer fills up, and a large number of radio blocks is lost.

This is confirmed by the right plot of Figure 4 which shows the average number of radio blocks in the base station buffer conditional on the number of active voice calls, for 5, 6, or 7 active voice calls, and 3 data sessions. The buffer fills when all traffic channels are busy with voice calls; this is the main source of the base station buffer overflow, hence of the radio block loss. Moreover, the fact that the voice call dynamics are much slower than the radio block generation dynamics makes the buffer overflow periods quite long in relative terms. As a result, although the average number of radio blocks in the base station buffer is quite low, the radio block loss probability is rather high.

A further proof of this phenomenon comes from the left plot of Figure 5, which reports the radio block loss probability for different values of the buffer size. Observe that the buffer size has a minor impact on loss probability. This is again due to the different time scales of voice call and radio block generation dynamics. The time interval during which radio blocks are lost is much larger than the time needed to fill the buffer so that radio block loss probability results to be almost independent from the buffer capacity. We notice a significant increase of the radio block loss probability only when the buffer size reduces to 25 radio blocks.

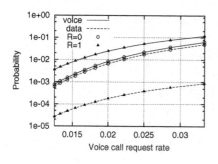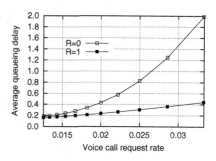

Fig. 6. Voice call blocking probability and radio block loss probability (left) and average queueing delay of radio blocks (right) when either 0 or 1 traffic channels are reserved to GPRS traffic, with 4 packet sessions.

This is due to the data traffic burstiness: the probability that two packets of 16 radio blocks each are generated close to each other is quite large. The right plot of Figure 5 reports the average radio block buffer occupancy for different values of buffer capacity. Observe again that the average buffer occupancy is always quite small.

Finally, the left plot of Figure 6 compares the radio block loss probability and the voice call blocking probability estimates when either zero or one traffic channels are reserved to GPRS traffic, assuming that 4 packet sessions are in progress. Note that the presence of one reserved channel produces a drastic reduction of the radio block loss probability (over one order of magnitude), with a modest increase of the voice call blocking probability (about a factor 3). A similar gain can be observed for the average radio block queueing delay in the right plot of the same Figure.

5   Complexity of the Model Solutions

The CPU times required for the solution of the DSPN models are reported in Table 5 together with the state space size of the underlying stochastic model. All models were solved on a Pentium III running at 600 Mhz with 512 Mbytes of main memory. The models were solved by using the software package DSPNexpress [11].

6   Conclusions

In this paper we have described a DSPN model for the performance analysis of mobile packet data services based on GPRS, and exploiting a common cellular infrastructure with GSM mobile telephony services.

The DSPN model is quite flexible and computationally parsimonious, allowing the accurate estimation of a variety of interesting performance metrics for the considered system.

Table 5. CPU times for the DSPN solution.

N	State space size	DSPN sol.[s]
1	58176	172
2	87264	522
3	116352	1136
4	145440	2046
5	174528	4281

Simple and accurate models of GSM/GPRS systems, like the one that we presented here, are a must for the effective design and planning of multiservice cellular systems.

References

1. M. Mouly, M.-B. Pautet, "The GSM System for Mobile Communications", Cell&Sys, 1992.
2. B.H. Walke, "Mobile Radio Networks: Networking and Protocols", John Wiley, 1999.
3. "Digital Cellular Telecommunications System (Phase 2+), General Packet Radio Service, Service Description, Stage 1", GSM 02.60 v.6.2.1, August 1999.
4. "Digital Cellular Telecommunications System (Phase 2+), General Packet Radio Service, Service Description, Stage 2", GSM 03.60 v.6.3.0, April 1999.
5. G. Brasche, B.H. Walke, "Concepts, Services, and Protocols of the New GSM Phase 2+ General Packet Radio Service", *IEEE Communications Magazine*, Vol. 35, N. 8, pp. 94-104, August 1977.
6. J. Cai, D. Goodman, "General Packet Radio Service in GSM", *IEEE Communications Magazine*, Vol. 35 N. 10, pp. 122-131, October 1977.
7. R. Kalden, I. Meirick, M. Meyer, "Wireless Internet Access Based on GPRS", *IEEE Personal Communications*, Vol. 7, N. 2, pp. 8-18, April 2000.
8. J.P. Castro, "Air Interface Techniques for Future Mobile Networks: The UMTS Radio Access Technology", John Wiley, 2001.
9. A. Samukic, "UMTS Universal Mobile Telecommunications System: Development of Standards for the Third Generation", *IEEE Transactions on Vehicular Technology*, Vol. 47, N. 4, pp. 1099-1104, November 1998.
10. M. Ajmone Marsan, G. Chiola, "On Petri Nets with Deterministic and Exponentially Distributed Firing Times," in: G.Rozenberg (editor), *Advances in Petri Nets 1987*, Lecture Notes in Computer Science n.266, Springer Verlag, 1987.
11. C. Lindemann, "Performance Modelling with Deterministic and Stochastic Petri Nets", John Wiley, 1998.
12. M. Ajmone Marsan, G. De Carolis, E. Leonardi, R. Lo Cigno, M. Meo, "How Many Cells Should Be Considered to Accurately Predict the Performance of Cellular Networks?", *European Wireless'99 and 4th ITG Mobile Communications*, Munich, Germany, October 1999.
13. "Universal Mobile Telecommunications System (UMTS); Selection procedures for the choice of radio transmission technologies of the UMTS (UMTS 30.03 version 3.2.0)", ETSI TR 101 112 V3.2.0 (1998-04).

Author Index